Comparing Nanofiltration and Reverse Osmosis for Treating Recycled Water

Awwa Research Foundation

About the Awwa Research Foundation

The Awwa Research Foundation (AwwaRF) is a member-supported, international, nonprofit organization that sponsors research to enable water utilities, public health agencies, and other professionals to provide safe and affordable drinking water to consumers.

The Foundation's mission is to advance the science of water to improve the quality of life. To achieve this mission, the Foundation sponsors studies on all aspects of drinking water, including supply and resources, treatment, monitoring and analysis, distribution, management, and health effects. Funding for research is provided primarily by subscription payments from approximately 1,000 utilities, consulting firms, and manufacturers in North America and abroad. Additional funding comes from collaborative partnerships with other national and international organizations, allowing for resources to be leveraged, expertise to be shared, and broad-based knowledge to be developed and disseminated. Government funding serves as a third source of research dollars.

From its headquarters in Denver, Colorado, the Foundation's staff directs and supports the efforts of more than 800 volunteers who serve on the board of trustees and various committees. These volunteers represent many facets of the water industry, and contribute their expertise to select and monitor research studies that benefit the entire drinking water community.

The results of research are disseminated through a number of channels, including reports, the Web site, conferences, and periodicals.

For subscribers, the Foundation serves as a cooperative program in which water suppliers unite to pool their resources. By applying Foundation research findings, these water suppliers can save substantial costs and stay on the leading edge of drinking water science and technology. Since its inception, AwwaRF has supplied the water community with more than $300 million in applied research.

More information about the Foundation and how to become a subscriber is available on the Web at **www.awwarf.org.**

Comparing Nanofiltration and Reverse Osmosis for Treating Recycled Water

Prepared by:
Jörg E. Drewes, **Christopher L. Bellona**, and **Pei Xu**
Colorado School of Mines, Environmental Science and Engineering Division,
Golden, CO 80401-1887

Gary L. Amy
UNESCO-IHE, Institute for Water Education, Westvest 7, 2601 DA Delft, The Netherlands

Gerald Filteau
Separation Processes, Inc., 2386 Faraday Avenue, Suite 100, Carlsbad, CA 92008
and

Gregg Oelker
United Water, West Basin Water Recycling Plant, 1935 Hughes Way, El Segundo, CA 90245

Jointly sponsored by:
Awwa Research Foundation
6666 West Quincy Avenue, Denver, CO 80235-3098

WateReuse Foundation
1199 N. Fairfax Street, Suite 410, Alexandria, VA 22314
and

Tailored Collaboration Partner
West Basin Municipal Water District
17140 S Avalon Blvd, Ste 210, Carson, CA 90746-1218

Published by:

Awwa Research Foundation

Distributed by:

American Water Works Association

IWA Publishing

DISCLAIMER

This study was jointly funded by the Awwa Research Foundation (AwwaRF), the WateReuse Foundation (WRF) and Tailored Collaboration partner West Basin Municipal Water District. AwwaRF, WRF and West Basin Municipal Water District assume no responsibility for the content of the research study reported in this publication or for the opinions or statements of fact expressed in the report. The mention of trade names for commercial products does not represent or imply the approval or endorsement of AwwaRF, WRF or West Basin Municipal Water District. This report is presented only for informational purposes.

CONTENTS

TABLES

FIGURES

FOREWORD

The Awwa Research Foundation is a nonprofit corporation that is dedicated to the implementation of a research effort to help utilities respond to regulatory requirements and traditional high-priority concerns of the industry. The research agenda is developed through a process of consultation with subscribers and drinking water professionals. Under the umbrella of a Strategic Research Plan, the Research Advisory Council prioritizes the suggested projects based upon current and future needs, applicability, and past work; the recommendations are forwarded to the Board of Trustees for final selection. The foundation also sponsors research projects through the unsolicited proposal process; the Collaborative Research, Research Applications, and Tailored Collaboration programs; and various joint research efforts with organizations such as the U.S. Environmental Protection Agency, the U.S. Bureau of Reclamation, and the Association of California Water Agencies.

This publication is a result of one of these sponsored studies, and it is hoped that its findings will be applied in communities throughout the world. The following report serves not only as a means of communicating the results of the water industry's centralized research program but also as a tool to enlist the further support of the nonmember utilities and individuals.

Projects are managed closely from their inception to the final report by the foundation's staff and large cadre of volunteers who willingly contribute their time and expertise. The foundation serves a planning and management function and awards contracts to other institutions such as water utilities, universities, and engineering films. The funding for this research effort comes primarily from the Subscription Program, through which water utilities subscribe to the research program and make an annual payment proportionate to the volume of water they deliver and consultants and manufacturers subscribe based on their annual billings. The program offers a cost-effective and fair method for funding research in the public interest.

A broad spectrum of water supply issues is addressed by the foundation's research agenda: resources, treatment and operations, distribution and storage, water quality and analysis, toxicology, economics, and management. The ultimate purpose of the coordinated effort is to assist water suppliers to provide the highest possible quality of water economically and reliably. The true benefits are realized when the results are implemented at the utility level. The foundation's trustees are pleased to offer this publication as a contribution toward that end.

David E. Rager
Chair, Board of Trustees
Awwa Research Foundation

Robert C. Renner,P.E.
Executive Director
Awwa Research Foundation

ACKNOWLEDGMENTS

The authors of this report are very thankful to the West Basin Municipal Water District, Carson, CA and the City of Scottsdale, Scottsdale, AZ for their cooperation and participation in this project. We especially appreciate the help from the many individuals, who assisted in the multiple facets of this research:

- Richard Nagel and Jennifer Bender, West Basin Municipal Water District
- Temitayo Abegunde, Michelle Rapanut, Chris Scardino, and Tina Tran with the United Water Laboratory and Kevin Tirado and Dennis Le with the United Water Process Control Group at the West Basin Water Recycling Plant in El Segundo, California
- Binga Talabi and Michael Helton with the Scottsdale Water Campus
- Eric Owens with Separation Processes, Inc.
- John Luna, Christiane Hoppe, Matthew Oedekoven and Gary Wang with the Water Technology Group at the Colorado School of Mines

We also acknowledge the technical support we received from membrane manufacturers, especially Koch Membrane Systems, Dow/Filmtec, Hydranautics, and Toray America. We are thankful to the members of the Project Advisory Committee for their guidance and critique:

- Robert C. Cheng, Ph.D., P.E. with the Long Beach Water Department
- Steven J. Duranceau, Ph.D., P.E. with Boyle Engineering Corporation
- Mehul Patel, P.E. with the Orange County Water District
- Richard Sakaji, Ph.D., P.E. with the California Department of Health

Finally, we would like to thank Jason Allen, My-Linh Nguyen, Ph.D., and Alice Fulmer with the Awwa Research Foundation (AwwaRF) for their support in managing this project.

EXECUTIVE SUMMARY

INTRODUCTION

An increasing number of municipalities are using surface water under the influence of wastewater effluents or recycled wastewater as a source to augment drinking water supplies. Today membrane treatment of source water of impaired quality by an integrated membrane system (IMS) such as microfiltration (MF) pretreatment followed by reverse osmosis (RO) represents the industry standard for subsurface injection projects leading to groundwater recharge or for surface water augmentation projects. Such an IMS is primarily driven by high removal efficiencies for unregulated and unidentified organic contaminants and total nitrogen. However, the rejection of organic micropollutants, which are associated with potential adverse human health effects, such as pesticides, disinfection by-products (DBPs), endocrine disrupting compounds (EDCs), pharmaceutically active compounds (PhACs), and organic compounds appearing on the candidate contaminant list (CCL) is of increasing interest for membrane applications. While the current concept of an IMS is effective from a performance standpoint, it involves the use of non-selective, higher pressure RO membranes. An alternative IMS involving nanofiltration (NF) membranes and a new generation of ultra-low pressure RO (ULPRO) membranes in place of conventional RO membranes provide an opportunity for lower pressure/higher flux operating conditions and higher selectivity (e.g., targeting trace organics over monovalent salts). Although recent work has demonstrated the potential of NF and ULPRO membranes in rejecting trace organics, there is little experience about how these membranes would perform and operate under conditions representative of full-scale installations.

This research project was performed by a team of faculty, scientists, graduate students, and engineers from the Colorado School of Mines, United Water, Separation Processes, Inc., and the University of Colorado-Boulder in collaboration with the West Basin Municipal Water District and the City of Scottsdale.

RESEARCH OBJECTIVES

The Awwa Research Foundation (AwwaRF) and the WateReuse Foundation (WRF) funded this study to determine the viability of ULPRO and NF membranes in treating municipal wastewater effluent for water recycling projects in terms of water quality and operational performance. The following research objectives were identified to determine the viability of ULPRO and NF membranes:

- To explore whether NF and ULPRO membranes can consistently meet potable water quality requirements with respect to TOC, total nitrogen, and both regulated and unregulated trace organic compounds.
- To determine while meeting these water quality goals, whether or not operating characteristics of NF and ULPRO membranes (such as flux, fouling/scaling, and cleaning frequencies) are comparable to thin-film composite RO membranes currently employed at full-scale facilities. Are operating feed pressure requirements significantly lower than the pressure range currently employed at facilities using conventional RO?

- To develop findings from rejection studies on laboratory-, pilot- and full-scale into a model framework providing the basis to reliably predict – *a priori* – rejection of organic micropollutants by RO, NF and ULPRO membranes.

APPROACH

Since there is little guidance in the water industry in identifying and selecting potentially viable NF and ULPRO membranes to treat water of impaired quality, the work presented here involved the development and validation of a laboratory-scale membrane testing protocol to select viable membranes for pilot- and full-scale operation. This selection protocol balanced operational characteristics (such as pressure, flux and potential for fouling) with product water quality and allowed for a pre-selection of potentially viable candidate membranes. Membranes considered for selection were characterized as thin-film composite (TFC) polyamide membranes and included commercially available ULPRO (TFC-ULP, Koch Membrane Systems; XLE, Dow/Filmtec; TMG10, Toray America; RE-BLR, Saehan) and NF membrane products (NF-90, NF-270, NF-200 and NF-4040, Dow/Filmtec; TFC-S, Koch Membrane Systems; ESNA1-LF, Hydranautics; NE-90, Saehan). Three candidate membranes were selected and tested using a 70 L/min (18 gpm) membrane pilot skid that was designed, constructed and operated during this study. The three selected candidate membranes were each tested for at least 1,300 hours on microfiltered feed water at two full-scale facilities. The feed waters represented non-nitrified microfiltered effluent provided by the West Basin Water Recycling Plant (WBWRP) in El Segundo, CA and nitrified/denitrified microfiltered effluent provided by the Scottsdale Water Campus (SWC), AZ. Operational performance and rejection of trace organics, nutrients and total organic carbon was monitored during pilot-scale testing and compared to the performance of full-scale trains operating on the same feed water employing conventional RO membranes (i.e., TFC-HR, Koch Membrane Systems). The best performing ULPRO membrane was also selected for testing over 5,000 hours in a 6.6 m^3/min (2.5-mgd) full-scale RO train to further verify findings from pilot-scale and to determine rejection behavior and operating costs under full-scale conditions. The study was assisted through state-of-the-art membrane characterization to describe the fouling behavior of NF and ULPRO membranes and its role on operation (e.g., flux decline) and rejection. Findings of this study were compiled into a model framework to describe and predict the rejection of organic micropollutants during NF or RO treatment, which is taking into account physicochemical properties of solutes and membranes as well as key operational conditions affecting solute rejection.

CONCLUSIONS

Membranes tested during this study using the membrane testing protocol displayed a wide range of operating characteristics including fouling tendency, permeate productivity, and feed pressure while displaying a wide range of rejection behavior in terms of TOC, ammonium, nitrate and select trace organic contaminant removal. Based upon the laboratory-scale testing, the candidate NF and ULPRO membranes can be divided into two categories: 1) Permeate quality similar to the benchmark RO membrane (TFC-HR) while operating at significantly lower pressure (ULPRO membranes: ESPA2, TFC-ULP, TMG10, XLE, and one NF membrane: NF-90); and 2) Lower pressure than the category 1 membranes, effectively rejecting TOC, selective for most trace organic contaminants, but were limited in their removal of monovalent ions, including nitrate and ammonium (NF membranes: NF-270, NF-200, and NF-4040). Three NF membranes (TFC-S, ESNA1-LF, and NE-90) exhibited a flux decline between 50 and 80 percent diminishing the benefits of operating at lower feed pressure. The selection of membranes for pilot-scale testing, therefore, became an exercise in balancing the benefits of lower pressure and energy requirements with constituent rejection and permeate quality. The TMG10 (Toray America) and NF-90 (Dow/Filmtec) membranes were selected as candidate membranes for pilot-scale testing based upon high specific fluxes after a period of fouling as compared to the 'benchmark' RO membrane (TFC-HR) and efficient rejection of TOC (<0.5 mg/L), nitrogen species (<10 mg-N/L), and organic trace contaminants. Although the category 2 membrane NF-4040 (Dow/Filmtec) did not meet the nitrogen and TOC water quality criteria, this membrane was also selected for pilot-scale testing due to moderate flux decline, efficient trace organic rejection (usually greater than 90 percent), and a significantly higher specific flux than the 'benchmark' TFC-HR membrane.

The TMG10 and NF-90 membranes tested on microfiltered non-nitrified wastewater (WBWRP) exhibited a slightly lower specific flux (flux normalized to operating pressure) after initial membrane compaction and fouling of 0.03 L/m^2-hr-kPa (0.12 gfd/psi) as compared to the 'benchmark' RO membrane (0.022 L/m^2-hr-kPa (0.09 gfd/psi)). Similar to observations made during membrane selection on laboratory-scale, the NF-90 exhibited a larger flux decline as compared to the TMG10 during pilot-scale testing. During these test periods, the TMG10 and NF-90 achieved permeate TOC and ammonia concentrations of less than 0.3 and 3 mg-N/L, respectively. Comparing pilot- and full-scale testing of the TMG10 to a period of full-scale operation indicated that the TMG10 performed similarly or better than the RO membrane TFC-HR in terms of TOC and ammonia permeate concentrations and rejection. Rejection of regulated organics, such as THMs, HAAs and NDMA, was also found to be similar to the 'benchmark' TFC-HR membrane with a poor to moderate rejection of chloroform and NDMA. Both, the TMG10 and NF-90 tested at pilot-scale and the TFC-HR and ESPA2 employed on full-scale were highly efficient in rejecting unregulated organics with non detectable concentrations or only traces of the target compounds in the permeates. The NF-4040 membrane was tested at pilot-scale on microfiltered nitrified/denitrified wastewater (SWC) and exhibited a relative small initial flux decline followed by a very constant permeate specific flux of 0.08 L/m^2-hr-Pa (0.33 gfd/psi) over 1,200 hours. As expected, during the 1,200 hours of testing the specific flux of the NF-4040 was more than four times higher as compared to the 'benchmark' RO membrane TFC-HR. At all times, TOC permeate concentrations of the NF-4040 were below 0.5 mg-C/L. Nitrate, however, was not rejected by the NF-4040 and negative rejections were observed in the NF-4040 membrane permeate. Of the unregulated trace organic compounds targeted, only traces of chlorinated flame retardants, carbamazepine and bisphenol-A were detected in the permeate of the NF-4040 membrane immediately after start-up. Rejection of trace organics, however, improved with membrane fouling and subsequent sampling campaigns revealed no detections in permeate samples. This was an interesting finding and especially important for NF membranes, because initial (organic) fouling will always occur where membranes are employed to treat surface water or secondary and tertiary treated effluents. Foulant precipitation and cake-layer formation resulted in a considerable change of membrane surface characteristics with respect to membrane hydrophobicity, surface charge and surface morphology, hence affecting transport mechanisms of contaminants as compared to virgin membranes. The transport of hydrophilic ionic organic contaminants was especially hindered in NF membranes as a result of improved electrostatic exclusion, pore clogging and an increased adsorption capacity of fouled polyamide membranes.

Due to the relatively high fouling potential of secondary effluent feed water, initial high specific flux rates of the TMG10 and NF-90 as observed during both laboratory- and pilot-scale testing were reduced to values similar to the commonly employed ULPRO membrane, ESPA2, but exceeded the conventional RO membrane, TFC-HR. For a full-scale installation operating at a feed pressure of 896 kPa (130 psi) and 85 percent recovery, this relates to an electrical consumption of 0.37 kWh/m^3 (1.40 kWh/kgal) of permeate produced. Based on an example electrical cost of \$0.12/kWh, the operating cost of producing one m^3 of permeate utilizing the TMG10 or ESPA2 membrane would be approximately \$0.05. Pilot- and full-scale operational results appear to support the argument that the choice of membranes to achieve more significant cost savings associated with lower feed pressure is limited to certain low fouling *loose* NF membranes. During the testing period, the NF-4040 sustained pressures 3–4 times lower than the feed pressure of conventional RO membranes currently employed at full-scale facilities. These lower feed pressures would result in potential savings of \$0.03–0.08 per m^3 (\$0.1–0.3 per kgal) as compared to current installations using conventional RO membranes while treating reclaimed water with an IMS.

Findings of this research clearly demonstrated that a single model does not exist which is capable of reliably predicting the mass transport of organic micropollutants during high-pressure membrane treatment. The physicochemical properties of the solutes are the key factors determining rejection and if properly considered and put in context with relevant membrane properties, rejection can be qualitatively predicted. This qualitative model, a solute rejection diagram, highlights the complexity of membrane rejection performance while considering solutes possessing different physicochemical properties. Although operational conditions, such as flux or recovery, are not considered, the rejection diagram allows utilities to estimated rejection of individual compounds semi-quantitatively. One approach was developed to quantitatively describe and predict the rejection of non-ionic and ionic compounds of concern (Spiegler-Kedem model). With the exception of highly hydrophobic compounds, the Spiegler-Kedem model provided a high to moderate degree of accuracy for the rejection of non-ionic solutes by a conventional RO, an ULPRO and an NF membrane. For ionic solutes, the Spiegler-Kedem model underpredicted rejection as expected since electrostatic exclusion is not properly considered in the model.

RECOMMENDATIONS TO UTILITIES

This report provides guidance for utilities regarding the viability of low-pressure RO (ULPRO) and NF membranes to replace conventional RO membranes employed in integrated membrane systems (IMS) to treat source water of impaired quality. Findings of this study reveal that ULPRO membranes can consistently meet potable water quality requirements for treating source water of impaired quality with respect to TOC, total nitrogen, and both regulated and unregulated trace organic compounds. ULPRO membranes (such as ESPA2, TMG10, XLE, TFC-ULP) exhibited specific flux values that were slightly higher as compared to conventional RO membranes (TFC-HR). Because purchase cost and expected life of these membranes are similar, ULPRO membranes potentially offer lower operating expenses and therefore should be considered in membrane procurement or replacement programs of an IMS. Achieving significant cost savings associated with lower feed pressure, however, is limited to certain low fouling *loose* NF membranes (such as NF-200, NF-4040 or NF-270). While these membranes demonstrated efficient rejection of TOC and a high selectivity for a wide range of trace organics, they will result in lowered permeate water quality in terms of nitrogen species such as ammonia and nitrate. These membranes will likely not be capable of meeting stringent nitrogen standards and would require a tailored wastewater pretreatment to reduce ammonia and nitrate concentrations below the Federal Drinking Water Standard of 10 mg-N/L. Providing the nitrogen standard can be met prior to membrane treatment and given the selectivity and low operating cost of *loose* NF membranes, these membranes are attractive for the design of multiple barrier treatment trains, consisting of high-pressure membranes followed by an advanced oxidation process (such as UV/hydrogen peroxide or ozone) or groundwater recharge of impaired water followed by membrane treatment. Findings of this study underline the fact that proper pre-treatment of ULPRO and NF membranes is as critical as it is for conventional RO membranes. If not properly maintained, the benefit of operating NF and ULPRO membranes at lower feed pressure is quickly lost, where severe biofouling is not prevented and a constant permeate flux is maintained. The laboratory-scale testing protocol developed to select viable candidate ULPRO and NF membranes was a useful tool in separating the performance of different membranes. The candidate membranes selected using this protocol performed as expected during pilot- and full-scale testing. This testing protocol can, therefore, assist the water industry in better pre-selecting viable membranes for applications using a given impaired water source such as recycled water. Findings of this research clearly demonstrated that a single model does not exist which is capable of describing the mass transport of organic micropollutants during high-pressure membrane treatment. The physicochemical properties of the solutes are the key factors determining rejection and if properly considered and put in context with relevant membrane properties, rejection can be qualitatively predicted.

CHAPTER 1
INTRODUCTION

PROJECT OVERVIEW

An increasing number of municipalities are using surface water under the influence of wastewater effluent or recycled wastewater as a source to augment drinking water supplies. Today membrane treatment of source water of impaired quality by an integrated membrane system (IMS) such as microfiltration (MF) pre-treatment followed by reverse osmosis (RO) represents the industry standard for subsurface injection projects leading to groundwater recharge or for surface water augmentation projects (National Research Council, 1998, Drewes et al. 2003). Such an IMS is primarily driven by high removal efficiencies for unregulated and unidentified organic contaminants and total nitrogen. However, the rejection of organic micropollutants, which are associated with potential adverse human health effects, such as pesticides, disinfection by-products (DBPs), endocrine disrupting compounds (EDCs), pharmaceutically active compounds (PhACs), and organic compounds appearing on the candidate contaminant list (CCL) is of increasing interest for membrane applications. While the current concept of an IMS is effective from a performance standpoint, it involves the use of non-selective, higher pressure RO membranes. An alternative IMS involving nanofiltration (NF) membranes and a new generation of ultra-low pressure RO (ULPRO) membranes in place of conventional RO membranes provide an opportunity for lower pressure/higher flux operating conditions and higher selectivity (e.g., targeting trace organics over monovalent salts). Although recent work has demonstrated the potential of NF and ULPRO membranes in rejecting trace organics, there is little experience about how these membranes would perform and operate under conditions representative of full-scale installations.

Research Objectives

The objectives of this study are three-fold and formulated in the following questions:

(1) Can nanofiltration (NF) and ultra-low pressure reverse osmosis (ULPRO) membranes consistently meet potable water quality requirements for treating source water of impaired quality with respect to total organic carbon (TOC), total nitrogen, and both regulated and unregulated trace organic compounds?
(2) Are operating characteristics of NF and ULPRO membranes (such as flux, fouling/scaling, and cleaning frequencies) comparable to thin-film composite RO membranes currently employed at full-scale facilities? Is the operating feed pressure of NF and ULPRO membranes significantly lower than the pressure range currently employed at facilities using conventional RO?
(3) Can findings from rejection studies at the laboratory-, pilot-, and full-scale be developed into a model framework providing the basis to reliably predict - *a priori* - rejection of organic micropollutants by RO, NF and ULPRO membranes?

Research Approach

Given the plethora of commercial NF, ULPRO, and RO product availability and the lack of guidance of how to select potentially viable membranes for pilot- and full-scale testing justified the development and validation of a laboratory-scale approach that can assist the water industry in selecting viable NF and ULPRO membranes. Three candidate membranes selected were tested using a 70 L/min (18 gpm) membrane pilot skid that was designed, constructed and operated during this study. The three selected candidate membranes were each tested for at least 1,300 hours on microfiltered feed water at two full-scale facilities. The feed waters represented non-nitrified microfiltered effluent provided by the West Basin Water Recycling Plant in El Segundo, CA and nitrified/denitrified microfiltered effluent provided by the Scottsdale Water Campus, AZ. Operational performance and rejection of trace organics, nutrients and total organic carbon was monitored during pilot-scale testing and compared to the performance of full-scale RO trains operating on the same feed water. During several occasions, challenge tests were conducted at pilot-scale to examine rejection of trace organics during periods of elevated feed concentrations. The best performing ULPRO membrane was also selected for testing over 5,000 hours in a 6.6 m^3/min (2.5 mgd) full-scale RO train to validate findings from pilot-scale and to determine rejection behavior and operating costs under full-scale conditions. The study was assisted through state-of-the-art membrane characterization to describe the fouling behavior of NF and ULPRO membranes and the role of fouling on operation (e.g., flux decline) and rejection. Findings of this study were compiled into a model framework to predict the rejection of organic micropollutants during NF and RO treatment, which takes into account physicochemical properties of solutes and membranes as well as key operational conditions affecting solute rejection.

BACKGROUND

Viability of Nanofiltration Membranes in Treating Water of Impaired Quality

In past research, it has been demonstrated that some micropollutants such as pesticides (e.g., atrazine) can be effectively removed by RO and, to a lesser extent, by NF membranes (Duranceau et al. 1992, Van der Bruggen et al. 1998, Taylor et al. 2000, Crozes and Seacord 2001). Duranceau et al. (1992) conducted pilot-scale membrane experiments investigating the rejection efficiencies of NF membranes (NF-70, Dow/Filmtec) for selected uncharged, hydrophobic pesticides. In this study groundwater was spiked with target compounds and findings indicated that rejection of these compounds by the membrane depended upon the molecular weight (MW). Sieving was identified as the key mechanism of rejection for uncharged organics that have an MW greater than the nominal molecular weight cut-off (MWCO) of the membrane. Taylor et al. (2000) evaluated a suite of RO and NF membranes for pesticide removal at bench- and pilot-scale and reported rejection rates for different pesticides varying between 58 and 82 percent both for RO and NF membranes. Physical-chemical solute properties and operational conditions in these studies were more important for rejection than MWCO of NF or RO membranes.

Recent studies of NF and RO treatment report a partial rejection of compounds of concern with molecular weight below the MWCO of the membranes. These compounds include

1,4-dioxane and disinfection by-products (DBPs), such as N-nitrosodimethylamine (NDMA), trihalomethanes (THMs) and haloacetic acids (HAAs) (Jacangelo et al. 1994, Najim and Trussell 2001, Vernon et al. 2001, Deeb et al. 2003, Mitch et al. 2003, Schippers et al. 2004).

Reinhard et al. (1986) studied the removal of trace organics by RO using cellulose acetate and polyamide membranes; all membranes rejected branched, complex molecules but varied in their rejection characteristics for smaller molecules, such as chlorinated solvents, base neutrals, and low molecular weight acids. Levine et al. (2000) demonstrated that low-molecular base neutrals showed the poorest rejection during RO treatment. Schäfer and Waite (2002) evaluated nanofiltration membranes (TFC-S and TFC-SR1, Koch Membrane Systems) for removal of steroids and reported rejections higher than 92 percent. Drewes et al. (2003) conducted pilot-scale experiments comparing RO (TFC-HR, Koch) and NF (TFC-S, Koch; NF-90, DOW/Filmtec) membranes fed with tertiary treated effluent to assess their viability for indirect potable reuse applications. During this test, RO and NF membranes were operated at an average flux of 23.7 L/m^2 hr (14 gpd/ft^2) and the NF operation required significantly less feed pressure (approximately 250 kPa) compared to RO treatment. Findings of this study revealed that NF treatment of tertiary treated effluent can achieve permeate TOC concentrations below 0.5 mg/L, nitrate nitrogen concentrations below 5 mg/L as N, and target trace organic compound concentrations (such as ethylenediaminetetraacetic acid (EDTA), alkylphenol-polyethoxycarboxylates (APECs), and pharmaceutical residues) below the level of detection. Disinfection by-products, such as bromoform and bromodichloromethane, were only partially rejected (i.e., 9 to 65 percent) by both membranes with the RO membrane exhibiting a slightly better removal.

Rejection of Organic Solutes by NF and RO Membranes

Rejection of organic compounds is a complex interaction of steric hindrance, electrostatic repulsion, solution effects within the membrane, and solute properties. Some interactions are fairly well understood; for example, the major mechanism of solute rejection by NF is physical sieving of solutes larger than the membrane MWCO (Duranceau et al. 1992). Pesticide rejection has been shown to be influenced by the physical-chemical properties of the compound (e.g., molecular weight, solubility, diffusivity, polarity, hydrophobicity, and charge), membrane properties (e.g., permeability, pore size, hydrophobicity, and charge), and membrane operating conditions (e.g., flux, transmembrane pressure, and recovery) (Taylor et al. 2000). Several studies point to the size of the molecule represented by molecular weight or molecular diameter as the most important structural property for retention (Tödtheide et al. 1997, Van der Bruggen et al. 1998). Kiso et al. (2000) reported a positive correlation between rejection and molecular weight for solutes with a molecular weight larger than 250 g/mol, whereas smaller molecules did not show a positive correlation. In a previous study, Kiso et al. (1992) reported a good relationship between molecular width and rejection of alcohols using cellulose acetate RO membranes. Hofman et al. (1993) reported that increasing size of the side chain results in higher rejection of organic compounds by an RO membrane. Steric hindrance of branched molecules was also observed in membrane studies by Reinhard et al. (1986) and Berg et al. (1997). Ozaki and Li (2002) showed that the percentage removal of undissociated organic compounds increased linearly with the molecular weight as well as with the molecular width. However, neither of these parameters served as a definitive factor for rejection. Ionic strength and pH are also recognized to influence the rejection behavior of charged compounds by affecting the double layer thickness surrounding a charged membrane (Braghetta et al. 1997). The rejection of undissociated organic compounds, however, did not show a close relationship with the feed pH (Ozaki and Li 2002). Membrane surface charge was identified as one key property affecting solute rejection. For the majority of commercially available thin-film composite NF and RO membranes, the surface charge of the membrane becomes increasingly negatively charged as the pH of the feed water increases (Xu et al. 2005). The extent of this surface charge is dependent upon the water quality matrix. Zander and Curry (2001) pointed out that the presence of divalent cations in solution greatly affects the performance of NF membranes studied. Berg et al. (1997) demonstrated that rejection of non-charged organics during NF treatment was mainly determined by steric hindrance. No evidence was found in this study that the transmembrane transport could be explained by the solution-diffusion model. Studies conducted by Van der Bruggen et al. (1998) using NF membranes indicated that a higher dipole moment resulted in a lower retention, which points to a higher retention of nonpolar substances. Therefore, the retention of a compound with a high dipole moment is lower than expected on the basis of molecular size. The high rejection of negatively-charged organics appeared to be determined by electrostatic repulsion between the solute and a negatively charged membrane and was also influenced by molecular size (Berg et al. 1997). It is the nature of the organic compounds that influences solute adsorption including water solubility and hydrogen-bonding ability (Ozaki and Li 2002). Studies conducted by Kiso et al. (2000) with pesticides and NF membranes indicated that the rejection was influenced mainly by hydrophobicity. A strong positive correlation was found between the affinity of a solute for the membrane as expressed by the octanol-water partitioning coefficient (K_{OW}) and rejection for *loose* NF membranes. A higher K_{OW} led to a better rejection which is most likely caused by solute adsorption onto the membrane and lower (hindered) diffusion. Since molecular size and dipole moment determine the retention of a non-charged organic molecule, these parameters could be used to predict the retention qualitatively and possibly quantitatively. In contrast to hydrophobic neutral compounds, many polar trace organic compounds (e.g., many PhACs) are negatively charged (Heberer 2002), and thus electrostatic rejection may be dominant.

Although the rejection of wastewater-derived micropollutants during membrane treatment is extremely complex, findings of a comprehensive literature review by Bellona et al. (2004) were structured in terms of interactions of molecular mass, acidity, functionality, and spatial characteristics of the solutes and the charge and hydrophobicity of the membrane surface. Rejection of non-charged solutes is mainly dependent upon steric exclusion and molecular size descriptors such as molecular width, molecular length, and molecular mean size may serve as good predictors of the influence of steric hindrance on rejection. The molecular size is less important for charged solutes, but electrostatic repulsion between solute and membrane surface is the main factor for rejection. Many studies published previously are characterized as single solute experiments using virgin membrane specimens with limited characterization of operational parameters and frequently limited experimental run times. Membrane and solute properties can change with different feed water composition, such as pH, solute and membrane surface charge, ionic strength, and presence of effluent organic matter (EfOM). It is well known that hydrophobic micropollutants (e.g., polycyclic aromatic hydrocarbons, PAHs) can bind to humic substances (HS, humic and fulvic acids), an important component of EfOM. This phenomenon can also influence solute interactions with membranes as the original micropollutant properties are masked by the humic macromolecules. Micropollutant binding to HS becomes important when its log K_{OW} is greater than 3. While the log K_{OW} for atrazine is 2.61, the log K_{OW} for bisphenol A is 3.32. The implication is that rejection of EfOM will be accompanied by rejection of an EfOM-bound micropollutant. These humic acid/solute interactions may be more important than details of molecular structure and thus may mask effects predicted by quantitative structure-property relationship (QSPR) or other predictive algorithms. Adsorption of hydrophobic compounds onto membrane surfaces and partitioning into membrane pores may also be an important factor in solute rejection. Studies characterized by short experimental runs may not fully account for solute adsorption and tend to report higher rejection performance than would have been achieved after the membrane is saturated with the hydrophobic solute. Members of the research team previously demonstrated that hydrophobic compounds can partition through NF and RO membranes leading to a significantly lower rejection as compared to hydrophilic compounds using virgin membrane specimens (Kimura et al. 2003). EfOM fouling has the potential to modify membrane surfaces to increase electrostatic repulsion as a rejection mechanism, favor size exclusion, and can also affect hydrophobic/hydrophobic solute-membrane interactions. The research team demonstrated during previous research that the transport of ionic organic micropollutants was hindered as a result of improved Donnan exclusion (electrostatic repulsion) due to a more negative surface charge (Xu et al. 2006a). Membrane fouling also resulted in an increased rejection of non-ionic compounds as a result of enhanced sieving effect due to pore clogging, and hindered partition/diffusion of solutes through the fouling layer of the membrane (Xu et al. 2006a, Xu et al. 2006b).

Rejection of Nitrogen Compounds by NF/RO Membranes

Nitrogen present in water of impaired quality is a concern regarding drinking water supply due to potential human health effects associated with specific nitrogen species. Nitrogen exists in various chemical forms including inorganic compounds such as ammonia (NH_3), nitrate (NO_3^-), and nitrite (NO_2^-) as well as dissolved organic nitrogen (DON) compounds. The chemical form of the nitrogen compounds significantly affects the rejection in NF/RO membranes. Small organic compounds, such as urea, have lower rejection than ionic compounds such as ammonium, nitrate, and nitrite or large organic compounds such as creatine. This suggests that electrical interaction as well as size exclusion play a role in rejection. The use of RO and NF membranes for the removal of some nitrogen species (i.e., nitrate) is well documented, indicating that in general *tighter* membranes are required to achieve a significant rejection (> 75 percent) of nitrate (Van der Bruggen et al. 2001). Drewes et al. (2001) conducted pilot-scale studies with RO and NF membranes fed with tertiary treated effluent (nitrified/denitrified). Findings of this study indicated that nitrate was rejected by 77 percent and 90 percent by two different NF membranes (TFC-S, Koch; NF-90, Dow/Filmtec), respectively, and more than 94 percent by a RO membrane (TFC-HR, Koch).

Less work is reported on the rejection of ammonia usually present in non- or partly nitrified secondary effluents at concentrations between 10 and 40 mg/L as N, with the dominant species being ammonium ion (NH_4^+) at ambient pH. Lee and Lueptow (2001) conducted rejection experiments with various nitrogen species using RO, ULPRO and NF membranes. The results of this study indicated that if either the ratio of the solute radius to pore radius was greater than about 0.8 or the membrane charge was greater than 0.05V, the rejection was quite good. However, if neither of these conditions was met, the rejection was poor. Thus, electrical effects are critical to rejection of small ions and molecules. Consequently, urea is quite difficult to reject, while smaller ammonium ions are quite easy to reject. Koyuncu (2002) reported over 80 percent rejection of ammonia ion using the TFC-S (Koch) nanofiltration membrane and over 90 percent rejection using TFC-ULP (Koch) ULPRO membrane in a pilot-scale study. An earlier study by Koyunco et al. (2001) using Dow/Filmtec RO membranes reported 90 percent removal of 100 mg/L ammonium ion concentration in the feed water. Lower concentrations yielded higher removal with a maximum of 99 percent removal at 8.5 mg/L as feed concentration. Kurama et al. (2002) presented results using NF membranes (N30F and NF-PES-10, Celgard) and RO membranes (Desal-3LP and Desal-3B SE, Osmonics) in flat sheet experiments. The feed to the membranes contained 6.5 mg/L NH_4^+. The N30F and NF-PES-10 NF membranes achieved 27 percent and 12 percent rejection, respectively, and the RO membranes Desal-3LP and Desal-3B SE removed 92 percent and 97 percent ammonium ion, respectively. Nitrate rejections of 97 to 99 percent have been reported for three ULPRO membranes (Toray Membranes, Japan, personal communication); corresponding rejections of urea (pK_a = 13.9) were 11 to 63 percent. These urea results suggest that non-ionized ammonia will be more difficult to reject than ammonium ion, however, given the pK_a (9.3) of ammonia, ammonium ion will be the dominant form at pH < 9.3. These findings indicate that *tighter* NF membranes with lower MWCO may be required for meeting stringent total nitrogen requirements.

Despite the numerous research studies attempting to relate physicochemical properties of solutes and membranes to organic solute rejection, a systematic and comprehensive approach is still needed in order to identify key parameters that could be used effectively for the prediction of solute separation by considering different water quality matrices and membrane operational conditions. An improved knowledge base is also required to guide the water industry in identifying and selecting NF and ULPRO membranes, which are capable of meeting stringent water quality standards while treating water of impaired quality at lower costs as currently achieved by conventional RO membranes.

SIGNIFICANCE OF THE PROJECT

The occurrence of unregulated organics (such as DBPs, PhACs, EDCs, or CCL compounds) in the aquatic environment and in drinking water has drawn public attention to the issue of receiving safe, aesthetically pleasing water today and in the future. These emerging contaminants are representative of the kind of problem the water industry will be confronting at an accelerating rate. Regardless of the size of a utility, the water supply industry must respond to environmental and health concerns, which are becoming a permanent part of the water industry landscape (Trussell 2001). There is a consensus within the water industry to be proactive about these issues. In order to be proactive, water utilities have to build confidence in their existing practice of treatment processes to assure proper removal of regulated and unregulated trace contaminants. This confidence relies on the capability to predict the rejection behavior of compounds of concern *a priori*. While committed to provide the highest quality water while using source water of impaired quality, utilities are also constantly exploring avenues to reduce both capital and operational costs when high-pressure membranes are employed.

Findings of this study provide the water industry with a knowledge base about the capability of NF and the new generation of RO membranes (ULPRO) to remove organic trace pollutants of concern and nitrogen species while treating source water of impaired quality. This knowledge was gathered through both laboratory- and pilot-scale studies to examine removal processes and full-scale studies supported by participating utilities. Findings of this study allow an assessment of the economic benefits of NF and ULPRO membranes under various operating conditions by considering cleaning and operating costs, while meeting the same water quality standards achieved with RO membranes while treating reclaimed water.

CHAPTER 2
EXPERIMENTAL METHODS

SELECTION OF TARGET COMPOUNDS

Target parameters and compounds chosen for this study were selected based upon characteristics relevant for drinking water augmentation projects using water of impaired quality, such as membrane performance (e.g. conductivity), rejection behavior (e.g., molecular geometry and charge), resistance to biodegradation, and associated potential adverse human health effects. Compounds and/or parameters were ranked into four tiers based upon their monitoring frequency and scope of study as outlined in Table 2.1. Tier 1 parameters consisted of conductivity, pH, total organic carbon (TOC), UV absorbance at 254 nm, ammonium and nitrate and represented constituents that were monitored frequently during laboratory-, pilot-, and full-scale membrane evaluations. Tier 2 and 3 target compounds represent endocrine disruptors, pesticides, disinfection by-products, pharmaceutically active compounds, a stimulant, chlorinated flame retardants, aliphatic amines, and selected compounds appearing on the CCL. Tier 2 and 3 target compounds required complex analytical methodologies for detection and quantification and were monitored less frequently than Tier 1 parameters during membrane testing. Tier 4 compounds were only used during bench-scale studies and consisted of surrogate compounds such as straight-chain aliphatic, branched-chain aliphatic, aromatic and ring- structured organic chemicals. Tier 5 compounds were pharmaceuticals and CCL compounds used during bench-scale studies for solute transport modeling.

During this study, Tier 2 and 3 target compounds were classified into bins representing similar physicochemical properties (based on hydrophobicity/hydrophilicity and solute charge) to further elucidate removal mechanisms during membrane treatment. The hydrophobicity/hydrophilicity is provided by the value of log D at pH 6 (log D_{pH6}), which is the octanol-water partitioning coefficient (log K_{OW}) at a specific pH. In this study, trace organic compounds with a log D_{pH6} greater than 3 are operationally defined as “hydrophobic”. Compounds with a log D_{pH6} value smaller than 3, and are defined as “hydrophilic” in this study. The charge of each compound is defined by relating the acid dissociation constant, pK_a, to the operational pH (pH 6-7.5) of high-pressure membrane applications. Accordingly in this study, trace organic contaminants with a pK_a smaller than 5 are considered as ionic and those with a pK_a greater than 5 are considered non-ionic. Target compounds sorted into these bins are considered indicator compounds representing similar physicochemical properties.

A summary of the physical and chemical properties of Tier 2, 3, 4 and 5 compounds is presented in Appendix A, Table A.1.

Table 2.1
Target compounds selected for this study

Tier	Scope of Study	Compound/ CAS No.	Representative of	Physicochemical Properties
Tier 1	Rejection performance using laboratory-, pilot-, and full-scale installations	Conductivity	Total dissolved solids	Hydrophilic, ionic
		pH	Acidity	N/A
		Total organic carbon (TOC)	Bulk organic carbon	N/A
		UV absorbance (UVA 254 nm)	Aromatic moieties	N/A
		Ammonium	Nutrient	Hydrophilic, ionic
		Nitrate	Nutrient	Hydrophilic, ionic
Tier 2	Rejection performance using laboratory-, pilot-, and full-scale installations	Bisphenol-A (80-05-7)[1]	Endocrine disrupting compounds	Hydrophobic, non-ionic
		Fenofibrate (49562-28-9)[2]	Blood lipid regulator	Hydrophobic, non-ionic
		Gemfibrozil (25812-30-0)[2]	Blood lipid regulator	Hydrophilic, ionic
		Trichloroacetic acid (76-03-9)[2]	Disinfection by-product	Hydrophilic, ionic
		Dichloroacetic acid (79-43-6)[2]	Disinfection by-product	Hydrophilic, ionic
		Mecoprop (93-65-2)[2]	Pesticide	Hydrophilic, ionic
		Dichlorprop (120-36-5)[2]	Pesticide	Hydrophilic, ionic
		Ibuprofen (15687-27-1)[2]	Analgesic	Hydrophilic, ionic
		Naproxen (22204-53-1)[2]	Analgesic	Hydrophilic, ionic
		Ketoprofen (22071-15-4)[2]	Analgesic	Hydrophilic, ionic
		Clofibric acid (882-09-7)[2]	Blood lipid regulator	Hydrophilic, ionic
		Salicylic acid (69-72-7)[2]	Analgesic	Hydrophilic, ionic
		Diclofenac (15307-86-5)[2]	Analgesic	Hydrophilic, ionic
		Chloroform (67-66-3)[4]	Disinfection by-product	Hydrophilic, non-ionic
		Bromoform* (75-25-2)[4]	Disinfection by-product	Hydrophilic, non-ionic
		Carbamazepine (298-46-4)[1]	Anticonvulsant, antimanic agent	Hydrophilic, non-ionic
		Phenacetine (62-44-2)[2]	Analgesic	Hydrophilic, non-ionic
		Primidone (125-33-7)[1]	Antiepileptic	Hydrophilic, non-ionic
		Caffeine (58-08-2)[2]	Stimulant	Hydrophilic, non-ionic
		Propyphenazone (479-92-5)[2]	Analgesic	Hydrophilic, non-ionic
		Tris(2-chloroethyl)phosphate (TCEP) (115-96-8)[2]	Chlorinated flame retardants	Hydrophilic, non-ionic
		Tris(1-chloro-2-propyl)phosphate (TCPP) (13674-84-5)[2]	Chlorinated flame retardants	Hydrophilic, non-ionic
		Tris(1,3-dichloro-2-propyl) phosphate (TDCPP) (13674-87-8)[2]	Chlorinated flame retardants	Hydrophilic, non-ionic

(continued)

Table 2.1 (Continued)

Tier	Scope of Study	Compound/ CAS No.	Representative of	Physicochemical Properties
Tier 3	Rejection performance using laboratory-, pilot-, and full-scale installations	17β–Estradiol (50-28-2)[3]	Endocrine disrupting compounds	Hydrophobic, non-ionic
		Estriol (50-27-1)[3]	Endocrine disrupting compounds	Hydrophobic, non-ionic
		Testosterone (58-22-0)[3]	Endocrine disrupting compounds	Hydrophobic, non-ionic
		N-nitrosodimethylamine (NDMA) (62-75-9)[5]	Disinfection by-product	Hydrophilic, non-ionic
		Methylamine (74-89-5)[7]	Aliphatic amine	Hydrophilic, ionic
Tier 4	Bench-scale studies for membrane characterization	Methanol (67-56-1)[6]	Alcohol	Straight aliphatic
		Ethylene glycol (107-21-1)[6]	Alcohol	Straight aliphatic
		1,3-Propanediol (504-63-2)[6]	Alcohol	Straight aliphatic
		1,4-Butanediol (110-63-4)[6]	Alcohol	Straight aliphatic
		Diethylene glycol (111-46-6)[6]	Alcohol	Straight aliphatic
		Triethylene glycol (112-27-6)[6]	Alcohol	Straight aliphatic
		Diethylene glycol monobutyl ether (112-34-5)[6]	Alcohol	Straight aliphatic
		Tetraetyhylene glycol (112-60-7)[6]	Alcohol	Straight aliphatic
		Tetra ethylene glycol dimethyl ether (143-24-8)[6]	Alcohol	Straight aliphatic
		Isopropanol (67-63-0)[6]	Alcohol	Branched aliphatic
		1,2-Propanediol (57-55-6)[6]	Alcohol	Branched aliphatic
		1,3-Butanediol (107-88-0)[6]	Alcohol	Branched aliphatic
		Glycerin (56-81-5)[6]	Alcohol	Branched aliphatic
		Neopentyl glycol (126-30-7)[6]	Alcohol	Branched aliphatic
		Phenol (108-95-2)[6]	Aromatic alcohol	Aromatic/ringed
		Benzyl alcohol (100-51-6)[6]	Aromatic alcohol	Aromatic/ringed
		Dextrose (50-99-7)[6]	Sugar	Aromatic/ringed

(continued)

Table 2.1 (Continued)

Tier	Scope of Study	Compound/ CAS No.	Representative of	Physicochemical Properties
Tier 5	Bench-scale studies for transport modeling	Primidone (125-33-7) [8]	Antiepileptic	Hydrophilic, non-ionic
		Phenacetine (62-44-2) [8]	Analgesic	Hydrophilic, non-ionic
		Naphthalene (91-20-3) [8]	Moth repellent	Hydrophobic, non-ionic
		Carbamazepine (298-46-4) [8]	Anticonvulsant, antimanic agent	Hydrophilic, non-ionic
		2-Naphthol (37350-58-6) [8]	Dye and antioxidant	Hydrophilic, non-ionic
		Ibuprofen (15687-27-1) [8]	Analgesic	Hydrophilic, ionic
		Naproxen (22204-53-1) [8]	Analgesic	Hydrophilic, ionic
		Diclofenac (15307-86-5) [8]	Analgesic	Hydrophilic, ionic

N/A – not applicable
1 – quantified using solid phase extraction, derivatization (MTBSTFA) and GC/MS
2 – quantified using solid phase extraction, derivatization (PFBBr) and GC/MS
3 – quantified using solid phase extraction and HPLC-ELISA
4 – quantified using GC/ECD
5 – quantified using solid phase extraction and GC/MS
6 – quantified using a Sievers 800 TOC analyzer
7 – quantified using HPLC-DAD and fluorescence
8 – quantified using HPLC-DAD

ANALYTICAL METHODS

Standard Methods

Conductivity

Conductivity is an estimate of the total dissolved solids content of water, and is measured in microSiemens per centimeter (μS/cm). Conductivity measurements were taken with a handheld Cole Parmer EC Meter (model 1481-61) (Standard Method 2510). Prior to use, the conductivity meter was calibrated with a 1,000 micromho per centimeter conductivity/TDS calibration solution obtained from RICCA Chemical Company.

pH

The pH was determined using a Beckman 260 portable pH meter with combination of a gel-filled electrode (Beckman, Fullerton, CA) (Standard Method 4500-H^+). Prior to use, the pH meter was calibrated with calibration solutions of pH 4, 7, 10 obtained from RICCA Chemical Company.

Total Organic Carbon

TOC is a measure of the concentration of total organic carbon in a water sample. For dissolved organic carbon (DOC), the sample was first filtered using a 0.45 μm filter to remove particulate organic carbon. Organic carbon concentrations were determined using a Sievers 800 Total Organic Carbon Analyzer (Ionics, Boulder, CO) according to Standard Method 5310 B. The Sievers 800 measurements are based on the oxidation of organic carbon to form carbon dioxide using ammonium persulfate and UV radiation. Carbon dioxide is measured using a selective membrane-based conductometric detection technique. The detection limit for TOC using the Sievers 800 was 0.06 mg/L. Fresh standard solutions were run with samples for each measurement.

UV Absorbance

UVA is a measure of the quantity of compounds present in a sample that absorb ultra-violet (UV) light at a wavelength of 254 nm. Compounds that absorb UV light at 254 nm are aromatic in structure. UVA measurements at CSM were conducted with a Beckman DU800 UV/Vis Scanning Spectrophotometer (Fullerton, CA) and at WBWRP with a HACH 4000 (Loveland, CO) according to Standard Method 5910 B. The absorbance was first zeroed using type I water created by processing in house deionized water through a Millipore Milli-Q™ water system. Cuvettes used are 1 cm and 4 cm, rectangular, quartz glass cuvettes. Once the absorbance was zeroed, the absorbance of each sample is read two times, and the average of the two readings is recorded. The specific UV absorbance (SUVA) was calculated as the ratio between UV absorbance and DOC.

Ammonia and Nitrate

Ammonia was measured using a Hach DR/2500 spectrophotometer at CSM (Hach method 8038) and using an ion selective probe (Standard Methods 4500-NH3) at WBWRP and SWC. Nitrate was measured with a Dionex DS600 Ion Chromatograph (IC) (Dionex, Sunnyvale, CA) using an AS14 column and a sodium hydroxide eluent according to Standard Method 4110C. Samples were 0.45 μm-filtered prior to analysis.

Other Inorganic ions

With the exception of ammonium, inorganic cations were measured with an Optima 3000 Inductive Coupled Plasma (ICP) Spectrometer (Perkin Elmer, Norwalk, CT) according to Standard Method 3120B. Prior to analysis, samples were pretreated by 0.45 μm membrane filtration and acidified to pH 2.

DBP Analysis

The analysis of chloroform and bromoform in this study followed EPA method 551.1. The analysis of trichloroacetic acid and dichloroacetic acid followed EPA method 552.2. Water samples were collected in 40 ml glass bottles with Teflon lines caps, with care being taken to completely fill the bottles and eliminate all air bubbles within the sample. EPA methods 551.1 and 552.2 use liquid-liquid extraction and analysis by gas chromatography electron capture detection (GC-ECD). DBP analysis was performed using a Hewlett Packard 6890 GC (Agilent, Palo Alto, CA) equipped with an ECD detector, a HP 7683 auto injector, and a DB-1 column. Method reporting limits for the DBP analytes employed in this study are presented in Appendix A, Table A.2.

THM Analysis

Bromoform and chloroform were quantified according to EPA method 551.1 for trihalomethanes. A 25 μL surrogate standard (30,000 μg/L 1,2,3-trichloropropane in acetone) was added in 30 mL sample. After addition of 8.0 gram of sodium chloride and 3.0 mL of methyl t-butyl ether (MtBE), the samples were mixed for about 30 seconds until most of the salts were dissolved. Then the samples stood for about 15 minutes for separation. Two milliliters of the top organic layer was transferred to a GC vial and analyzed using the GC-ECD.

HAA Analysis

Trichloroacetic acid and dichloroacetic acid were quantified according to EPA method 552.2. A 30 μL surrogate standard (25,000 μg/L 2-bromoacetic acid in methanol) was added in a 30 mL sample that had been adjusted to pH 0.5 using sulfuric acid. Once acidified, 13 g of sodium sulfate and 3 mL MtBE was added to the samples. The samples were then mixed using an orbital mixer for about 30 seconds and subsequently allowed to stand for about 15 minutes for separation. Exactly 2 mL of the top organic layer was transferred to a 2 mL volumetric flask with a Pasteur pipette and placed into the freezer for 30 minutes. Samples were then derivatized with diazomethane by transferring 250 μL of diazomethane into the 2 mL volumetric flask containing the sample, and mixing gently by inverting once. After placing samples in the refrigerator for 15 minutes to allow for esterification, the sample was allowed to reach room temperature for 15 minutes. Approximately 10 mg of silica gel was added to sample to stop the reaction. The organic layer of the sample was transferred to a GC vial and analyzed by using the GC-ECD.

NDMA Analysis

N-nitrosodimethylamine (NDMA) samples were analyzed according to EPA Method 1625M with a reporting limit of 2 ng/L by Weck Laboratories (City of Industry, CA). N-nitrosamine samples were shipped on ice via overnight courier to Weck Laboratories.

Methods for Unregulated Organics

Analytical Method for the Analysis of Pharmaceutical Residues, Pesticides and Personal Care Products

The analysis of the selected trace organics (Tier 2, Table 2.1), not including DBPs, and steroidal hormones, was performed at Colorado School of Mines (CSM) by adopting a method published by Reddersen and Heberer (2003). Methods employed for individual target compounds are indicated in Table 2.1. For the analysis of the target compounds, 1 L of each sample was collected and acidified to pH 2 using residue free hydrochloric acid. Two internal standards, 100 ng of 10,11-dihydrocarbamazepine and 100 ng of 2-(m-chlorophenoxy) propionic acid (100 μL of a 1 ng/μL solution in methanol), were spiked into each sample. A 10 mL volume of methanol was then added as a modifier for solid phase extraction (SPE). SPE was carried out by using 1 g of RP-C-18 material (Bakerbond Polar Plus, Mallinckrodt-Baker, Phillipsburg, NJ) filled in a 6 mL polyethylene cartridge. The cartridges were conditioned by applying 5 mL of acetone, 10 mL of methanol and 10 mL of type I water (adjusted to pH 2.0). After conditioning, a vacuum was applied to a PreSep 12-port manifold (Fisher Scientific Inc. Pittsburgh, PA) and the water samples were passed through the cartridges at a flow rate of 3 to 5 mL/min. The C-18 cartridges were then dried overnight with a gentle stream of medical grade nitrogen.

PFBBr method. The analytes were eluted from the cartridges three times with 1 mL of acetone directly into 2 mL auto-sampler vials (elution was stopped at an elution volume of approximately1.5 mL). Afterwards the eluate was dried again and the eluate dissolved in 100 μL of a pentafluorobenzyl bromide (PFBBr) solution (2 percent in toluene). A volume of 4 μL of triethylamine was added as a catalyst into the sample vial which was then placed in a drying cabinet for one hour at 100°C. The residue was dissolved again in toluene (100 μL), transferred into 200 μL glass inserts and analyzed by a HP 6890 gas chromatograph equipped with a HP 5973 quadrupole mass spectrometer from Agilent Technologies (Palo Alto, CA).

MTBSTFA method. The analytes were eluted from the cartridges three times with 1 mL of methanol directly into sampler vials (the process was stopped at an elution volume of approximately1.5 mL). Afterwards the eluate was dried again and subsequently dissolved in 50 μL of acetonitrile and 50 μL of N-(t-butyldimethylsilyl)-N-methyl-trifluoraacetamide (MTBSTFA). Sealed vials were placed into a drying cabinet for one hour at 80 °C. The remaining solution was transferred into 200 μL glass inserts and analyzed by a HP 5890 Series II gas chromotograph and a HP 5971 quadrupole mass spectrometer from Agilent Technologies (Palo Alto, CA).

Detection and Quantification Limit Establishment. The team conducted a verification of the limit of detection (LOD) and limit of quantification (LOQ) for all target pharmaceutical residues, pesticides and personal care products and both derivatization methods considering the two water matrices important for membrane selection and operation (i.e., type II water and secondary treated effluent). This sample matrix consisted of six different concentrations for every compound in type II water with six duplicates for each concentration. For secondary effluent, five concentrations for every compound with six duplicates were prepared. Method detection and quantification limits for individual Tier 2 analytes are presented in Table A.2 (Appendix A).

LOD values were computed by determining the average concentration for each compound over the replicate samples in which the signal to noise ratio was approximately 3:1. LOQ values were computed by determining the average concentration for each compound over the replicate samples in which the signal to noise ratio was approximately 11:1. Because LOD and LOQ determination experiments were performed on the 5971 MS system and a 5973 MS system was acquired in the middle of the study, LOD and LOQ values were lowered due to the increased sensitivity of the newer GC/MS system. In addition, since noise and target ion abundances can change from one GC/MS run to another, LOD and LOQ values determined during this study are an estimation of analytical sensitivity.

QA/QC Protocol. Due to method detection and quantification limits in the low nanogram per liter range, several QA/QC protocols were adopted in order to assure the validity of analytical results.

To assure that sample recoveries were sufficient through the various steps of the method, two surrogate standards were spiked into samples at 100 ng/L. The first surrogate standard, 2-(m-chlorophenoxy) propionic acid was employed to evaluate the recovery of all analytes with the exception of carbamazepine. The chemical 10,11-dihydrocarbamazepine was used to evaluate the recovery of carbamazepine. Samples which resulted in surrogate standard recoveries of less than 60 percent or greater than 120 percent were not considered for reporting.

To address the possibility of contamination, one blank sample was included in every GC/MS sample run. At least 30 blank samples were analyzed. Different results regarding contamination were observed for the two GC/MS systems employed during the study. For the HP 5971 GC/MS system, less than 10 percent of blank samples showed detectable or quantifiable concentrations of any target analytes employed during the study. Analytes present in blank samples included the flame retardant TCPP and the plasticizer bisphenol-A occurring in the 20 to 30 ng/L range. For the newer GC/MS system, the HP 5973, approximately 75 percent of blank samples had detectable or quantifiable concentrations of the flame retardants TCEP and TCPP at concentrations of approximately 15 ng/L and 30 ng/L, respectively. Analytes detected or quantified in permeate samples that were also detected or quantified in blank samples were not included for reporting. However, if an analyte was quantified in a sample greater than ten times the concentration observed in a blank sample, the analyte was included in reporting. Blanks were also used to eliminate false positives as a result of environmental contamination. Field blanks consisting of type I water samples were collected during sampling campaigns at the utility laboratories of SWC and WBWRP.

In addition to standards and blanks, target analyte spike samples were prepared by spiking the Tier 2 analytes at 100 ng/L into type II water. These samples were employed to further investigate target solute recovery levels in environmental samples and generally showed that the surrogate standards were sufficient in estimating the recovery of the target analytes.

Analytical Method for the Analysis of Steroid Hormones

The analysis of the selected steroid hormones (e.g., 17β-estradiol, estriol, testosterone) was conducted following a protocol published in Mansell and Drewes (2004). The analytes were quantified using high pressure liquid chromatography coupled with enzyme linked immunosorbent assays (HPLC-ELISA). Samples were processed through a C-18 solid phase extraction disk. Organic matter was extracted from the C-18 disks using methanol/water solutions. The extracts were dried, resuspended in acetonitrile/water solutions, and subjected to two HPLC fractionation steps using a size-exclusion and a reversed phase column, which were used to isolate the target compounds from background organic matter. Quantification of the steroids in the samples was performed using ELISA. 17β-estradiol, estriol, and testosterone ELISA kits were obtained from Cayman Chemicals (Ann Arbor, MI). The recoveries for all three target compounds through the extraction procedure and the HPLC clean-up were 95±3 percent. Recoveries ranged from 45 to 95 percent when isolated compounds where quantified using ELISA. Recoveries of all compounds were verified using a standard addition of 30 ng/L added to a split sample prior to solid phase extraction. Blank samples were processed along with each batch of field or laboratory samples. The detection limit of the analytical method for a 1-liter sample for 17β-estradiol, estriol, and testosterone was calculated according to Standard Methods by considering three times the standard deviation of the blank samples and were 0.4, 0.6, and 0.5 ng/L, respectively (Table A.2).

Analytical Method for the Analysis of Aliphatic Amines

The method for methylamine follows Kutlan et al. (2002) with some modification. Methylamine was quantified as *o*-phthaldialdehyde/*N*-acetyl-L-cysteine (OPA/NAC) derivatives. Methylamine was quantified by a Hewlett Packard Series 1050 HPLC (Wilmington, DE) equipped with a Hypersil ODS column (Supelco, Bellefonte, PA, 150×4.6 mm, particle size 5 µm). DAD and fluorescence (1046 Hewlett Packard) detection was performed simultaneously, with the detectors connected in series. Blank and samples were evaluated at 334 nm as well as at the excitation/emission wavelengths of 337/454 nm. Solvent A was 0.05 *M* sodium acetate of pH 7.2, while solvent B was prepared from 0.1 *M* sodium acetate-acetonitrile-methanol (46:44:10) (mixed in volume ratios and titrated with gracial acetate acid or 1 *M* sodium hydroxide to pH 7.2). The eluent system consisted of 45 percent solvent A and 55 percent solvent B. The LOD was determined to be 15 µg/L and the LOQ was estimated to be 30 µg/L for a membrane permeate matrix.

Analytical Method for the Analysis of Tier 4 compounds

The Tier 4 compounds used for membrane characterization were measured as TOC using a Sievers 800 Total Organic Carbon Analyzer (Ionics, Boulder, CO) according to Standard Method 5310 B.

Analytical Method for the Analysis of Tier 5 compounds

A Hewlett Packard Series 1100 HPLC (Wilmington, DE) equipped with a reverse phase column (Supelco Discovery C-18, particle size 5 µm, pore size 180Å, 25 cm × 4.6 mm) and an diode-array (DAD) detector (at 220.2 nm) was employed to quantify eight indicator compounds at low microgram per liter levels. The method detection limits are listed in Table 2.2.

Table 2.2

HPLC method for the detection of 8 target compounds in aqueous solution

Compound	Retention Time (min)	Limit of Detection (µg/L)
Primidone	2.4	5
Phenacetine	3.7	5
Carbamazepine	4.8	5
2-Naphthol	6.2	5
Naproxen	7.0	<5
Diclofenac	9.3	5
Ibuprofen	9.7	5
Naphthalene	10.0	5

Solvent A: 25 mMol KH_2PO_4 pH=2.5 (adjusted with Phosphoric Acid)
Solvent B: Acetonitrile
Wavelength: t=0 min 205 nm bw 10 nm; ref 330 nm bw 60 nm
t=6.7 min 230 nm bw 24 nm; ref 330 nm bw 60 nm
t=7.7 min 205 nm bw 10 nm; ref 330 nm bw 60 nm
Gradient: t=0 min 70% Solvent A; 30% Solvent B
t=12 min 20% Solvent A; 80% Solvent B
t=15 min 70% Solvent A; 30% Solvent B
Run Time: 15 minutes
Post Time: 7 minutes

QA/QC for Unregulated Organic Analysis

All Tier 2 methods employed analytical grade chemicals and pesticide grade solvents. The collection of samples for Tier 2 target solute analysis, with the exception of DBPs and hormone analyses, were preferred in at least triplicate analyses. Replicate DBP and hormone sample collection was preferred, however, sample loads and schedules sometimes created sampling limitations where only one sample could be collected per location.

SELECTION OF CANDIDATE MEMBRANES FOR PILOT-SCALE TESTING

Feed Water Qualities

Three unique wastewater effluents collected at three different wastewater treatment plants were employed during the laboratory-scale membrane selection. Samples were collected at the West Basin WRP, the Scottsdale Water Campus (SWC), and the Denver Metro Wastewater Reclamation District's (Metro) plant. Key water quality constituents of these different water types are summarized in Table 2.3. Further information about the full-scale treatment processes employed at WBWRP and the Scottsdale Water Campus are provided in a subsequent section entitled "Full-Scale Membrane Testing and Monitoring" (p. 34). The South treatment train of

Metro's Plant 3 has an annual average operational capacity of 5.25 m^3/s (120 mgd) and consists of grit removal, primary clarification, aerobic (non-nitrified) conventional activated sludge treatment, secondary clarification, and chlorine disinfection. Hydraulic residence times for primary clarification range from 2.4 to 3.9 hours. Mixed liquor aeration (hydraulic residence) times average 2.1 hours, while solids retention times average 1.7 days. Sulfur dioxide is used after chlorine contact to dechlorinate treated wastewater prior to discharge. Solids are treated on-site through a combination of anaerobic digestion and centrifugation with subsequent land application for beneficial reuse (95 percent) or composting for home use (5 percent).

Table 2.3
West Basin WRP, Scottsdale Water Campus and Denver Metro Average Feed Water Qualities (after microfiltration)

Constituent	West Basin WRP*	Scottsdale WC*	Denver Metro
Lab pH	6.4	6.3	7.7
TDS (mg/L)	663	1000	520
Conductivity (μS/cm)	1318	1700	1300
Alkalinity (mg/L as $CaCO_3$)	169	98	80
Sulfate (mg/L)	211	446	87
Chloride (mg/L)	153	333	81
Nitrate (mg/L as N)	<0.1	6-11	<0.5
Nitrite (mg/L as N)	<0.1	NA	NA
Calcium (mg/L)	40	75	35
Magnesium (mg/L)	21	30	8
Hardness (mg/L as $CaCO_3$)	188	313	180
Sodium (mg/L)	137	210	93
Potassium (mg/L)	15	30	11
Aluminum	0.01	0.03	0.02
Barium	0.001	0.06	0.02
Iron	0.11	0.05	0.02
Manganese	0.06	0.05	0.02
Strontium	0.3	0.96	0.3
Zinc	0.02	0.02	1.1
Boron	0.72	0.43	0.2
Phosphorous	7	2	2
Silica (mg/L)	24	13	7
Ammonia (mg/L as N)	20-40	<3	19
Total nitrogen (mg/L)	20-40	6-14	20
TOC (mg/L)	8-13	5-8	9-11

Note: * Acidified using H_2SO_4 solution; NA: Not Available

Testing Protocol

In order to select candidate membranes for pilot-scale testing at the WBWRP and the SWC, a bench- and laboratory-scale membrane testing protocol utilizing flat-sheet and 4040 spiral wound membrane specimens was developed at CSM and implemented with commercially available NF and ULPRO membranes. Parameters relevant to indirect potable reuse projects were chosen as criteria for the evaluation and identification of membranes that could be used to efficiently treat wastewater effluents. The testing protocol evaluated membranes based on the rejection of relevant solutes and constituents present in wastewater effluents (e.g., TOC, nutrients such as ammonia and nitrate, conductivity, and selected trace organics), as well as operational performance (e.g., specific flux and flux decline due to fouling). A total of 15 different commercially available NF and ULRPO products were considered for this study. Only eight ULPRO and NF membranes, however, were evaluated according to the testing protocol and the performance compared with a benchmark RO membrane commonly employed at several full-scale water reuse facilities.

The testing protocol consists of three elements: 1. flat-sheet tests to determine pure water permeability (PWP) and salt rejection, 2. an assessment of fouling, and 3. rejection of bulk organics, nutrients, and selected organic target compounds using a 2-stage membrane test unit employing 4040 spiral wound elements. The 2-stage tests for every membrane were conducted using two operational modes (i.e., flow-through and internal recycle) simulating different recoveries (20 to 30 and 70 to 80 percent) and two different water matrices (i.e., type II water and secondary effluent) spiked with target compounds. Fouling experiments were performed initially utilizing secondary effluent from WBWRP, but due to difficulties transferring wastewater samples from California to Colorado were mostly conducted utilizing secondary effluent (not nitrified) from the local Denver Metro Reclamation Plant in Denver, Colorado. The characteristic of Denver Metro feed water has been compared to secondary effluent from WBWRP and was found to be very similar (Table 2.3). In order to investigate differences in fouling propensities of certain membranes as a function of feed water quality, fouling experiments were also performed employing tertiary treated effluent from the SWC. Two-stage membrane experiments were conducted using secondary effluent (not nitrified) from the Denver Metro Reclamation Plant in Denver, Colorado after microfiltration using a laboratory-scale MF skid (0.2 μm).

Relevant Equations

The feed water recovery is the percentage of feed water that becomes permeate and is given by

$$\text{Recovery (\%)} = \frac{Q_p}{Q_f} \times 100 \qquad (2.1)$$

Where Q_p is the permeate flow and Q_f is the feed flow. The removal of a component of a feed solution is often expressed as rejection which is given as

$$\text{Rejection (\%)} = 1 - \frac{C_p}{C_f} \times 100 \qquad (2.2)$$

Where C_p is the permeate concentration and C_f is the feed concentration. Another form commonly used in expressing the rejection of a feed water component is the feed/brine average rejection value which is used to compare systems that are operated at different operating conditions and is given by

$$\text{Rejection (Feed/Brine Concentrations)} = \frac{C_f * \left(\frac{\ln\left(\frac{1}{1 - \text{Recovery}}\right)}{\text{Recovery}} \right) - C_p}{C_f * \left(\frac{\ln\left(\frac{1}{1 - \text{Recovery}}\right)}{\text{Recovery}} \right)} \qquad (2.3)$$

This value takes into account the increased concentrations of a component at the surface of a downstream membrane in a system of a certain recovery, and offers a more useful comparison for two systems operated at different recoveries.

Several equations are used to calculate the productivity of a membrane in terms of permeate production. The temperature-corrected specific flux (TCSF) is defined by Equation 2.4.

$$\text{Temperature Corrected Specific Flux } (TCSF) = \frac{\text{Actual Flux} \times \text{TCF}}{NDP} \qquad (2.4)$$

Where Actual Flux = observed permeate flow divided by membrane area (L/m^2)
TCF = temperature correction factor
NDP = the net driving pressure.

The TCF is calculated by using Equation 2.5.

$$\text{TCF} = e^x \qquad (2.5)$$

The variable 'x' in Equation 2.5 is calculated by using Equation 2.6.

$$x = U \times \left(\frac{1}{T_a + 273} - \frac{1}{T_s + 273} \right) \qquad (2.6)$$

where T_a = actual temperature
T_s = the standard temperature
U = temperature-corrected flux coefficient that is membrane specific (2,300 for most thin film polyamide membranes).

The net driving pressure is the portion of the applied pressure which actually drives water through the membrane and is calculated using equation 2.7.

$$\text{Net Driving Pressure} = P_F - \frac{\Delta P}{2} - \Delta\Pi - P_P \tag{2.7}$$

where, P_F = feed pressure

ΔP = differential pressure

$\Delta\Pi$ = average osmotic pressure differential

P_P = permeate pressure

Target Membranes

Membrane products from seven different vendors were initially considered as potential candidates for pilot-scale testing. Initially, fifteen different NF, ULPRO, and RO membrane products, each in two configurations as flat sheet specimen and 4040 spiral wound elements, were obtained and considered for testing according to the membrane testing protocol. A list of the membrane products obtained and initially considered for this study is presented in Table 2.4. The ESPA2 membrane, a RO membrane with similar performance than the TFC-HR, was not considered for further study, since the TFC-HR membrane was selected as the "benchmark RO" membrane in this study. The Osmonics membrane MX07 was not considered for further testing due to withdraw of support from Osmonics to provide membrane samples for the study. A list of membranes employed in this study with relevant manufacturer specifications is provided in Table 2.5.

Table 2.4
Target membranes initially selected for study

Membrane	Vendor	Type	Flat-sheet	Spiral-wound elements
TFC-HR	Koch	RO	√	√
TFC-ULP	Koch	ULPRO	√	√
TFC-S	Koch	NF	√	√
CTA	Koch	RO	√	-
XLE	Filmtec	ULPRO	√	√
NF-90	Filmtec	NF	√	√
NF-200	Filmtec	NF	√	√
NF-270	Filmtec	NF	√	√
NF-4040	Filmtec	NF	√	√
TMG10	Toray America	ULPRO	√	√
NE-90	Saehan	NF	√	√
RE-BLR	Saehan	RO	√	-
ESPA2	Hydranautics	RO	√	-
ESNA1-LF	Hydranautics	NF	√	√
MX07	Osmonics	NF	√	-

Table 2.5
General, manufacturer and project specifications of 4040/8040 membranes employed during membrane experiments

General Specifications			Manufacturer Specifications					Project Specifications
Membrane	Type	Manufacturer	Salt Rejection (catalog)	pH Range	Pure Water Flux (L/d)	4040/8040 Element Area (m^2)	Material	4040/8040 Element Employed on
TFC-HR	Brackish Water RO	Koch Membranes	99.5% NaCl[1]	4-11	7950	7.2	Polyamide	Lab-scale
TFC-HR-400	Brackish Water RO	Koch Membranes	99.5% NaCl[1]	4-11	39750	37	Polyamide	Full-scale
TFC-ULP	Low Pressure RO	Koch Membranes	98.5% NaCl[1]	4-11	6625	7.2	Polyamide	Lab-scale
XLE	Low Pressure RO	Dow/Filmtec	99% NaCl[2]	2-11	9085	7.6	Polyamide	Lab-scale
TMG10	Low Pressure RO	Toray America	99.5% NaCl[2]	2-11	9085	8.1	Polyamide	Lab- and pilot- scale
TMG20-430	Low Pressure RO	Toray America	99.5% NaCl[2]	2-11	41640	40	Polyamide	Full-scale
NF-90	Nanofiltration	Dow/Filmtec	85-95% NaCl[3], 97.7% $MgSO_4$[4]	2-11	7000	7.6	Polyamide	Lab- and pilot-scale
NF-200	Nanofiltration	Dow/Filmtec	45-65% $CaCl_2$[5], 97% $MgSO_4$[4]	3-10	5110	7.6	Polyamide	Lab-scale
NF-4040	Nanofiltration	Dow/Filmtec	99% $MgSO_4$[4]	2-11	7380-11350	7.6	Polypiperazine	Lab- and pilot-scale
NF-270	Nanofiltration	Dow/Filmtec	40-60% $CaCl_2$[5], 97% $MgSO_4$[4]	2-11	11000	7.6	Polyamide	Lab-scale
TFC-S	Nanofiltration	Koch Membranes	85% NaCl[1], 98.5% Hardness Total[1]	2-11	7570	7.2	Polyamide	Lab-scale
ESNA1-LF	Nanofiltration	Hydranautics	80% NaCl[2]	3-10	6435	7.9	Polyamide	Lab-scale
ESPA2	RO	Hydranautics	99.5% NaCl[7]	3-10	34100	37	Polyamide	Full-scale

Test Conditions: [1]700 ppm (>45% monovalent), [2]500 ppm NaCl, [3]2000 ppm NaCl, [4]2000 $MgSO_4$, [5]500 ppm CaCl2, [6]1,000 ppm $MgSO_4$, [7]1,500 ppm NaCl

Bench-scale Membrane Fouling Assessment

Membrane fouling propensity was evaluated by employing a common experimental set-up for studying fouling mechanisms and fouling characteristics in controlled laboratory studies (Zhu et al. 1997, Cho et al. 2000, DiGiano et al. 2000, Balannec et al. 2002, Seidel and Elimelech 2002, Lee et al. 2004, Xu et al. 2006a). Flat-sheet membrane specimens were fouled in parallel using two cross-flow flat-sheet membrane units (Sepa II, Osmonics) consisting of rectangular plate-and-frame cells and having a membrane surface area of 139 cm^2 and a cross-sectional flow area of 0.90 cm^2. Membrane specimens were pre-compacted for a period of 24 hours using deionized water. For each membrane, feed water was prepared by collecting approximately 50 L of microfiltered wastewater effluents from the WBWRP, SWC, or the Denver Metro Wastewater facilities, and pH adjustment to 6 to 6.3 with sulfuric acid and sodium hydroxide solutions. Applied feed pressure during fouling experiments was kept at 60 psi, and the feed water flow rate for each membrane unit was kept at 1,000 mL/min equaling a superficial crossflow velocity of approximately 0.19 m/s. The experiments were operated in recycling mode in which concentrate and permeate were circulated into the feed water tank. Feed water temperature was kept at 22.5±1.5°C by a stainless steel water cooling loop immersed in the feed solution. The duration of fouling experiment lasted nine days (220 hours) for all membranes to ensure that the membranes reached a relatively stable extent of fouling before experiment was terminated. A schematic of the bench-scale membrane testing unit is presented in Figure 2.1.

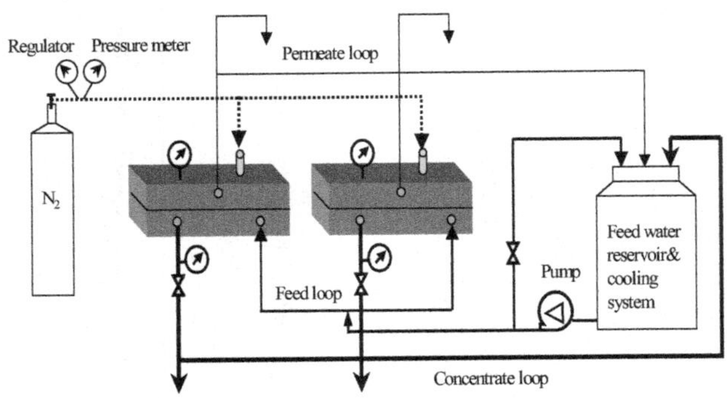

Figure 2.1 Schematic of the flat sheet membrane unit (CSM)

The extent of membrane fouling is described by permeate flux decline and delivered DOC. Permeate flux decline is defined as the percentage of reduced permeate flux compared to initial permeate flux, and is calculated by using Equation 2.8.

$$\text{Permeate flux decline (\%)} = (1-J/J_o) \times 100 \qquad (2.8)$$

where J_o = initial permeate flux taken at filtration time of 30 minutes
J = permeate flux at filtration time t

Although the fouling experiments did not simulate actual hydrodynamic or pressure conditions (0.19 m/s and 60 psi) typical for pilot- or full-scale systems, the membranes were tested under the same conditions, and it is assumed that the measured flux decline is relative, allowing the use of this data for comparison of the candidate membranes.

The fouling experiments were initially conducted using microfiltered non-nitrified effluent shipped from the WBWRP to CSM. Eight candidate membrane specimens were assessed including two RO membranes TFC-HR (Koch) and RE-BLR (Saehan), two ultra low-pressure RO membranes XLE (Dow/Filmtec) and TFC-ULP (Koch), and four NF membranes TFC-S (Koch), NF-90 (Dow/Filmtec), NE-90 (Saehan) and ESNA (Hydranautics). Because it was costly to ship water from the WBWRP, a local effluent source was utilized (Denver Metro non-nitrified secondary effluent) to substitute the WBWRP effluent for fouling experiments. Comprehensive water quality analyses showed that the two non-nitrified waters were comparable having similar water quality parameters (Table 2.3). Since water quality parameters are gross and may not be sensitive enough to detect subtle differences that might have a profound impact on membrane performance, we conducted control experiments to compare the fouling potential of NF-90 using the WBWRP effluent and the Denver Metro effluent. The NF-90 membranes exhibited similar permeate flux decline using the WBWRP effluent and the Denver Metro effluent (Figure 2.2). We therefore considered the data obtained from fouling experiments using the Denver Metro microfiltered non-nitrified effluent to be comparable with the data derived from the WBWRP effluent.

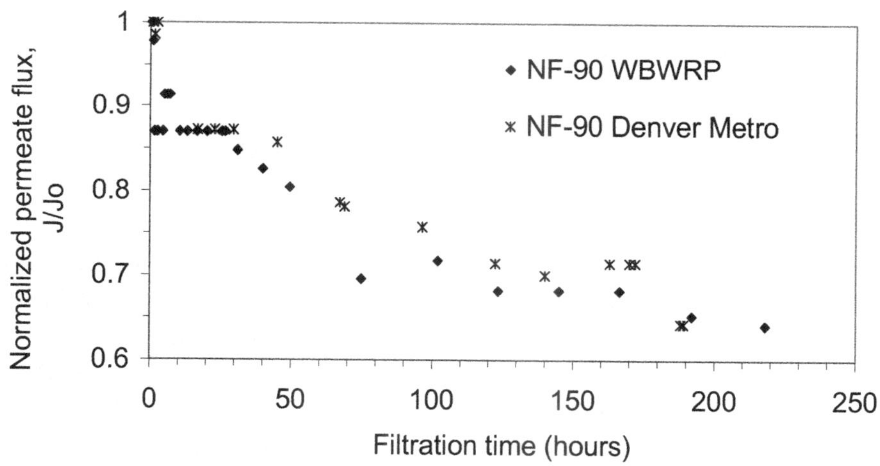

Figure 2.2 Control experiment to compare flux decline of NF-90 membrane using microfiltered non-nitrified wastewater effluents from the WBWRP and the Denver Metro Reclamation Facility

2-stage Membrane Laboratory-scale Membrane Testing Unit

A 35 L/min (9.2 gpm) two-stage membrane laboratory-scale unit was employed for testing all spiral wound membrane specimens. The membrane unit employed two single element (4040 spiral wound) vessels arranged in a two-stage array. A LabView (National Instruments) customized SCADA system was used to acquire data for the following parameters: flow (feed, permeate and concentrate), conductivity (permeate and concentrate), pressure (feed and concentrate), and temperature (feed). Data collected by the SCADA system was used to compare operational performance among candidate membranes. A baffled stainless steel feed tank (200 L) was used to supply the feed water to the high-pressure pump. Before membrane performance testing, spiral wound membrane specimens were tested in accordance to manufacturer specification sheets to validate element integrity. Figure 2.3 presents a flow schematic of the two-stage membrane testing unit.

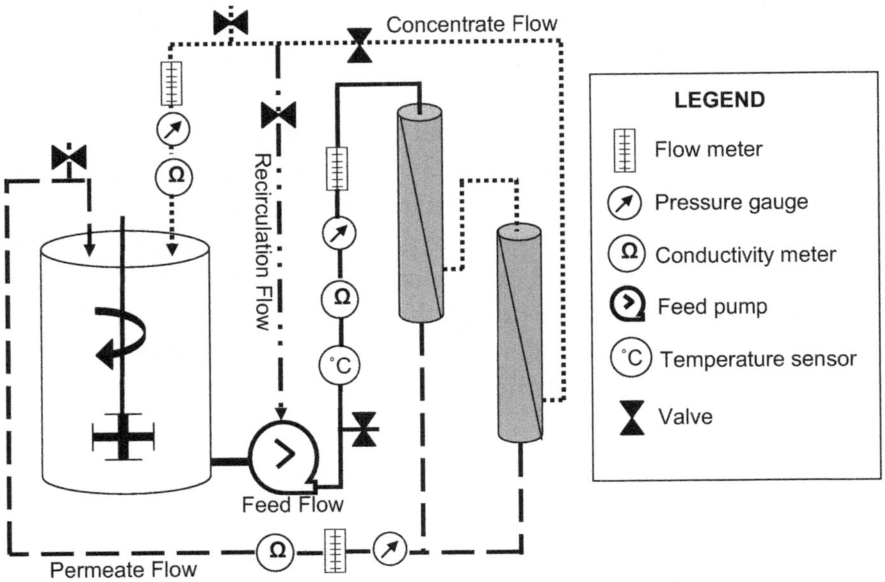

Figure 2.3 Schematic of the 2-stage membrane testing unit (CSM)

Membrane performance experiments conducted with candidate membranes on the 2-stage testing unit were carried out with 2 feed water matrices, deionized water and secondary effluent. For deionized feed water matrices, NaCl and $CaSO_4$ were added to achieve approximately 200 mg/L of hardness as $CaCO_3$ and a conductivity of 1,400 µS/cm. For all 2-stage membrane experiments a feed water pH of 6.1 to 6.3 was maintained using concentrated hydrochloric acid. Secondary effluent used for feed water was 0.2 µm microfiltered prior to membrane experiments. During all 2-stage membrane experiments, a vertical mixer and a tank recycle pump was used to insure proper mixing. During operation, combined permeate and concentrate flows from the membrane unit were recycled to the stainless steel tank. A stainless steel cooling loop was used to maintain a constant feed water temperature (23±1.5°C) during membrane experiments.

Membrane performance was evaluated separately in two flow regimes termed 'flow through' and 'internal recycle'. The flow through regime was designed to mimic the first stage of a full-scale membrane treatment unit by maintaining a permeate flux of 30.6 to 37.4 L/m^2-d (18 to 22 gfd) and a system recovery (equation 2.1) of 19 to 21 percent. The internal recycle regime utilized an internal concentrate recycle loop to simulate higher system recoveries (75 to 85 percent) and solute concentrations while maintaining lower permeate flux 20.4 to 23.8 L/m^2-d (12 to 14 gal/ft^2•day) found in the second stage of a full-scale membrane treatment plant. High system recoveries were simulated by returning a large portion of the second stage concentrate flow to the feed pump, thereby reducing the flow from the feed container.

Initial Membrane Integrity Verification

Each candidate membrane was subjected to two operational performance experiments with two different matrices using the two-stage membrane laboratory scale unit. Immediately following the installation of virgin elements onto the two-stage membrane unit, verification experiments were conducted in order to verify that the performance of the membrane specimens used during testing matched manufacturer performance data sheets. During these experiments, feed water was prepared to replicate the feed water used during manufacturer tests.

Permeate flux and conductivity rejection data obtained during operational performance verification experiments and performance data reported by the membrane manufacturer can be found in Table 2.6. Specific flux and conductivity rejection values were obtained during two-stage flow through regime experiments, using type II water with NaCl and $CaSO_4$ concentrations that replicated manufacturer experiments, and were computed with the permeate flux from the first stage at 13 to 15 percent recovery. Manufacturer reported specific flux values were found to be significantly higher than specific flux values observed during two-stage experiments (Table 2.6). It was determined that manufacturer tests achieved higher permeate flux at 15 percent recovery due to significantly higher feed flow rates (40 to 44 L/min) and feed water temperature (25°C) than were used during two-stage experiments (35 L/min and 23.9°C, respectively). During two-stage experiments, recovery values were kept below 15 percent for the first element (approximately 13 percent), which would lead to a lower specific flux value. In addition, individual element flow rates can vary +/-25 percent according to manufacturer specification sheets. Conductivity rejection observed during two-stage experiments was similar to manufacturer reported conductivity rejection and the membrane specimens provided by the manufacturer were deemed of high enough quality to proceed with the selection process.

Table 2.6
Comparison of manufacturer reported and two-stage specific flux and conductivity rejection

Membrane	Manufacturers Specific Flux (gfd/psi) [2]	CSM Two-Stage Specific Flux (gfd/psi) [1]	Manufacturers Reported Percent Salt Rejection [2]	CSM Two Stage Conductivity Percent Rejection [1]
TFC-HR	0.146	0.123	99.4	99.6
TFC-ULP	0.265	0.2	99	99
XLE	0.293	0.26	99	98.1
TMG10	0.197	0.22	99.4	99.3
TFC-S	0.29	0.24	80	92.0
ESNA1-LF	NA	0.15	77.5 – 90.5	92.6
NF-270	0.51	0.49	40-60	42.2
NF-200	0.26	0.22	50-65	56.4
NF-90	0.283	0.3	85-95	94

1. Salt rejection tests for TFC-HR, TFC-S, NF-90, NF-270 and NF-200 spiral wound membranes conducted in a two-stage configuration with 700 mg/L solution consisting of 40 percent $CaSO_4$ and 60% NaCl at pH of 6, with 15 percent recovery. Salt rejection tests for ESNA1-LF spiral wound membranes conducted in a two-stage configuration with 500 ppm NaCl at pH of 6, with 15 percent recovery.
2. Test conditions for salt-rejection reported by manufacturers:
   TFC-HR: 700 mg/L TDS (45 % monovalent), at 200 psi, 15 % recovery, 77°F and pH 7
   TFC-S: 700 mg/L TDS (45 % monovalent), at 110 psi, 15 % recovery, 77°F and pH 7
   NF-90 and TMG10: 500 ppm NaCl, at 100 psi, 15 % recovery, 77°F

NF-200:	500 ppm $CaCl_2$ at 70 psi, 15% recovery, 77°F
NF-270:	500 ppm $CaCl_2$ at 70 psi, 15% recovery, 77°F
ESNA1-LF:	500 ppm NaCl at 75 psi, 15% recovery, 77°F

Performance Parameters: Flux, Fouling and Adjusted Specific Flux

During 2-stage membrane experiments, permeate flow, feed and concentrate pressure, and feed water temperature data were collected in order to calculate permeate flux (L/m^2-hr (gal/ft^2-d)) and specific flux (L/m^2-hr-kPa (gal/ft^2-d-psi)) for a specific membrane. These parameters were measured at the end of the flow-through two-stage unit experiments using non-nitrified microfiltered secondary effluent as feed water.

For a particular membrane, flux decline measured during flat sheet fouling experiments was used to correct the temperature-corrected specific flux value (equation 2.4) obtained during two-stage experiments in order to characterize the fouling potential of a specific membrane. The adjusted specific flux for a particular membrane is calculated with equation 2.9.

$$\text{Adjusted TCSF} = (\text{TCSF}) \times (1 - J/J_o) \qquad (2.9)$$

where J/J_o = the flux decline that was measured during fouling experiments.

The adjusted specific flux is defined as the difference between the specific flux and the flux lost due to fouling (equation 2.8) and was used to assess the operational performance of a specific membrane.

Nitrogen and TOC

Membranes were evaluated for rejection of inorganic nitrogen species, relevant to differing wastewater operational conditions for nitrogen conversion and removal. The rejection of both ammonia and nitrate was used to determine a membrane's potential for meeting federal drinking water quality requirements, as well as defining membrane specificity for rejection of nitrogen species in nitrified or non-nitrified effluent. Secondary effluent feed water was spiked with ammonia and nitrate to achieve concentrations of 40 mg/L of each constituent as nitrogen. Experiments were run in both flow through and internal recycle flow regimes for approximately three hours after which samples were taken from the feed, concentrate and permeate lines in duplicate and analyzed for nitrate, ammonia and TOC. The percent rejection for a specific membrane was then calculated.

Nitrogen rejection. Membranes were evaluated for rejection of nitrogen species with different wastewater operational conditions for nitrogen conversion and removal. Membrane rejection for both ammonia and nitrate was used to determine a membranes potential for meeting federal drinking water quality requirements, as well as defining membrane specificity for rejection of nitrified or non-nitrified waters. In this study, a rejection of 80 percent was considered to be acceptable for indirect potable use application, given that a nitrogen concentration of 40 mg/L-N would be reduced to 8 mg/L-N prior to surface spreading or direct injection. Membrane selectivity for the rejection of ammonia or nitrate was considered as a precaution for the use of a membrane with differing feed water qualities.

TOC. The California Department of Public Health (CDPH) has selected total organic carbon (TOC) as a parameter to be monitored in groundwater recharge projects using reclaimed water in order to address and limit potential human health risks associated with unidentified and unregulated organics present in reclaimed water (CDPH 2007). In this study, TOC rejection experiments were used to specify a membrane's potential for meeting the CDPH Draft Groundwater Recharge regulation. TOC rejection, however, was a selection criteria where membranes that meet the CDPH Draft regulations are preferred over membranes that could not, however, was not grounds for disqualification.

Trace Organic Rejection. Two-stage membrane experiments were conducted in order to assess the removal of selected trace organics by candidate membranes. These experiments were performed with ionic strength adjusted deionized feed water matrix in order to reduce matrix interference during analysis. Experiments were run in both flow through and internal recycle flow regimes. Before two-stage membrane rejection experiments were started, selected trace organic solutes were spiked to the feed water at a nominal concentration of 400 to 600 ng/L. Permeate and feed samples were taken in triplicate after one hour and analyzed by GC-MS to determine rejection performance.

Candidate Membrane Selection

Operational performance was assessed by comparing flux decline and "adjusted specific flux values" for the membranes tested. In terms of ranking membrane performance, the adjusted TCSF value was given the most weight when selecting a membrane as long as it achieved the rejection performance parameters specified above. For example, a highly permeable ultrafiltration membrane with low fouling propensity would have a high adjusted TCSF value, but would most likely fail all rejection performance parameters and would not be chosen for pilot-scale testing.

TOC and ammonium rejection was assessed using the laboratory-scale experimental data from 15 and 80 percent recovery experiments. For the purpose of this study, it was necessary to select a membrane that had low permeate concentrations of TOC and ammonium, explicitly equal to or less than 0.5 mg-C/L and equal to or less than 10 mg-N/L for both 15 and 80 percent recovery experiments. However, as specified above, membranes that achieved the TOC permeate requirement would be preferred over membranes that did not, but would not be grounds for disqualification. Trace organic rejection was assessed by summing the number of individual compound detections in membrane permeates (more than two samples) from both 15 and 80 percent recovery laboratory-scale membrane experiments.

Membranes having adjusted specific flux values similar to or less than the adjusted specific flux value for the TFC-HR (0.029 L/m^2-hr-kPa (0.116 gfd/psi)) were not considered for pilot-scale testing. Membranes displaying TOC permeate concentrations less than 0.5 mg-C/L, and/or ammonia or nitrate permeate concentrations less than 10 mg-N/L as nitrogen during rejection experiments were preferred over candidate membranes that did not achieve these criteria. Selection weight was given to membranes that efficiently rejected trace organic contaminants by summing the permeate detections and quantifications and calculating the percentage of compounds present in the feed water that were detected in three permeate samples. Ranking was performed using whole number integers, with the best performing membrane receiving a value of 1, the second receiving a value of 2, and continuing with the next whole number integer in series for the remaining membranes. Membranes that performed similarly to one another were given the same ranking, with the next best membrane receiving the next available ranking.

The membrane selection protocol was developed in order to discern one membrane from another and allow for the selection of membranes for pilot-scale testing. Because the number of membranes for pilot-scale testing was limited to two or three, distinguishing the performance of one membrane from another became an important component of the study and the selection protocol became an important tool for the final selection of candidate membranes for pilot-scale testing. It should be noted, however, that most membranes performed very well during testing and could be considered viable as candidate membranes for pilot-scale testing with wastewater effluent as feed water.

PILOT-SCALE TESTING OF CANDIDATE MEMBRANES

Pilot-scale Membrane Testing Unit

Design of Pilot-scale Membrane Testing Unit

CSM, with assistance from Separation Processes, Inc., designed a mobile 4-stage pilot-scale membrane skid using 4040 spiral wound NF/ULPRO elements capable of simulating the hydraulics and recovery of a full-scale two-stage RO train. In order to meet this goal, a 2:2:1:1 pressure vessel array with each pressure vessel holding 4, 3, 4, and 3 elements, respectively was utilized. This design has been proven to simulate a two-stage full-scale train employing seven 8040 elements per vessel. The pilot skid was constructed by Applied Membranes, Inc., CA and delivered to WBWRP for the initial membrane testing. The pilot is equipped with a customized SCADA system to monitor and log flux, pressure, and selected water quality parameter online (e.g., pH, temperature, conductivity). The pilot skid also has a standalone clean-in-place (CIP) system. Figure 2.4 is a schematic of the pilot-scale membrane testing unit, and Figure 2.5 is a photograph of the 2-stage membrane testing unit employed at the SWC.

Operation of pilot-skid

A summary of pilot-skid operation is presented in Table 2.7 that provides the membrane product tested, type of feed water and pretreatment, length of operation and membrane cleanings (CIP) performed during operation. The pilot-skid was initially installed at the WBWRP, where two membrane products were tested on microfiltered non-nitrified secondary effluent. After testing at the WBWRP was completed, the pilot-skid was transferred to and installed at the SWC for testing two membrane products on microfiltered nitrified/denitrified tertiary effluent.

Sampling and Monitoring of Pilot-scale Testing Unit

Pilot-skid Sampling. Sampling campaigns were performed for the monitoring of Tier 1, 2 and 3 analyte rejection performance by candidate membranes tested on pilot-scale. Samples for Tier 1 analyte analysis were collected weekly from the feed, permeate and concentrate streams of the pilot-skid. Samples for Tier 2 analyte analysis were collected monthly from the feed, permeate and concentrate streams. Samples for Tier 3 analyte analysis were collected one to two times per candidate membrane operation from feed and permeate streams.

Pilot-skid Monitoring. The pilot-skid SCADA system was programmed to download online measurement signals into a database every five minutes in order to monitor system performance during operation. As a backup to the online measurements, manual readings were taken daily by United Water staff at the WBWRP and by City of Scottsdale staff at the SWC.

Table 2.7
Summarized pilot-skid operation

Product Tested	Feed Water	pH Adjustment	Anti-scalant Addition	Duration of Operation (hours)	CIP (after hours of operation)
TMG10	Non-nitrified microfiltered secondary effluent	Yes	Yes	2,965	No
NF-90	Non-nitrified microfiltered secondary effluent	Yes	Yes	1,300	No
TMG10	Nitrified/denitrified microfiltered tertiary effluent	Yes	Yes	2,025	Yes (2,000)
NF-4040	Nitrified/denitrified microfiltered tertiary effluent	Yes	Yes	1,670	No

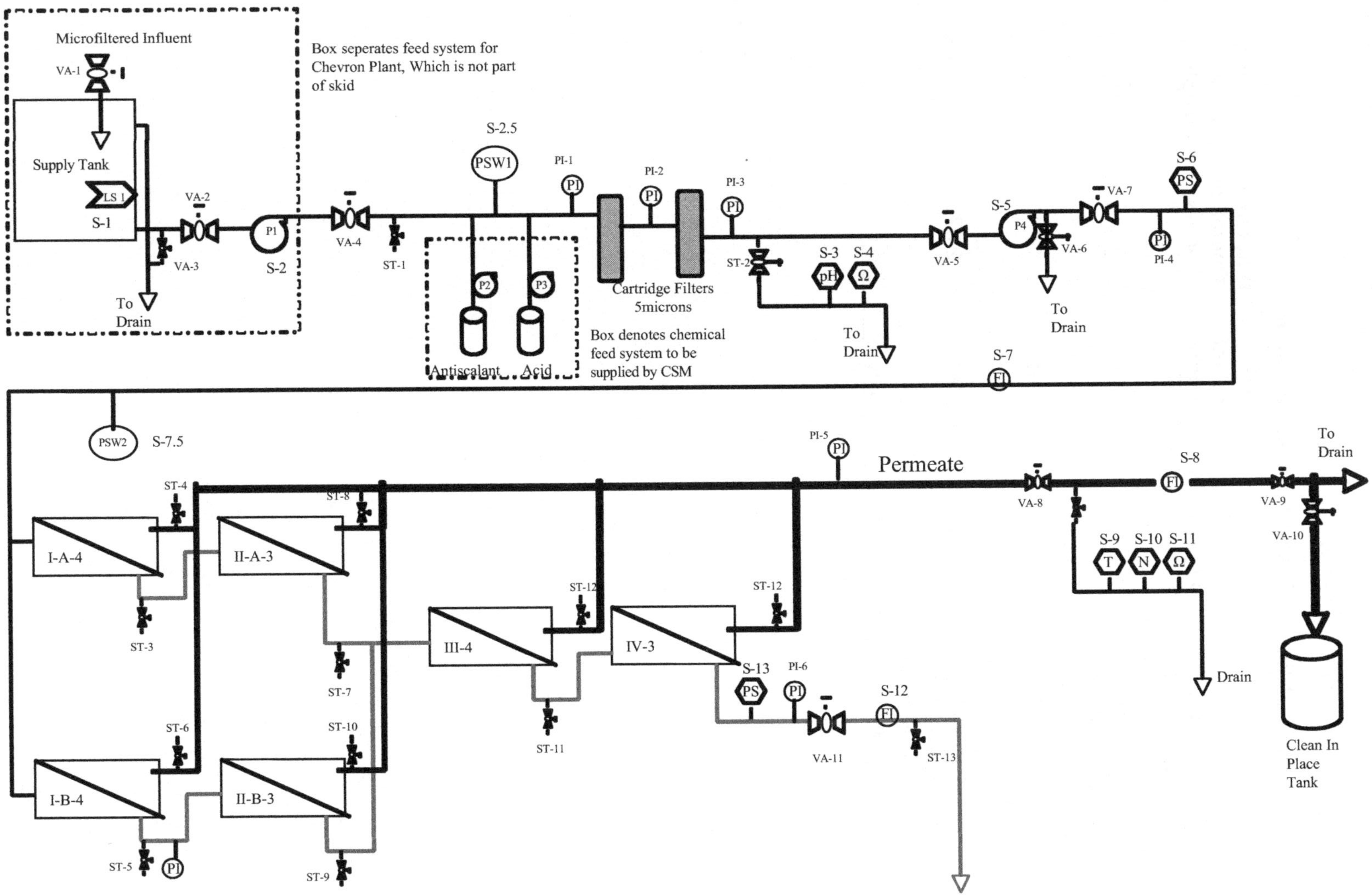

System array is 2:2:1:1, with a 4:3:4:3 element configuration.

Figure 2.4. Schematic of the pilot-scale membrane test unit

Figure 2.5 Pilot-scale membrane unit operational at the Scottsdale Water Campus

Challenge Tests of Pilot-scale Testing Unit

Preparation of Spiking Solution. A spiking solution was created using standards prepared for Tier 2 analysis including the pharmaceutical residues, personal care products, flame retardants and hormones (Appendix A, Table A.1). Individual compound stock solutions (approximately 90 mL of 1 mg/mL concentration solutions) were added to a single 125 mL bottle and dried down under a gentle stream of nitrogen. Once all desired components of the mixture were added to the bottle, the volume of the solution was adjusted to approximately 90 mL with methanol to achieve a stock solution concentration of approximately 1 mg/mL. Due to differences in initial stock solution volumes, the final concentration varied between individual compounds by approximately 10 percent.

Challenge Test Method. A pilot-skid spike feed solution was prepared by adding approximately 45 mL of the spike solution to 15 L of deionized water in a covered 20 L plastic bucket. A LMI pump (Milton Roy, Ivyland, PA) was used to deliver the spike solution into the feed line of the pilot-skid at approximately 1 L/hr in order to achieve a concentration of approximately 600 to 700 ng/L in the feed of the pilot-skid. This level represents 200 to 300 percent of environmental concentrations observed in the microfiltered feed water. Spiking was performed for 1.5 hours prior to sample collection for analysis. Samples were collected from samples taps on the feed, permeate 1, permeate 2, combined permeate and concentrate lines. Eight samples for Tier 2 analysis were collected (3 samples for each GC/MS method employed at CSM, 1 sample for hormones and 1 back-up sample) from each of the five sampling locations. Single samples were collected for all other trace organic analyses.

FULL-SCALE MEMBRANE TESTING AND MONITORING

West Basin Water Recycling Plant

The West Basin Water Recycling Plant receives secondary effluent from Los Angeles' Hyperion Waste Water Plant, pre-treats this water with microfiltration membranes and then utilizes reverse osmosis membranes to desalinate the water. Prior to entering the MF system, the secondary effluent from Hyperion is dosed with chlorine to form chloramines with excess ammonia already in the water. The MF system employs US Filter/Memcor CMF 90M10C units.

Sulfuric acid and antiscalant are added to the MF filtrate prior to being fed to the reverse osmosis system. The RO feed is pretreated with sulfuric acid to reduce the feed pH and prevent precipitation of calcium carbonate. The RO feed is pretreated with proprietary antiscalants (threshold inhibitors) to prevent the precipitation of sparingly soluble salts such as calcium sulfate, calcium phosphate, and silica.

RO Train 3 at the Water Recycling Plant is one of three trains known as Barrier Trains. Train 3 operates with Koch Membrane Systems TFC-HR membranes, 85 percent recovery, and approximately 17.0 L/m^2-hr (10 gfd) permeate flux. RO permeate from the Barrier Trains is blended with Metropolitan Water District potable water and injected into the groundwater aquifer to protect against sea water intrusion. Figure 2.6 includes a process flow diagram representative of the West Basin WRP treatment system.

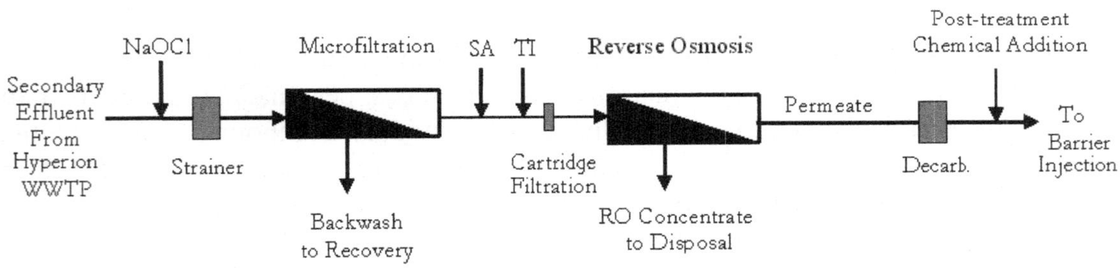

Figure 2.6 West Basin WRP MF/RO Process Flow Diagram

Scottsdale Water Campus

The City of Scottsdale Water Campus contains two treatment facilities: a Water Reclamation Plant and an Advanced Water Treatment (AWT) facility. The WRP treats raw wastewater to a quality suitable for turf irrigation. When irrigation reuse demand is less than the Water Reclamation Plant effluent flow, excess effluent from the Water Reclamation Plant is further treated at the AWT facility and injected into vadose zone recharge wells. The AWT process consists of chloramination, microfiltration, reverse osmosis, decarbonation and chemical stabilization as the principal treatment steps. The AWT facility currently produces up to 45,000 m^3/d (11.9 mgd) of RO permeate.

The tertiary effluent from the Water Reclamation Plant is dosed with ammonia and chlorine to create chloramines prior to entering the MF system. The MF system uses US Filter/Memcor CMF 90M10C units. Similar to West Basin, sulfuric acid and antiscalant are added to the MF filtrate prior to being fed to the reverse osmosis system. The RO trains operate with Koch Membrane Systems TFC-HR membranes, 85 percent recovery, and approximately 18.2 L/m^2-hr (10.7 gfd) permeate flux. Figure 2.7 includes a process flow diagram representative of the Scottsdale Water Campus treatment system.

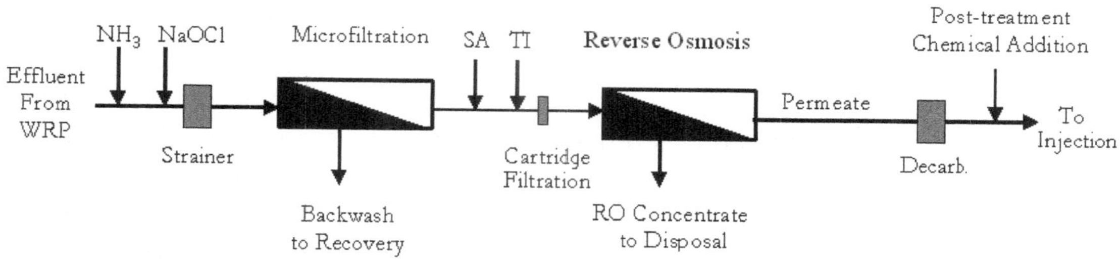

Figure 2.7 City of Scottsdale Water Campus MF/RO Process Flow Diagram

PILOT-SCALE MEMBRANE AUTOPSY AND FOULING CHARACTERIZATION

Membrane Autopsy

Upon completion of the pilot-scale investigation for each membrane, one lead element of the first stage and one tail-end element of the last stage were removed from the skid and soaked in 1 percent sodium metabisulfite solution for preservation. After shipping to the CSM laboratory, the membrane elements were prepared for autopsy analyses. The unrolled membranes were cut into small pieces, and rinsed carefully with DI water (type II) to remove sodium metabisulfite solution. The membrane specimens used for the biomass assay were put into sealed plastic bags immediately to avoid contamination and were preserved at 4°C pending analysis. For other characterizations, the membrane specimens were placed in glass sample bottles to preserve the fouling layer or any deposition on the membrane. All samples were dried at room temperature for two days before characterization, or stored in a desiccator for 3 days prior to FTIR measurement. For comparison and control, the same characterization protocol was applied to both virgin (unfouled) and fouled membrane specimens.

Fouling Characterization

Contact Angle Measurement

The wetting and adhesion properties of membranes are characterized by contact angle measurement using a NRL contact angle Goniometer-Model 100-00 (Ramé-hart, Inc. Surface Science Instrument, Landing, NJ). For each membrane sample, two specimens (about 1.5 cm × 5 cm each) were cut from the membrane for duplicate measurement. A 5.0 µL water droplet (Milli-Q water) was applied on the specimen surface and the contact angle was measured immediately after the droplet deposited on the membrane. To minimize the interference from roughness and morphology of membrane surface, at least ten droplets were applied onto membrane specimen and contact angle was measured from both sides of the droplet.

Membrane Functionality

Functional group characteristics of membrane specimens were measured using a Nicolet Nexus 870 Fourier transform infrared (FTIR) spectrometer (Nicolet, Madison, WI). Membrane specimens (two pieces 1 cm × 4 cm for each sample) were placed in close contact with a ZnSe flat plate crystal. Using a liquid nitrogen cooled mercury cadmium telluride (MCT) detector, the spectra were recorded by the attenuated total reflection (ATR) method with 500 scans and a wave number resolution of 2.0 cm^{-1}.

Membrane Surface Structure and Morphology

Scanning electron microscopy (SEM) has emerged as a powerful tool to examine the morphological features of a membrane and membrane foulants. It has been used to characterize membrane porosity (Meng et al. 2005), fouling and defects (Her et al. 2000; Beverly et al. 2000), and deposition of colloids (Vrijenhoek et al. 2001, Hoek et al. 2003). During SEM, characteristic X-rays are also emitted as a result of electron bombardment. The analysis of the characteristic X-radiation emitted from samples can yield both qualitative identification and quantitative elemental information from regions of a specimen nominally 1 µm in depth under normal operating conditions. EDS screen scanning can identify elemental information for the entire area shown on the SEM screen while spot scanning can focus on the point of interest nominally 1 µm in diameter. EDS has been used to provide rapid qualitative analysis of elemental composition for membrane fouling characterization (Schneider et al. 2005, Xu and Drewes 2006). When standards are not available, EDS is not a technique of choice to quantify the elemental composition of polymer membranes because the intensities of carbon and oxygen atoms are often erratic and misleading to calculate the weight percentage of other elements. However, an EDS spectrum gives excellent elemental analysis of fouling components by identifying the peaks shown in the EDS scan.

Membrane surface structure and morphology were imaged by an environmental scanning electron microscopy (ESEM) Quanta 600 (FEI Company). ESEM operates by scanning an electron probe across a membrane specimen, high resolution electron micrographs of the specimen morphology can be obtained at very low or very high magnifications. Membrane specimens were cut into approximately 1 cm × 1 cm pieces and attached to a carbon tape on an aluminum holder. The membranes were then coated with a thin layer of gold in a Hummer VI sputtering system (Technic Inc., Providence, RI). The plasma discharge current was 20 milliampere and the chamber vacuum was adjusted to 50 to 100 millitorr. Sputtering time was approximately 2 minutes. The coated membrane samples were examined with ESEM at accelerating voltage of 20 to 30 kV, spot size of 2.0 to 2.5, and working distance of 15 mm. Membrane specimens were imaged from low magnifications to high magnifications. Initial imaging at low magnifications allowed observation of flaws, cracks, precipitates grains, fibers and other foulants on membrane surface. It also assisted in selecting representative spots observing scalants and foulants at high magnifications and resolutions. Because of the cost and time for analysis, ESEM was conducted on only one specimen per membrane sample.

Elemental Composition

Elemental composition of virgin and fouled membrane specimens was quantified by energy dispersive spectroscopy (EDS) mounted in the ESEM. Prior to EDS analysis, the membrane specimens were coated with a thin carbon layer by Denton DV-502 Vacuum Evaporator (Moorestown, NJ). With ESEM imaging, EDS can be performed on scanning membrane areas of interest and spots like specific grains and precipitates.

In the EDS analysis, only peaks that are statistically significant should be considered for identification. The minimum size of the peak after background subtraction should be three times the standard deviation of the background at the peak position. The peak height can be approximately estimated directly on the EDS display from the statistical scatter in the background on either side of the peak (Goldstein et al. 2003).

Phospholipid Analysis

The phospholipid analyses were based on the method developed by Findlay et al. (1989) with some modification. Phospholipids extraction estimates the amount of viable microbial biomass as foulants regardless of its physiological activity, e.g. in terms of heterotrophic decay, nitrification, or denitrification. Phospholipids are components of biological cell membranes (about 6 percent of dry cell mass) and are degraded within minutes to hours of cell death by cellular enzymes (Vestal and White 1989).

All glassware, ampules and pipets used were cleaned and free of phosphate and lipids. For each sample, triplicates of 20 to 50 cm^2 wet membranes were extracted with a chloroform:methanol:water solution (0.8:2:1 by volume) for 24 hours. A defined volume of the lipid containing chloroform phase was separated from the aqueous phase and evaporated under a purified air stream at 39°C. A volume of 0.9 mL of 5 percent potassium persulfate (in 0.36 N sulfuric acid) was added before digestion at 96°C for at least 8 hours to separate inorganic phosphate from fatty acid tailing. After cooling, the ampule was opened and 0.2 mL of ammonium molybdate reagent (2.5 percent $(NH_4)_6Mo_7O_{24}\cdot 4H_2O$ in 5.72 N sulfuric acid) and 0.9 mL of malachite green reagent were added. The sample was allowed to sit for 30 minutes for color development and then analyzed with a Beckman Coulter DU 800 spectrophotometer (Fullerton, CA) at 610 nm. Blanks were prepared in triplicates without addition of membrane sample following the same procedure. The stock solution was prepared by dissolving β-glycerophosphate disodium salt, hydrate in methanol.

All chemicals used were of reagent or HPLC grade and purchased from J. T. Baker, Inc. (Philippsburg, NJ) and Sigma-Aldrich (St. Louis, MO).

CHAPTER 3
CANDIDATE MEMBRANE SELECTION FOR PILOT-SCALE TESTING

INTRODUCTION

In order to select candidate membranes for pilot-scale testing at two full-scale facilities, a laboratory-scale membrane testing protocol utilizing flat-sheet and 4040 spiral wound membrane specimens was developed at CSM and implemented with commercially available NF and ULPRO membranes. The testing protocol evaluated membranes based on the rejection of relevant solutes and constituents present in impaired water sources (e.g., TOC, nutrients such as ammonia and nitrate, conductivity, and selected trace organics), as well as operational performance (e.g., specific flux and flux decline due to fouling). A total of 15 different commercially available NF and ULRPO products were considered for this study. Only eight ULPRO and NF membranes, however, were evaluated according to the testing protocol and the performance compared with a benchmark RO membrane commonly employed at several full-scale water reuse facilities.

The testing protocol consists of three components: 1. flat-sheet tests on pure water permeability (PWP) and salt rejection (results are presented in Chapter 2), 2. an assessment of fouling, and 3. rejection of bulk organics, nutrients, and selected organic target compounds using a 2-stage membrane test unit employing 4040 spiral wound elements. The 2-stage tests for every membrane were conducted using two operational modes (i.e., flow-through and internal recycle) simulating different recoveries (20 to 30 and 70 to 80 percent) and two different water matrices (i.e., type II water and secondary effluent) spiked with target compounds. Fouling experiments were performed initially utilizing secondary treated effluent from full-scale wastewater treatment facilities.

MEMBRANE SELECTION PROTOCOL

Pressure, Specific Flux and Fouling

As an example of results generated during bench-scale fouling experiments, flux decline curves for the TFC-HR, TMG10, TFC-ULP, XLE, TFC-S and ESNA1-LF membranes are presented in Figure 3.1. The TFC-HR reverse osmosis membrane was observed to be the most resistant to fouling among the membranes tested, while the ESNA1-LF was the least resistant, exhibiting more than 80 percent flux decline by the termination of the experiment. Flux decline values, initial specific flux values, adjusted specific flux values and relevant membrane characteristics for the candidate membranes are presented in Table 3.1. Out of the candidate ULPRO and NF membranes tested, the NF-4040, TMG10, TFC-ULP and NF-4040 exhibited the least amount of flux decline, maintaining approximately 92, 87, 78, and 76 percent of initial flux by the termination of the experiment (approximately 220 hours), respectively. The NF-270 and NF-90 membranes exhibited final flux decline values of approximately 34 and 39 percent, respectively while the TFC-S membrane was slightly higher at approximately 50 percent.

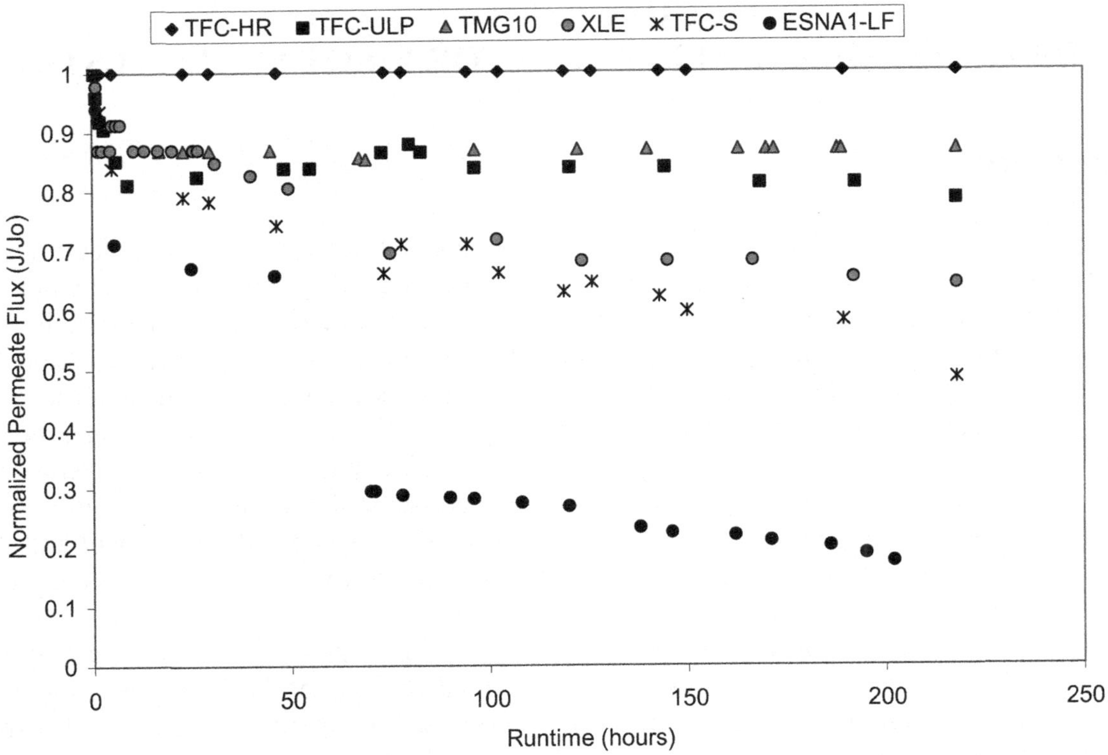

Figure 3.1 RO and ULPRO flux decline (normalized flux J/Jo) versus filtration time during fouling experiments

Table 3.1
Unadjusted and adjusted specific flux values for candidate membranes

Membrane	Flux Decline	Initial Specific Flux		Adjusted Specific Flux		Contact Angle	Mean Roughness	Zeta Potential*
	%	L/m^2 hr kPa	gfd/psi	L/m^2 hr kPa	gfd/psi	degrees	nm	mV
TFC-HR	0	0.030	0.12	0.030	0.12	35	64	-7
TFC-ULP	21.6	0.042	0.17	0.032	0.13	38	41	-7.6
TMG10	13.1	0.057	0.23	0.049	0.20	55	44	-42
XLE	35.9	0.071	0.29	0.047	0.19	66	73	-14
ESNA1-LF	80	0.037	0.15	0.007	0.03	30	24	-41.3
TFC-S	48.4	0.057	0.23	0.030	0.12	57	73	-11
NE-90**	80	-	-	-	-	N/A	N/A	N/A
NF-90	34.3	0.074	0.30	0.049	0.20	63	64	-26.2
NF-200	8	0.054	0.22	0.049	0.20	45	5	-20.2
NF-270	39	0.111	0.45	0.066	0.27	31	5	-35.5
NF-4040	24	0.069	0.28	0.054	0.22	30	29	-12.9

*zeta potential at pH 7
**given the extensive flux decline, the NE-90 was not considered for further testing using the 2-stage membrane skid

Of the membranes tested using spiral-wound elements in a 2-stage configuration, the TFC-HR membrane required the highest feed pressure to achieve a flux of 30.6 to 37.4 L/m^2-hr (18 to 22 gfd) and subsequently displayed the lowest specific flux (SF) value (0.03 L/m^2 hr (0.12 gfd/psi), Table 3.1). Three Dow/Filmtec products, the XLE, NF-90 and NF-270 membranes had significantly higher SF values than the other membranes tested with the NF-270 being highly permeable, requiring less than 517 kPa (75 psi) of feed pressure and resulting in a SF value of 0.11 L/m^2 hr kPa (0.45 gfd/psi). The SF values determined for each membrane, however, reflect the flux of a virgin un-fouled membrane, which changes significantly with membrane fouling. SF values obtained during two-stage membrane experiments with secondary effluent (Table 3.1) were subsequently adjusted using the percentage of the initial flux remaining after fouling obtained during fouling experiments. After flux decline adjustment the ULPRO membrane with the highest adjusted SF was the TMG10, followed closely by the XLE. The NF membrane with the highest adjusted SF was the NF-270, followed by the NF-4040, NF-200 and NF-90 which had values of approximately 0.05 L/m^2 hr kPa (0.2 gfd/psi) (Table 3.1).

Membranes with adjusted SF values equal to or less than the TFC-HR membrane, were not considered for pilot-scale testing, since, they would yield similar operating pressures as a 'traditional' RO membrane after a period of compaction and fouling. Therefore, the ESNA1-LF and TFC-S membranes were not considered viable candidate membranes for pilot-scale testing using municipal wastewater effluents as feed water based on operational results.

Total organic carbon, ammonium and nitrate

Candidate membranes were evaluated for the rejection of total organic carbon (TOC) during experiments using secondary treated effluent followed by MF as feed water. Spiral-wound elements in a 2-stage configuration were employed during this assessment. Secondary effluent serving as feed water was collected multiple times at a local wastewater treatment facility during the execution of these rejection experiments. The TOC concentrations of the feed waters varied between 8.7 and 11.7 mg/L as carbon. TOC rejection values for candidate membranes obtained during 80 percent recovery experiments are presented in Figure 3.2. All candidate membranes rejected TOC present in the feed water greater than 90 percent, with the RO and ULPRO membranes achieving TOC rejection values of 95 percent and above. While the NF-90 was found to achieve a TOC rejection similar to the RO and ULPRO membranes, the NF membranes NF-200, NF-270, and NF-4040 exhibited permeate TOC concentrations two times greater (approximately 0.8 mg-C/L) resulting in rejection values close to 90 percent. TOC concentrations during 80 percent recovery experiments were similar to permeate concentrations observed during 30 percent recovery experiments indicating that recovery and feed concentrations have little effect on TOC rejection.

Because the ESNA1-LF and TFC-S membrane failed operational testing due to significant flux decline, the remaining NF membranes NF-200, NF-270 and NF-4040 were still considered viable. However, it is noteworthy that permeate TOC concentrations greater than 0.5 mg-C/L during 80 percent recovery experiments exhibited by these membranes did not meet the CDPH draft regulations for TOC.

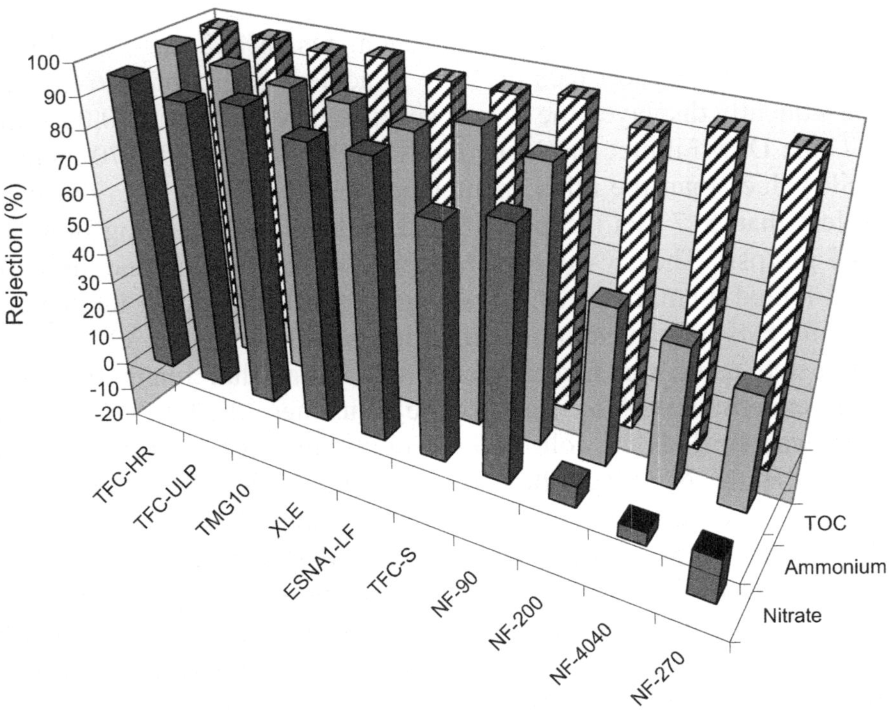

Figure 3.2 Average nitrate, ammonium and TOC rejection for the candidate membranes during 80 percent recovery laboratory-scale testing experiments

Candidate membranes were evaluated for ammonium rejection performance due to the fact that nitrogen in a non-nitrified wastewater effluent is primarily composed of ammonium. Secondary effluent feed water employed in ammonium rejection experiments was adjusted when necessary with ammonium chloride to achieve ammonium concentrations of 35 to 40 mg-N/L. The performance of the TFC-HR, TMG10 and TFC-ULP membranes in terms of ammonium rejection during 30 (not shown) and 80 percent recovery experiments was similar and all three membranes exhibited ammonium rejection of 95 percent or above resulting in permeate concentrations below 5 mg-N/L (Figure 3.2). The XLE and NF-90 membranes were only slightly less efficient in rejecting ammonium, with rejection values of 93 and 89 percent, respectively (Figure 3.2). The NF-200, NF-4040, and NF-270, however, were found to only moderately reject ammonium (30 and 50 percent, respectively) and could only be used for applications where the rejection of nitrogen species is of secondary importance.

Nitrate experiments with the candidate membranes highlighted the differences between the membranes that were selected for this study. The highly permeable NF-270 and NF-4040 membranes exhibited no rejection of nitrate during both 30 and 80 percent recovery experiments (Appendix B, Figure B.9, 30 percent recovery results not presented). In contrast, the NF-90, TFC-S and ESNA1-LF displayed nitrate rejection similar to the ULPRO and RO membranes chosen, exhibiting nitrate rejection in the 80 to 90 percent range. Permeate nitrate concentrations, however, exceeded the threshold of 10 mg-N/L during most NF experiments, with the exception of 30 percent recovery experiments for the NF-90, TFC-S and ESNA1-LF membranes. Therefore, a limitation of NF membranes for water reuse applications arises when the rejection of nitrate is desired and feed water concentrations exceed 10 mg-N/L. Nitrate ions have been shown to be much more difficult to reject than ammonium ions and that rejection was highly membrane dependent with membrane electrical properties being the most important (Lee and Lueptow 1999, Van der Bruggen et al. 2001). The observed poor rejection of nitrate by certain NF membranes is likely the result of the more mobile nitrate anion permeating in order to balance the transport of cations (i.e. sodium), which has been observed in various studies investigating the transport of ions in NF systems (Tsuru et al. 1991, Bowen et al. 2002).

Trace organic compounds

Candidate membranes were evaluated for the rejection of trace organic compounds in an ion strength type II water feed solution using spiral-wound elements in a 2-stage configuration. Summarized trace organic contaminant 80 percent recovery experimental results are presented in Figure 3.3 for the TFC-HR, TFC-ULP, NF-90 and NF-270 representing an RO membrane, an ULPRO membrane, a *tight* NF membrane and a *loose* NF membrane, respectively. For nonionic trace organic contaminants, where rejection depends upon size exclusion and solute-membrane interactions, increasing permeate concentrations were observed with increasing permeate productivity and decreasing salt rejection, which are traits generally considered to separate *tight* membranes from *loose* membranes (Van der Bruggen et al. 2001, Bellona et al. 2004, Nghiem et al. 2006). Ionic trace organic contaminants, however, were well rejected by all membranes tested since the rejection of theses solutes depends upon both size exclusion and electrostatic exclusion (Kimura et al. 2003b, Bellona et al. 2004, Bellona and Drewes 2005, Nghiem et al. 2005, Braeken et al. 2006).

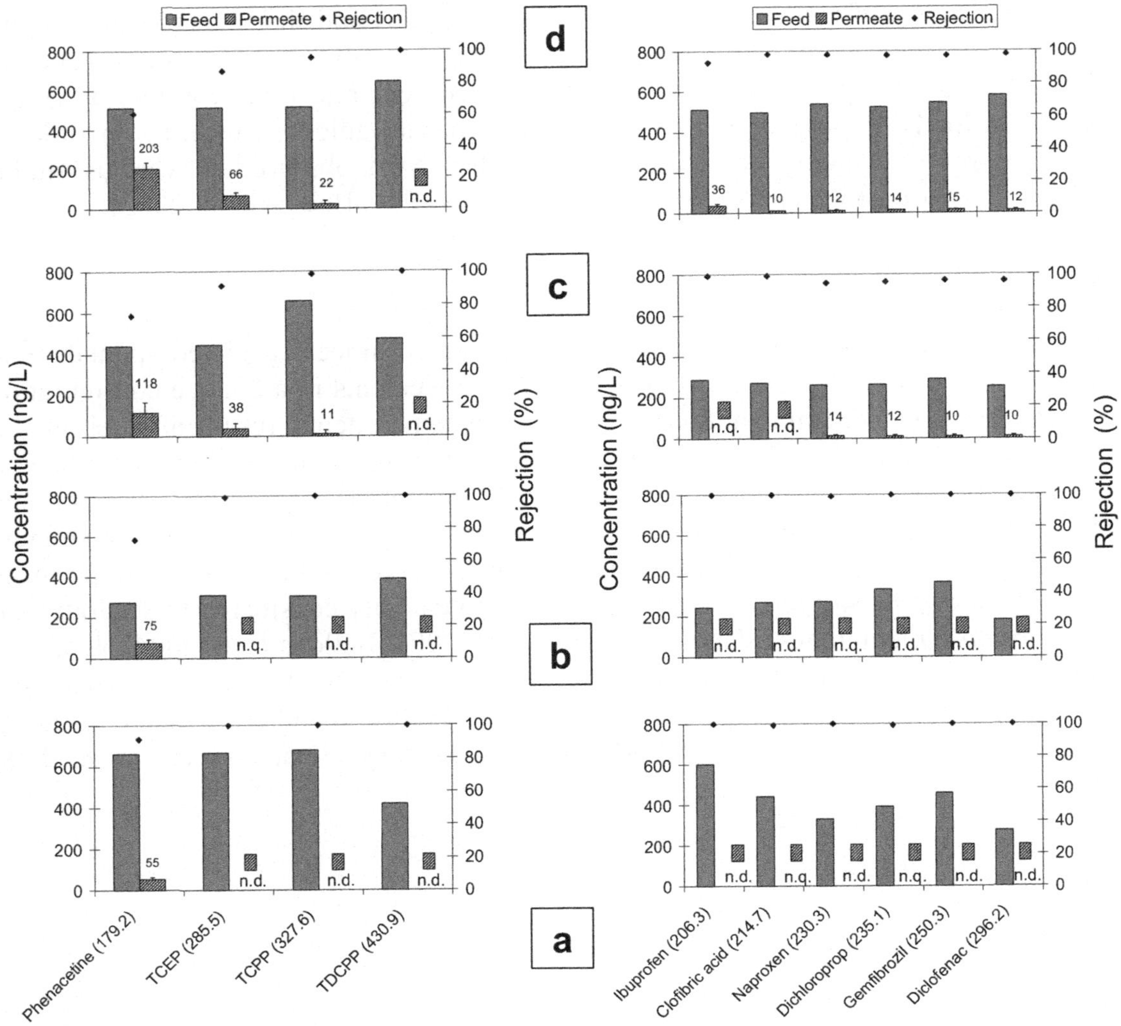

Figure 3.3 Feed and permeate concentrations of select nonionic (left side) and ionic (right side) trace organic contaminants and subsequent rejection values for the: a. TFC-HR, b. TFC-ULP, c. NF-90, and d. NF-270 (molecular weight of compound given in parenthesis)

The rejection results for four nonionic trace organic contaminants by the candidate membranes employed in the study are presented in Figure 3.4. The nonionic pharmaceutical residue phenacetine was quantified in all permeate samples taken during 80 percent recovery experiments with all of the candidate membranes. The chlorinated flame retardants TCEP and TCPP were more frequently detected in permeate samples from membranes with high permeate productivity. The largest chlorinated flame retardant and nonionic trace organic contaminant employed in the study, TDCPP (molecular weight = 430.9 g/mol), was not detected in any permeate sample. The rejection of nonionic solutes was found to increase with increasing molecular weight. Although the presence of actual pores in ULPRO and NF membranes is debatable, multiple studies have determined that the rejection of nonionic solutes is highly dependent upon the size of a solute in relation to the 'effective' pore size, and both descriptors are often employed as parameter used in modeling the rejection of uncharged solutes (Bellona et al. 2004, Nghiem et al. 2006, Braeken et al. 2006).

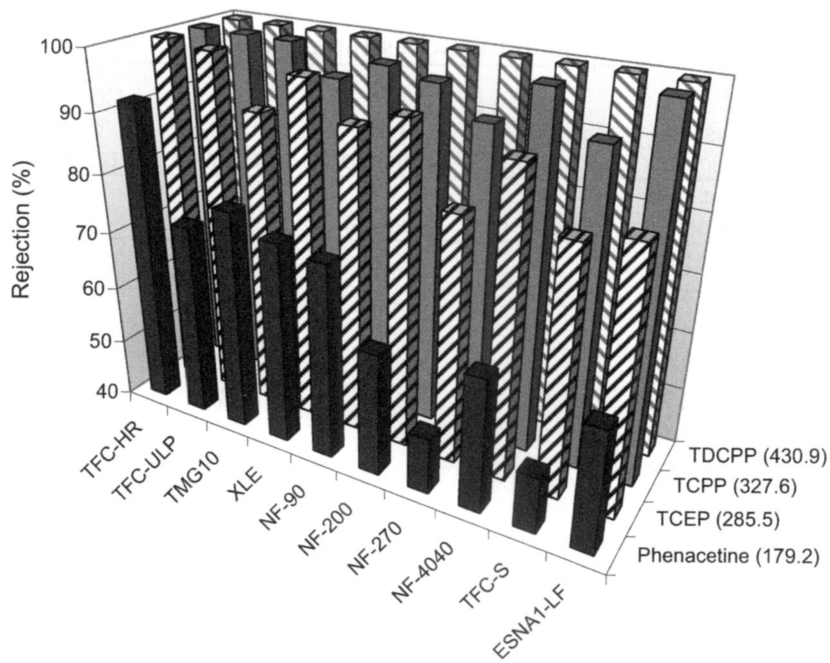

Figure 3.4 Average rejection of select nonionic trace organic contaminants by the candidate membranes during 80 percent recovery experiments (molecular weight of solutes given in parenthesis)

The RO and ULPRO membranes TFC-HR, TFC-ULP, TMG10, and XLE were found to efficiently reject trace organics employed in this study and permeate concentrations were generally below 10 ng/L or non-quantifiable with exception of the low molecular weight pharmaceutical residue phenacetine. Ionic trace organic contaminants were found to be well removed by the RO and ULPRO membranes due to contributions of steric hindrance and electrostatic exclusion to the rejection of these solutes. Although ionic trace organic solutes were more often detected in NF permeate samples, concentrations were below 40 ng/L, resulting in rejection values greater than 90 percent. As expected, the limitations of NF membranes were observed to be the removal of small molecular weight non-ionic organic solutes such as phenacetine and TCEP.

Ranking. A summary of ULPRO and NF membrane experiments discussed in previous sections in terms of operation and TOC and nutrient rejection is presented in Table 3.2. All of the ULPRO membranes tested had adjusted specific flux values greater than the TFC-HR membrane, and also met the CDPH draft regulation requirement of 0.5 mg-C/L permeate TOC concentration. The XLE membrane had permeate concentrations exceeding 10 mg-N/L during 80 percent recovery nitrate rejection experiments, however, met the nitrogen requirement during the

30 percent recovery experiments. Of the NF membranes tested, all of the membranes achieved better than expected TOC and ionic trace organic rejection, but differed considerable in flux decline, permeability and ammonia and nitrate rejection.

Table 3.2
Summary of membrane testing results

Membrane	Flux Decline	Adjusted Specific Flux	TOC Rejection*	NH_4^+ Rejection*	NO_3^- Rejection*	Compounds detected perm/feed*
	percent	L/m^2 hr kPa (gfd/psi)	Percent	percent	percent	percent
TFC-HR	0	0.30 (0.12)	99	99	94	24
TFC-ULP	21.6	0.032 (0.135)	99	97	92	35
TMG10	13.1	0.049 (0.198)	97	92	95	43
XLE	35.9	0.047 (0.186)	99	88	88	64
NF-90	34.3	0.049 (0.198)	98	88	80	75
TFC-S	48.4	0.030 (0.119)	95	94	75	76
NF-200	8	0.049 (0.198)	92	51	7	67
NF-270	39	0.066 (0.275)	95	38	-15	78
NF-4040	24	0.054 (0.22)	96	46	-4	53
ESNA1-LF	80	0.007 (0.03)	96	97	89	71

*During 80 percent recovery experiments

The results of the ranking and selection process are summarized in Table 3.3 for the ULPRO membranes and Table 3.4 for the NF membranes. Of the ULPRO membranes, the XLE membrane was found to be the most prone to fouling during the laboratory-scale fouling experiments while the TMG10 and TFC-ULP membranes performed similarly. Although the XLE exhibited a higher fouling propensity, due to high initial permeability it had a similar adjusted specific flux value as the TMG10 membrane. Trace organic detections were similar between all three membranes and all three were given the same ranking. The TMG10 membrane had a lower ranking when the Tier parameters were summed indicating that it was the better performing membrane of the candidate ULPRO membranes during the laboratory-scale experiments. The results of summing Tier 2 and 3 parameters indicated that the TFC-ULP and TMG10 membranes performed similarly and the only difference between them was the adjusted specific flux value. Because the adjusted specific flux value was approximately 50 percent higher for the TMG10 than the TFC-ULP membrane, the TMG10 membrane was considered the top performing ULPRO membrane and chosen for pilot-scale testing.

Table 3.3
Rankings of candidate ULPRO membranes

Candidate Membranes	TFC-ULP	TMG10	XLE
Tier 1 Selection Criteria (x2)			
Flux Decline	1	1	2
Adjusted Specific Flux	3	2	2
Trace Organic Compounds	1	1	1
Sum	10	8	10
Tier 2 Selection Criteria (x1.5)			
Ammonia	1	1	2
Nitrate	1	1	2
Sum	3	3	6
Tier 3 Selection Criteria (x1)			
TOC	1	1	1
Sum	1	1	1
Total Sum	14	12	17

Best-performing membrane in category received a value of 1, the second receiving a value of 2, etc. Membranes that performed similarly were given the same ranking, with the next best membrane receiving the next available ranking.

The summarized NF rankings based on testing are presented in Table 3.4. Due to low adjusted specific flux values, the TFC-S and ESNA1-LF membranes were not considered viable due to high fouling propensity observed during flux decline experiments. The Dow/Filmtec membranes NF-200, NF-270 and NF-4040 were observed to only moderately reject ammonia and be extremely limited in the removal of nitrate during rejection experiments. Because the high adjusted specific flux values for these membranes would be advantageous in terms of reducing operating costs, certain scenarios where nitrogen rejection is of secondary importance (e.g. <10 mg-N/L already met in municipal wastewater effluent used as feed water) may allow for the application of these types of membranes. Based on the selection protocol, however, the NF-90 was determined to be the only NF membrane that met all of the criteria for pilot-scale testing.

Table 3.4
Rankings of candidate NF membranes

Candidate Membranes	NF-90	TFC-S	ESNA1-LF	NF-200	NF-270	NF-4040
Tier 1 Selection Criteria (x2)						
Flux Decline	2	3	4	1	2	2
Adjusted Specific Flux	2	3†	3†	2	1	2
Trace Organic Compounds	1	1	1	1	2	1
Sum	10	14	16	8	10	10
Tier 2 Selection Criteria (x1.5)						
Ammonia	2	1	1	3‡	3‡	3‡
Nitrate	2§	2§	1	3‡	3‡	3‡
Sum	6	4.5	3	9	9	9
Tier 3 Selection Criteria (x1)						
TOC*	1	2	2	2	2	2
Sum	1	2	2	2	2	2
Total Sum	17	20.5	21	19	21	21

Best-performing membrane in category received a value of 1, the second receiving a value of 2, etc. Membranes that performed similarly were given the same ranking, with the next best membrane receiving the next available ranking.

†Membranes that did not exceed the adj. SF of the TFC-HR were not considered

‡These membranes did not meet the nitrogen requirement

§These membranes only met the nitrogen requirement during 30 percent recovery experiments

*Only the NF-90 met the CDPH draft reg. for TOC of less than 0.5 mg-C/L in permeates

SUMMARY

The laboratory-scale testing protocol developed to evaluate candidate ULPRO and NF membranes was a useful tool in discerning the performance of one membrane from another. Membranes tested displayed a wide range of operating characteristics including fouling tendency, permeate productivity, and feed pressure while displaying a wide range of rejection behavior in terms of TOC, ammonium, nitrate and select trace organic contaminant removal. In general, however, the strict selection criteria favored ULPRO membranes due to high removal efficiencies of TOC, nitrogen species and trace organic contaminants, although candidate NF membranes often demonstrated significantly higher flux rates and lowered feed pressures. The selection of membranes for pilot-scale testing therefore became an exercise in balancing the benefits of lower pressure and energy with constituent rejection and permeate quality. Of the membranes tested, the TMG10 and NF-90 provided a good balance between permeate productivity and permeate quality, providing high rejection of TOC, nitrogen and trace organic contaminants while displaying moderate (30 percent) flux decline during fouling experiments and a significantly higher adjusted TCSF than the 'benchmark' RO membrane (TFC-HR). Although they did not meet the nitrogen and TOC criteria, the NF-200 and NF-4040 membranes were identified as promising candidates based on low (10 percent) to moderate flux decline (30 percent) values, efficient trace organic rejection (usually greater than 90 percent) and significantly higher adjusted TCSF than the 'benchmark' RO membrane (TFC-HR). From laboratory-scale testing results the TMG10 and NF-90 were identified as membranes for pilot-scale testing.

One large component of the testing protocol was that rejection experiments were conducted for a particular membrane at two different feed water recovery conditions by recycling a portion of concentrate stream into the feed stream. Since an important criteria of the testing protocol was that a particular membrane meet a nitrogen permeate concentration threshold of 10 mg-N/L and to a lesser extent the CDPH draft regulation TOC permeate requirement of less than 0.5 mg-C/L, it was necessary to examine the effect of recovery on permeate concentrations. Table 3.5 summarizes Tier 1 membrane testing results for three membranes to illustrate the effect of system recovery on permeate concentrations and rejection. The permeate concentrations and rejection of TOC and ammonium by the 'benchmark' RO membrane TFC-HR was found to be minimally affected by system recovery. Although nitrate rejection was unaffected by system recovery, permeate nitrate concentrations increased approximately 240 percent during 80 percent recovery experiments. Similarly, Tier 1 rejection by the NF-90 was largely unaffected by system recovery, however, permeate concentrations of TOC, ammonium and nitrate increased in proportion with increases in feed concentration as a result of increased recovery. NF-4040 permeate concentrations were less affected by recovery, however, nitrate rejection was decreased 180 percent and became negative. Due to the relative inability of the NF-4040 to reject ammonium and nitrate recycling the concentrate was found to minimally increase feed concentrations.

Researchers have shown that diffusion-controlled mass transfer in membrane systems is affected by feed concentrations and therefore system recovery while solutes controlled by sieving mechanisms are largely unaffected (Zhao and Taylor 2004). TOC, therefore, appears to be largely controlled by size exclusion since system recovery had a small effect on permeate concentrations by all three membranes. Therefore, TOC rejection and permeate concentrations is expected to be largely independent of system recovery and related to membrane characteristics.

The extreme opposite was found for nitrate in which permeate concentrations were largely affected by recovery. Nitrate mass transfer, however, may be related to ion mobility and pore and permeate electroneutrality during operation. Researchers have shown that during NF and to a lesser extent RO experiments with ternary and quaternary ion mixtures negative rejection of certain ions is often observed (Tsuru et al. 1991b, Hagmeyer and Gimbel 1998, Schaep et al. 1999, Bowen et al. 2002). This occurs generally when a more mobile ion (usually co-ion to the membrane charge) is introduced at much lower concentration than a less mobile counter-ion in solution (i.e. 20 mM NaCl and 1 mM $NaNO_3$). In this case, a small percentage of counter-ions (i.e. Na^+) in the feed water permeate (since no membrane rejects 100 percent of ions) forces a large percentage of co-ions (usually the mobile specie, NO_3^-) to permeate in order to maintain permeate electroneutrality. It is hypothesized that during 2-stage membrane experiments, the highly mobile nitrate ion is forced through the membrane to balance the charge of less mobile but highly concentrated cations (Tsuru et al. 1991, Bowen et al. 2002). The rejection of nitrate therefore is highly dependent upon the rejection of other ions, which in turn appears to be diffusion-controlled and a function of recovery and feed concentration.

Ammonium mass transfer, which would be assumed to be diffusion controlled, was unaffected by recovery conditions during TFC-HR tests, but affected during NF-90 experiments. It is possible that ammonium analytical methods are not sensitive enough to quantify the effect of system recovery on mass transfer during TFC-HR experiments. Ammonium permeate concentrations appear to be dependent upon system recovery and feed concentrations, but rejection should be largely unaffected since there appears to be a linear relationship between permeate and feed concentrations.

Finally, trace organic solute detections were observed to be more numerous in 80 percent recovery permeate samples as opposed to 30 percent recovery permeate samples (Tables 3.2 and 3.6). Trace organic solute mass transfer is the result of solute and membrane interactions and depending upon the solute has been shown to be governed both by diffusion and convection (Lee and Lueptow 2001, Yoon and Lueptow 2005). Permeate concentrations of trace organic solutes appears to be minimally dependent upon system recovery based on the increase in detections in 80 percent recovery experiments. Because environmental concentrations of trace organic contaminants are generally low (nanogram per liter range), system recovery is not expected to have a large impact on permeate concentrations especially when employing RO and ULPRO membranes.

Table 3.5
Effect of recovery on permeate concentrations and rejection for Tier 1 constituents

Membrane	Constituent	Concentration (mg/L)				Rejection (%)		Perm. Concentration	Rejection
		30 percent		80 percent		30 percent	80 percent	Percent change	Percent change
		Feed	Permeate	Feed	Permeate				
TFC-HR	TOC	10.1	0.2	22.2	0.2	98.0	99.1	0.00	1.10
	Ammonium	35	1	123.8	0.9	97.1	99.3	-10.00	2.19
	Nitrate	37.6	1.6	132.8	5.4	95.7	95.9	237.50	0.20
NF-90	TOC	10.4	0.3	21.2	0.5	97.1	97.6	66.67	0.54
	Ammonium	34.4	3.9	89.3	10.8	88.7	87.9	176.92	-0.85
	Nitrate	37.7	8.1	101	20.2	78.5	80.0	149.38	1.89
NF-4040	TOC	7.4	0.7	22	0.8	90.5	96.4	14.29	6.43
	Ammonium	31.5	16.3	36.5	19.8	48.3	45.8	21.47	-5.18
	Nitrate	36.2	34.2	35.8	37.4	5.5	-4.5	9.36	-180.89

CHAPTER 4
PILOT-SCALE TESTING OF CANDIDATE MEMBRANES

PERFORMANCE OF ULPRO AND NF MEMBRANES USING NON-NITRIFIED MICROFILTERED WASTEWATER EFFLUENT

Performance of TMG10 ULPRO and the NF-90 Nanofiltration Membranes

As the top performing membrane during the laboratory-scale membrane testing outlined in Chapter 3, the TMG10 ULPRO membrane was chosen for pilot-scale testing with microfiltered non-nitrified feed water at the WBWRP. Pilot-scale testing of the TMG10 membrane commenced on December 4, 2004 and was terminated on April 1, 2005 for a total testing period of 2,965 hours. As the top performing NF membrane during the laboratory-scale membrane testing outlined in Chapter 3, the NF-90 membrane was chosen for pilot-scale testing with non-nitrified feed water at the WBWRP. Pilot-scale testing of the NF-90 membrane with non-nitrified secondary effluent commenced at the WBWRP on April 7, 2005 and was terminated June 2, 2005 for a total testing period of approximately 1,300 hours.

Membrane Operational Performance.

The operational set-points for TMG10 and NF-90 testing at the WBWRP are summarized in Table 4.1. TMG10 and NF-90 pilot-scale testing was performed at a system recovery of approximately 82 percent and a flux rate of 20.4 (11.8 gfd) and 21.2 L/m^2-hr (12.4 gfd), respectively. These operational set points are representative of conditions employed at full-scale RO installations using recycled water. During the six months of testing, operational conditions were monitored with the customized SCADA system installed on the pilot-skid, with the exception of a period where the data logging computer malfunctioned during testing of both membrane products. The operational set-points were maintained by United Water staff at the WBWRP and adjusted whenever necessary.

Feed pressure and temperature corrected specific flux (TCSF) data for the TMG10 and NF-90 pilot-scale testing period are presented in Figures 4.1 and 4.2, respectively. The feed pressure at start-up for the TMG10 membrane (December 4, 2004) was approximately 900 kPa (130 psi), and increased over the first month (720 hours) of operation to approximately 1,100 kPa (160 psi) due to membrane compaction and fouling (Figure 4.1). During this membrane compaction and fouling period, the TMG10 membrane TCSF decreased from approximately 0.04 L/m^2-hr-kPa (0.17 gfd/psi) to 0.027 to 0.03 L/m^2-hr-kPa (0.11 to 0.12 gfd/psi) and remained relatively stable throughout the testing period (Figure 4.2). During the month of January of 2005 and corresponding to approximately 600 hours of runtime, there were two feed pressure spikes as the result of rain events that significantly decreased the temperature of the feed water and increased the driving pressure necessary to maintain the permeate flow set-point. After these rain events, however, the feed pressure of the TMG10 membrane was relatively stable and was maintained between 1,100 kPa and 1,170 kPa ((160 and 170 psi) over the testing period (Figure 4.2).

Table 4.1
Operational set points during pilot-scale WBWRP testing

Membrane Parameter	Units	Toray TMG10	Dow/Filmtec NF-90
Element Membrane Area	m^2	8.1	7.6
	ft^2	87	82
Pilot Membrane Area	m^2	170.1	160
	ft^2	1827	1722
Recovery	percent	82.5	82
Permeate Flux	L/m^2-hr	20.4	21.2
	gfd	11.8	12.4
Permeate Flow	L/min	56.8	56.4
	gpm	15	14.9
Concentrate Flow	L/min	12.1	12.5
	gpm	3.2	3.3
Feed Flow	L/min	68.9	68.9
	gpm	18.2	18.2
pH		6.3	6.3
Antiscalant Dose	ppm	3	3

The feed pressure of the NF-90 during the initial start-up period was approximately 760 kPa (110 psi) but increased within the first 24 hours to 970 kPa (140 psi) as a result of compaction and initial fouling. The initial feed pressure increase (and TCSF decrease, Figure 4.2) of the NF-90 membrane was observed to occur at a faster rate and to a greater extent than the TMG10 membrane (Figure 4.1). The difference in decreasing permeability rates between the two membranes appears to be related to membrane characteristics including pure water permeability, and surface roughness and charge, and was similarly observed during bench-scale fouling propensity studies. Initially, the temperature corrected specific flux decreased rapidly with compaction and fouling, but stabilized at 0.03 L/m^2-kPa-hr (0.12 gfd/psi) after 600 hours of operation. Although the NF-90 feed pressure remained relatively stable between 600 and 1,300 hours of operation, a slight but steady decrease in TCSF was observed between 900 and 1,300 hours of operation. The feed pressure required to achieve a constant permeate flux remained constant due to increased feed water temperature from an initial temperature of 23°C to a final temperature of 26°C. It appears, however, that during this time the actual permeability of the NF-90 decreased which is represented by the TCSF value.

As a comparison, the historic stabilized TCSF value for the TFC-HR and ESPA2 during operation on WBWRP's full-scale Train 3 is represented as a dashed line in Figure 4.2. The TFC-HR and ESPA2 have generally achieved a stabilized TCSF of approximately 0.022 L/m^2-hr-kPa (0.09 gfd/psi) (Appendix C, Figure C.1).

Due to the initial rapid increase in feed pressure in comparison to the TMG10, the advantage of employing an NF similar to the NF-90 (i.e. low feed pressure and reduced energy costs) diminished fairly quickly (approximately 500 hours) as compaction and fouling occurred. Testing of the NF-90 therefore, was abandoned after 1,300 hours since it appeared that the TMG10 was a superior product for this application regarding operational performance.

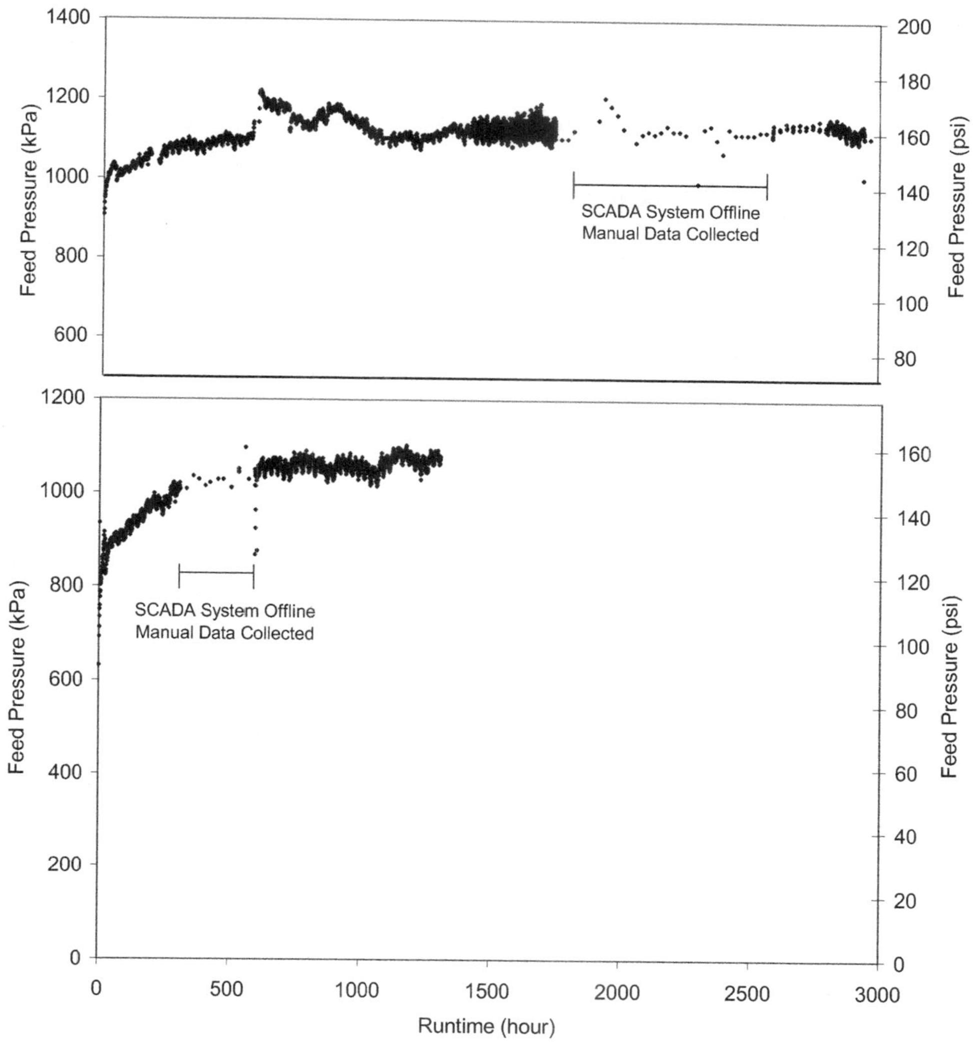

Figure 4.1 Feed pressure of the TMG10 (top graph) and NF-90 (bottom graph) membranes during pilot-scale testing

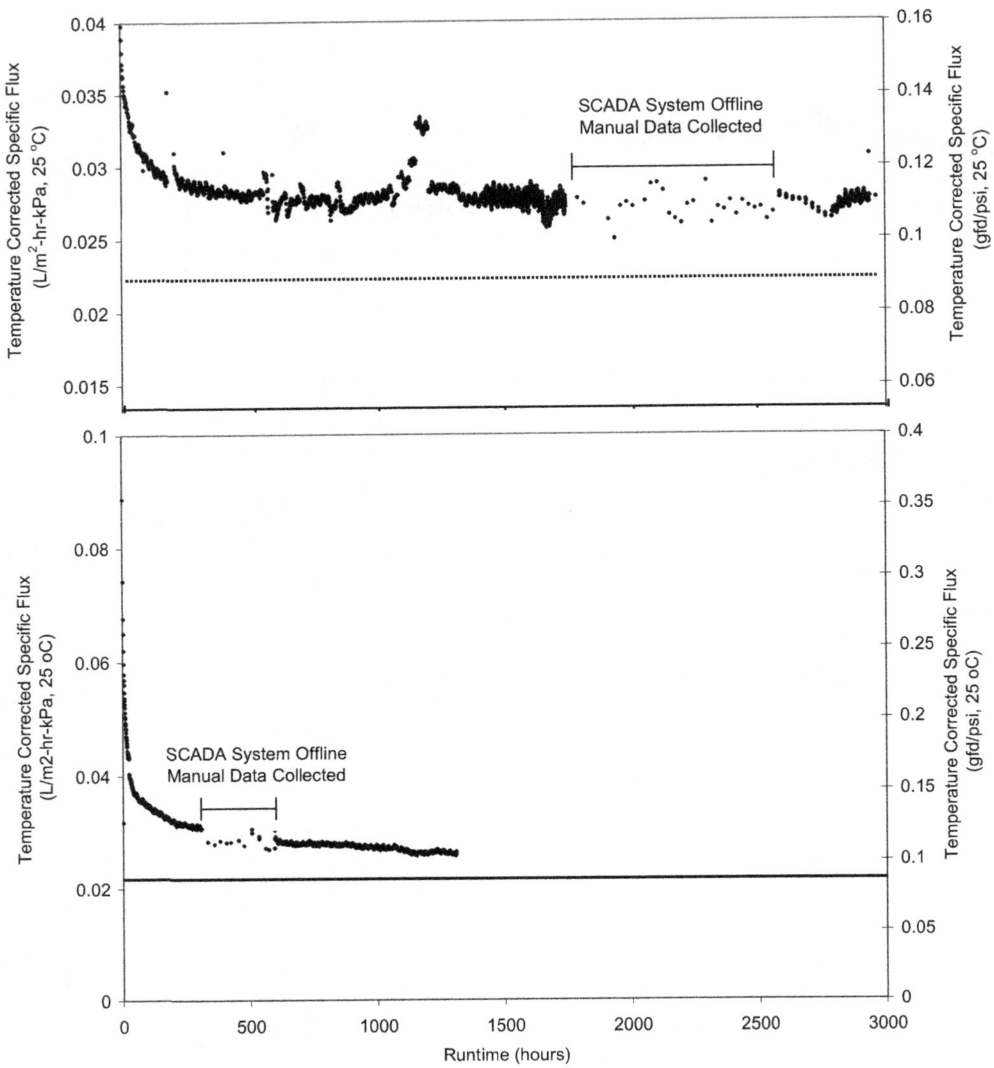

Figure 4.2 Temperature corrected specific flux for the TMG10 (top graph) and NF-90 (bottom graph) membranes during pilot-scale testing
(For comparison, black line in both graphs represents stabilized specific flux value of TFC-HR on full-scale)

Rejection of TOC and Nitrogen

Average TOC and ammonia concentrations for the TMG10 and NF-90 membrane testing on pilot-scale and the TFC-HR and ESPA2 membranes on full-scale are presented in Table 4.2. Combined permeate TOC concentrations were below 0.2 mg-C/L resulting in TOC rejection values above 97 percent during the TMG10 testing period. All of the permeate samples collected during NF-90 pilot-scale operation had TOC concentrations below 0.3 mg-C/L resulting in TOC rejection greater than 97 percent. TOC rejection by the NF-90 was comparable to TOC rejection by the TMG10 during pilot-scale testing and the TFC-HR and ESPA2 during full-scale operation (Table 4.2).

Table 4.2
Summary of TOC, ammonium and UV absorbance data for the TMG10 and NF-90 membranes on pilot-scale and the TFC-HR and ESPA2 membranes on full-scale

		Pilot-scale TMG10	Pilot-scale NF-90	Full-scale TFC-HR	Full-scale ESPA2
	Unit	Average	Average	Average (1/04 - 5/05)	Average (1/04 - 5/05)
Recovery	%	82	82	85	85
TOC					
Feed TOC	mg-C/L	11.8	13	11.3	11.6
Permeate TOC	mg-C/L	0.2	0.18	0.37	0.31
TOC Rejection	%	98.3	98.6	96.7	97.3
Ammonium					
Feed Ammonium	mg-N/L	31.6	37	33.2	31.8
Permeate Ammonium	mg-N/L	1.2	2.3	1.9	1.7
Ammonium Rejection	%	96.2	93.8	94.3	94.7
UV Absorbance					
Feed UVA	abs/cm	1.2	0.18	NA	NA
Permeate UVA	abs/cm	0.014	0.013	NA	NA
UVA Rejection	%	98.8	92.8	NA	NA

During the TMG10 membrane testing period, combined permeate ammonia concentrations were below 2 mg-N/L resulting in ammonia rejection values above 90 percent. Nitrate was detected in two feed samples taken from the pilot-skid, and was not detected in any combined permeate samples collected during the TMG10 membrane testing period. Nitrite was not detected in any sample taken from the pilot-skid. Ammonium rejection by the NF-90 was greater than 93 percent during testing resulting in permeate concentrations below 3 mg-N/L. Nitrate and nitrite were not detected in any feed samples during NF-90 testing. Ammonium rejection by the NF-90 during pilot-scale testing was comparable to the rejection of ammonium by the TMG10, TFC-HR and ESPA2 membranes and was better than expected based on the laboratory-scale testing results (Table 4.2).

Ultraviolet absorbance (UVA) measurements of feed and permeate samples were also collected during TMG10 testing and are included as a performance parameter (Table 4.2). Feed water UVA was significantly reduced in TMG10 membrane permeate samples and the average permeate UVA values was near the detection limit (0.01 absorbance/cm). In addition, UVA rejection by the NF-90 was similar to UVA rejection observed during TMG10 pilot-scale testing.

Comparing pilot-scale testing of the TMG10 membrane to a period of full-scale operation indicates that the TMG10 performed similarly or better than the ESPA2 and TFC-HR membranes in terms of TOC and ammonium permeate concentrations and rejection. All three membranes achieved high rejection and low permeate concentrations of ammonium and TOC. Somewhat unexpectedly, the nanofiltration membrane NF-90 achieved similar permeate concentrations of TOC and ammonium to the RO and ULPRO membranes examined during the testing period.

Rejection of Regulated Organics

Disinfection By-products. Feed and permeate samples were collected monthly from the pilot-skid (TMG10 and NF-90) and Train 3 (TFC-HR) and Train 4 (ESPA2) at the WBWRP for total halogenated methanes (i.e., chloroform and bromoform) and haloacetic acids (i.e, trichloroacetic acid and dichloroacetic acid) analysis. Initially, Train 3 (TFC-HR) was chosen for the full-scale sampling location, but went offline due to technical problems in February, 2005. Therefore, Train 4 (ESPA2) was subsequently selected as the full-scale sampling location for comparison with the pilot-skid.

Chloroform data generated from TMG10 and NF-90 membrane pilot-scale and full-scale sampling campaigns conducted between December 4, 2004 and June 2, 2005 are summarized in Figure 4.3. In general, feed water chloroform concentrations were between 5 and 22 μg/L during TMG10 and NF-90 membrane testing. TMG10 membrane combined permeate concentrations were between 4 and 9 μg/L, resulting in rejection values between 20 and 40 percent (Figure 4.3). Feed water chloroform concentrations were greater during NF-90 testing (between 15 and 24 μg/L) and NF-90 membrane combined permeate concentrations were between 10 and 15 μg/L during the testing period, which resulted in rejection values between 35 and 47 percent (Figure 4.3). While NF-90 membrane chloroform rejection was observed to be greater than TMG10 and full-scale TFC-HR and ESPA2 membrane chloroform rejection, chloroform permeate concentrations were similar among all four membranes evaluated.

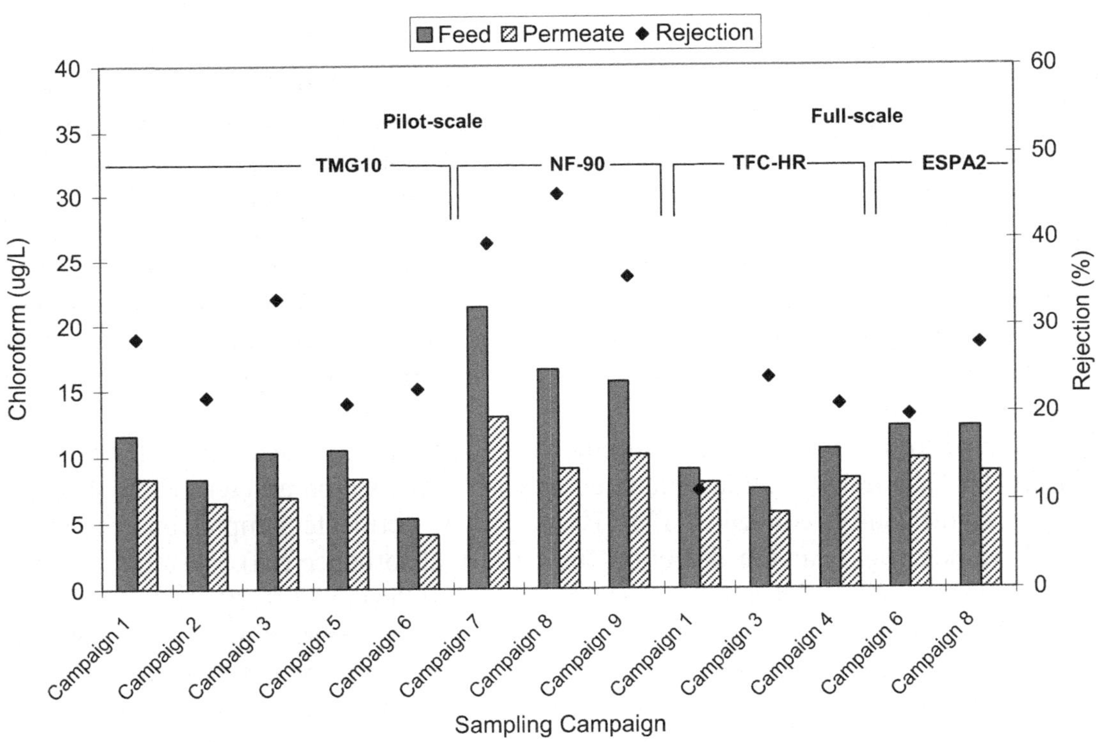

Figure 4.3 Feed and permeate concentrations of chloroform and rejection values for the TMG10 (pilot-scale), NF-90 (pilot-scale), TFC-HR (full-scale) and ESPA2 (full-scale) membranes

Chloroform, therefore, was found to be poorly (20 percent) to moderately (40 percent) rejected by all four membranes investigated during the study. Chloroform, a small molecular weight nonionic solute (117 g/mol) has been shown to be poorly removed by most RO and NF membranes due to adsorption to and partitioning across membranes materials (Xu et al. 2005).

Bromoform concentrations in pilot-skid and full-scale feed water were lower than the corresponding sample's chloroform concentration, and were often below the detection limit in feed samples taken during TMG10 and NF-90 membrane testing. When quantified in feed samples, bromoform permeate concentrations were below the detection limit indicating efficient rejection by all membranes evaluated.

Contrary to chloroform results presented in Figure 4.3, haloacetic acid rejection was found to be greater than 90 percent for all four membranes evaluated. Summarized results from sampling campaigns conducted on the pilot-skid (TMG10 and NF-90) and Train 3 (TFC-HR) and 4 (ESPA2) for trichloroacetic acid (TCAA) analysis are presented in Figures 4.4. Concentrations of TCAA in membrane permeate samples was less than 1 µg/L during testing of the NF-90 and TMG10, and similar results were observed for DCAA (results not presented). Although DCAA and TCAA have molecular weights similar to chloroform (128 and 163 g/mol versus 117 g/mol), both possess a negative charge at the operational pH range, which results in greater rejection (Braghetta et al. 1997, Ozaki and Li 2002, Bellona and Drewes 2005).

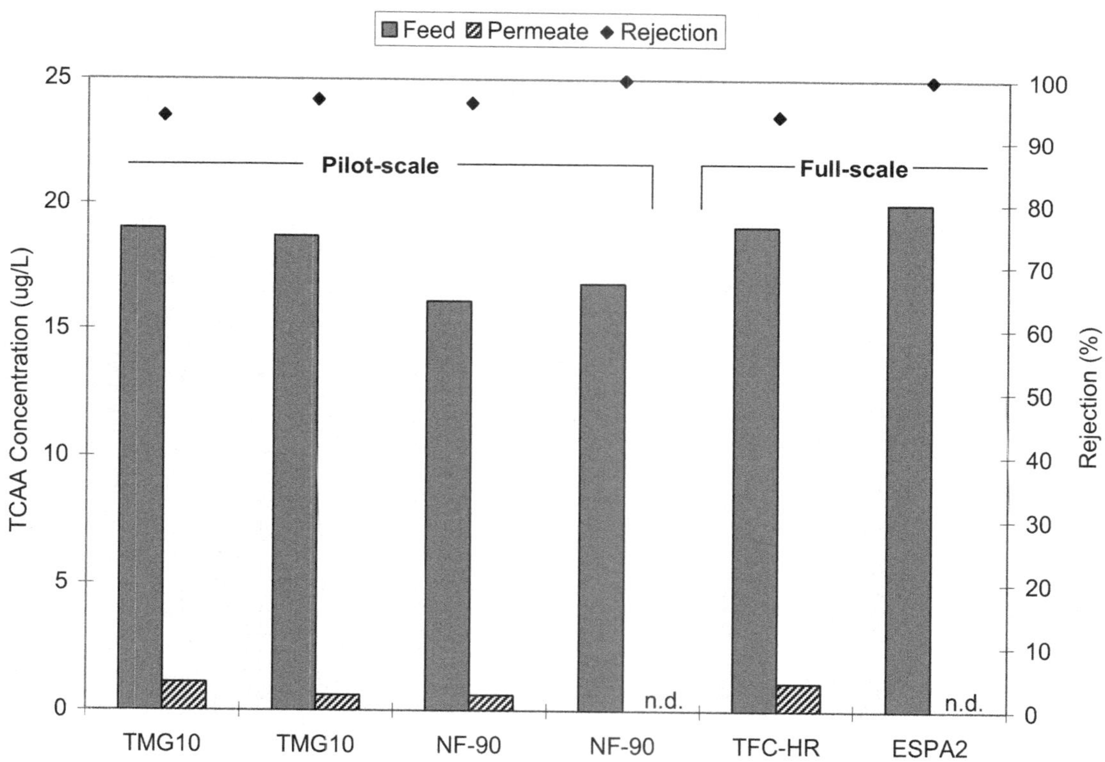

Figure 4.4 Feed and permeate concentrations of trichloroacetic acid and rejection values for the TMG10 (pilot-scale), TFC-HR (full-scale) and ESPA2 (full-scale) membranes

Two sampling campaigns for N-nitrosodimethylamine (NDMA) analysis (Tier 3) were conducted during pilot-scale testing, with samples taken from the feed and combined permeate of the pilot-skid (TMG10 and NF-90). Feed and combined permeate from Train 3 (TFC-HR) and 4

(ESPA2) were collected one time each for comparison. Summarized results from the two NDMA sampling campaigns are presented in Figure 4.5. Samples taken from the pilot-skid during sampling campaign #1 (December 4, 2004) and campaign #2 (March 22, 2005) showed NDMA concentrations in WBWRP feed water between 20 and 60 ng/L. Pilot-skid permeate concentrations were approximately 50 percent of feed concentrations and varied between 10 and 20 ng/L, which is equal to or greater than the CDPH drinking water reporting level of 10 ng/L for NDMA. The TFC-HR and ESPA2 membranes operating on full-scale treatment trains at the WBWRP were found to reject NDMA by approximately 30 percent, which resulted in permeate concentrations between 15 and 40 ng/L. The low rejection of NDMA shows the limitation of RO membranes in removing small molecular weight (74 g/mol for NDMA) nonionic polar solutes.

Samples taken from the pilot-skid during sampling campaign #1 (April 28, 2005) and campaign #2 (June 3, 2005) showed NDMA concentrations in WBWRP feed water between 10 and 25 ng/L which were generally lower than observed during pilot-scale TMG10 membrane testing. NDMA rejection by the NF-90, however, was found to be similar to the TMG10 membrane and was approximately 45 percent for both samples (Figures 4.7 and 4.14). The TFC-HR and ESPA2 membranes operating on full-scale treatment trains at the WBWRP were found to reject NDMA by approximately 30 percent, which resulted in permeate concentrations between 15 and 40 ng/L. Surprisingly, the NF membrane NF-90 was observed to reject NDMA at similar levels to the ULPRO membrane TMG10 and ESPA2 and RO membrane TFC-HR.

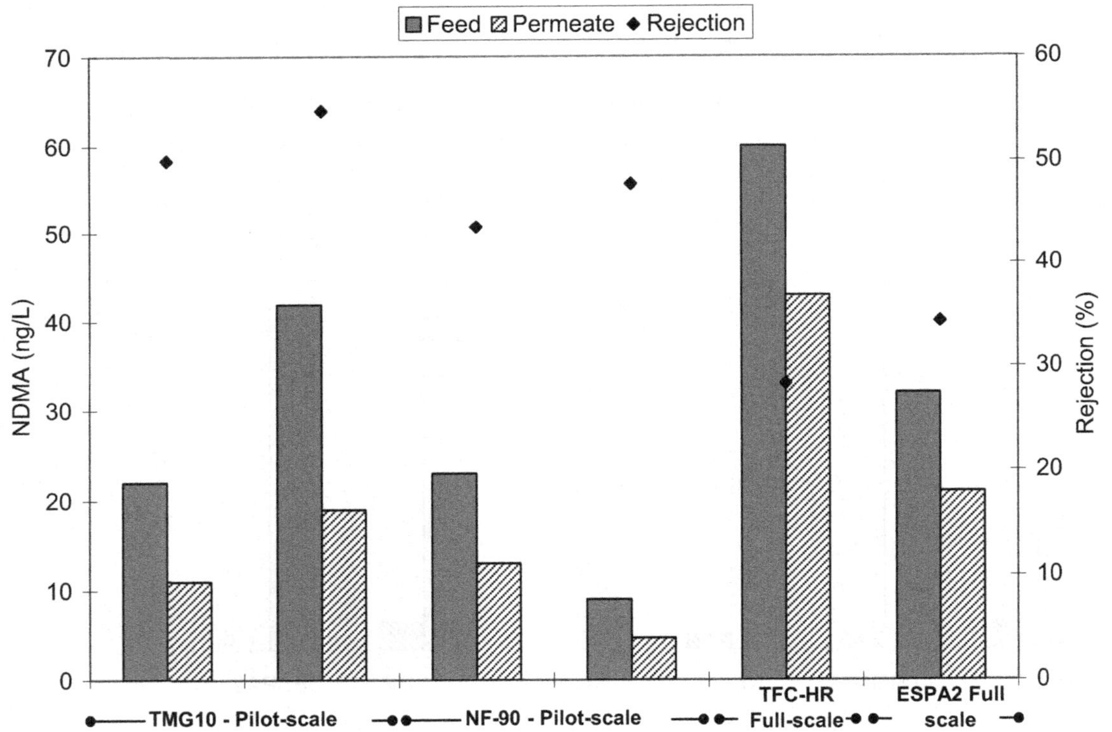

Figure 4.5 Feed and permeate concentrations of NDMA and rejection values for the TMG10 (pilot-scale), TFC-HR (full-scale) and ESPA2 (full-scale) membranes

Rejection of Unregulated Organics

During the total period of pilot-scale testing at the WBWRP, which includes the period of December 4, 2004 until June 6, 2005, samples were collected for the quantification of the Tier 2 unregulated trace organic contaminants listed in Table 2.1. WBWRP non-nitrified feed water was analyzed monthly by the GC/MS method outlined in Chapter 2 for the quantification of Tier 2 compounds including pharmaceutically active residues, personal care products and chlorinated flame retardants. The culmination of the sampling and analysis schedule was the generation of a large amount of data regarding trace organic occurrence in WBWRP feed water and corresponding concentrations. In order to effectively present trace organic contaminant data obtained through WBWRP feed water sampling, a box and whisker plot was generated and is presented in Figure 4.6. A graphical explanation of box and whisker plot presentation is presented as the legend in Figure 4.6 (Helsel 2004).

During pilot-scale testing at the WBWRP, approximately 30 feed water samples were analyzed by each GC-MS method for the Tier 2 unregulated compounds presented in Table 2.1. Sample results were not included into the data pool, however, when they exhibited surrogate standard recoveries less than 60 percent and greater than 120 percent. Due to matrix effects as a result of analyzing wastewater effluent samples, the results of approximately 50 percent of the samples analyzed by the MTBSTFA method were not included in Figure 4.8 due to poor surrogate standard recoveries. In general, the MTBSTFA method was affected more by the wastewater matrix than the PFBBr method, and therefore generated less acceptable data.

Non-nitrified feed water concentrations of the quantified trace organics were observed to be variable over the testing period and ranged from below quantification (Appendix A, Table A.1) for certain compounds to the low microgram per liter range for the chlorinated flame retardant TCPP and the pharmaceutically active compounds ibuprofen and gemfibrozil. Median concentrations of the pharmaceutically active compounds were generally between 50 and 400 ng/L with salicylic acid, ibuprofen, primidone, and carbamazepine most frequently quantified in feed water samples. Concentrations of the pharmaceutical residues ibuprofen, gemfibrozil and naproxen were highly variable, possibly due to fluctuations in wastewater treatment and/or usage patterns. The chlorinated flame retardants TCEP, TCPP, and TDCPP were quantified in more than 90 percent of the feed water samples collected with median concentrations increasing in the order of TDCPP, TCEP and TCPP. Bisphenol-A, a plasticizer that is also considered an endocrine disruptor was infrequently quantified in feed water samples in concentrations from 20 to 250 ng/L.

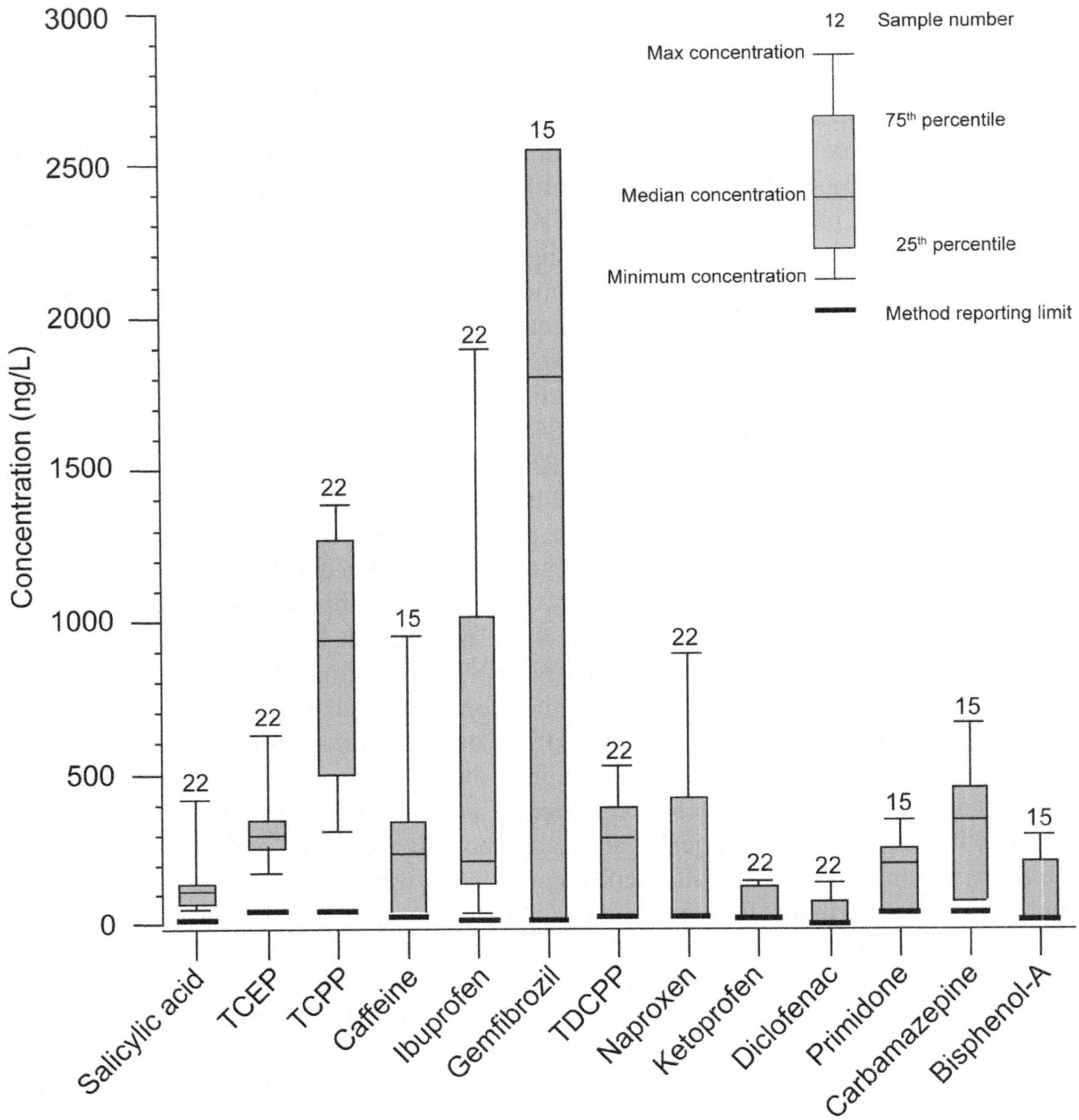

Figure 4.6 Unregulated trace organic contaminants identified in WBWRP microfiltered membrane feed water during TMG10 and NF-90 membrane testing
(Box and whisker plot explanation given in legend)

The summarized trace organic contaminant results for permeate samples taken from the TMG10 and NF-90 membranes on pilot-scale and the TFC-HR and ESPA2 membranes on full-scale are presented in Table 4.3. Bisphenol-A was detected in permeate samples taken from the TMG10, TFC-HR and NF-90 membranes. Ibuprofen was detected in one sample and quantified in the low nanogram per liter range is permeate samples taken from the TMG10 membrane on pilot-scale. The chlorinated flame retardants TCEP and TCPP were both detected and quantified in numerous permeate samples taken during testing of the TFC-HR and TMG10 membranes, but were also detected and quantified in two blank samples. Therefore, the occurrence of these compounds in permeate samples could not be verified and are not included in the Table 4.3.

However, other studies have reported the occurrence of the chlorinated flame retardant TCEP in RO permeate samples (Snyder et al. 2006). Due to concentrations present in feed water samples taken during testing (Figure 4.6) and the number of permeate samples without trace organic contaminants detections, the TMG10 and TFC-HR exhibited high removal efficiencies for compounds. Of the solutes quantified in WBWRP feed water, gemfibrozil, ketoprofen, naproxen and bisphenol-A were detected in one or more NF-90 permeate samples. In contrast, no permeate detections were observed for any of the compounds present in WBWRP feed water during ESPA2 sampling campaigns. Quantifiable concentrations of the organic contaminants found in NF-90 permeates were below 10 ng/L, with the exception of bisphenol-A, which was quantified in five permeate samples at an average of 25 ng/L. Bisphenol-A, however, was found previously to leach off of membrane material during laboratory-scale testing and therefore the origin of the bisphenol-A present in NF-90 permeate samples is unknown.

Sampling campaigns were also performed to determine the removal efficiency of the steroidal hormones estriol, 17β-estradiol and testosterone by the TMG10, TFC-HR, and NF-90 membranes. Although low nanogram per liter concentrations (<10 ng/L) of all three hormones were detected in WBWRP feed water, permeate concentrations were below the detection limit for all three hormones.

Table 4.3
Summarized results of unregulated organic concentrations observed in permeate samples

	TMG10 Pilot-Scale	TFC-HR Full-scale	NF-90 Pilot-scale	ESPA2 Full-scale
Samples (n)	14	6	10	5
Compound	ng/L	ng/L	ng/L	ng/L
Salicylic Acid	n.d. (14)	n.d. (6)	n.d.	n.d.
TCEP	n.d. (14)	n.d. (6)	n.d.	n.d.
TCIPP	n.d. (14)	n.d. (6)	n.d.	n.d.
Caffeine	n.d. (14)	n.d. (6)	n.d.	n.d.
Ibuprofen	n.d. (12) - detect (1) - 27 (1)	n.d. (6)	n.d.	n.d.
Gemfibrozil	n.d. (14)	n.d. (6)	n.d (3) - 9 (5)	n.d.
Primidone	n.d. (14)	n.d. (6)	n.d.	n.d.
TDCPP	n.d. (14)	n.d. (6)	n.d. (7) - 7 (3)	n.d.
Naproxen	n.d. (14)	n.d. (6)	n.d. (9) - Detect (1)	n.d.
Ketoprofen	n.d. (14)	n.d. (6)	n.d.	n.d.
Diclofenac	n.d. (14)	n.d. (6)	n.d.	n.d.
Carbamazepine	n.d. (14)	n.d. (6)	n.d.	n.d.
Bisphenol A	n.d. (9) - detect (4) - 25 (1)	n.d (4) - detect (2)	n.d. (5) - 25 (5)	n.d.

n.d. – not detected

To further validate unregulated trace organic rejection by the TMG10 and NF-90 membranes, challenge condition experiments (spiking experiment) were conducted on the TMG10 membrane after 2,950 hours of operation during testing at the WBWRP according to the protocol outlined in Chapter 2. The results of the TMG10 membrane spiking experiment are presented in Table 4.4. No spiked organic solutes in the feed water of the pilot-skid were detected in any permeate samples. Based on the concentrations of the spiked solutes that have never been detected in WBWRP feed water including phenacetine, clofibric acid, mecoprop and dichloroprop, it appears that the spike experiment resulted in an addition of 300 to 500 ng/L of each solute. Concentrations of certain solutes including diclofenac, naproxen, carbamazepine and primidone, however, were quantified at concentrations in the 100 ng/L range indicating that recoveries of these compounds during the analytical method was poor although surrogate standard recoveries were acceptable.

Table 4.4
Summary of average feed and permeate Tier 2 concentrations during the WBWRP TMG10 membrane spiking experiments

Sample Location	Feed	Combined Permeate
Number of Samples	3	6
Compound	ng/L	ng/L
Phenacetine	443	n.d.
Salicylic acid	111	n.d.
TCEP	993	n.d.
TCPP	2177	n.d.
Caffeine	238	n.d.
Clofibric acid	527	n.d.
Ibuprofen	590	n.d.
Mecoprop	348	n.d.
Dichloroprop	370	n.d.
Gemfibrozil	1159	n.d.
TDCPP	927	n.d.
Naproxen	115	n.d.
Fenofibrate	296	n.d.
Ketoprofen	523	n.d.
Diclofenac	164	n.d.
Carbamazepine	88	n.d.
Primidone	157	n.d.
Bisphenol-A	500	n.d.
Testosterone	275	n.d.

The results of the NF-90 spiking experiment are presented in Table 4.5. Similar to TMG10 membrane, feed water concentrations of spiked solutes were in the 200 to 400 ng/L range even for solutes never observed in WBWRP feed water (phenacetine, mecoprop and dichloroprop). Concentrations of salicyclic acid and fenofibrate were observed to be lower than expected based on other solutes. Concentrations of spiked solutes in permeate samples were non-detectable indicating that the NF-90 efficiently (90 percent) removes the trace organic contaminants employed during the spiking experiment.

Table 4.5
Summary of average feed and permeate Tier 2 concentrations during the WBWRP NF-90 membrane spiking experiments

Sample Location	Feed	Combined Permeate
Number of Samples	3	6
Compound	ng/L	ng/L
Phenacetine	374	n.d.
Salicylic acid	80	n.d.
TCEP	887	n.d.
TCPP	1386	n.d.
Caffeine	1288	n.d.
Clofibric acid	263	n.d.
Ibuprofen	432	n.d.
Mecoprop	208	n.d.
Dichloroprop	282	n.d.
Gemfibrozil	1789	n.d.
TDCPP	400	n.d.
Naproxen	574	n.d.
Fenofibrate	107	n.d.
Ketoprofen	200	n.d.
Diclofenac	218	n.d.
Primidone	396	n.d.
Carbamazepine	343	n.d.
BPA	209	n.d.

PERFORMANCE OF ULPRO AND NF MEMBRANES USING NITRIFIED/ DENITRIFIED MICROFILTERED WASTEWATER EFFLUENT

Performance of NF-4040 NF Membrane

By the time NF-90 membrane pilot-scale testing was terminated at the WBWRP it was determined that although the NF-90 membrane could match the permeate water quality of the ULPRO and RO membranes evaluated, the characteristics that made the membrane a top choice for pilot-scale testing (i.e. high permeate flux, low pressure) were reduced through compaction and fouling. A significant reduction in operating costs therefore would require the application of a low pressure and low fouling NF membrane. Subsequently, the laboratory-scale NF testing results (Chapter 3) were revisited in order to select a more 'typical' NF membrane in terms of flux, and salt rejection. The next best performing NF membrane, the NF-200, could not be procured and the next best performing NF membrane was the NF-4040 membrane. Therefore, the NF-4040 membrane was chosen for pilot-scale testing.

Membrane Operational Performance

Pilot-scale testing with 4040 NF elements (termed NF-4040) treating nitrified/denitrified tertiary treated effluent started at the Scottsdale Water Campus on August 1, 2005 and ended on October 20, 2005 for a total testing period of 1,700 hours. The pilot-skid operational set-points for the NF-4040 were identical to the NF-90 set-points during WBWRP testing and are presented in Table 4.6.

Table 4.6
Operational set points during pilot-scale NF-4040 membrane testing

Membrane Parameter	Units	Dow/Filmtec NF-4040
Element Membrane Area	m^2	7.6
	ft^2	82
Pilot Membrane Area	m^2	160
	ft^2	1722
Recovery	percent	82
Permeate Flux	L/m^2-hr	21.2
	gfd	12.4
Permeate Flow	L/min	56.4
	gpm	14.9
Concentrate Flow	L/min	12.5
	gpm	3.3
Feed Flow	L/min	68.9
	gpm	18.2
pH		6.3
Antiscalant Dose	ppm	3

The NF-4040 membrane feed pressure and net driving pressure during pilot-scale testing is presented in Figure 4.7. The start-up feed pressure was approximately 425 kPa (61 psi), after which the feed pressure stabilized at approximately 490 kPa (70 psi) after a period of approximately 200 hours until 1,200 hours of runtime. Around 1,200 hours of runtime the feed pressure increased by approximately 100 kPa (15 psi) as a result of the application of permeate backpressure (approximately 100 kPa) to the first-stage in order to equalize permeate flows between the first and second stage of the pilot-skid. After the adjustment of permeate flows, the NF-4040 membrane feed pressure and net driving pressure displayed a slight increase until approximately 1,400 hours when a significant increase in feed pressure began and lasted over the final 300 hours of testing.

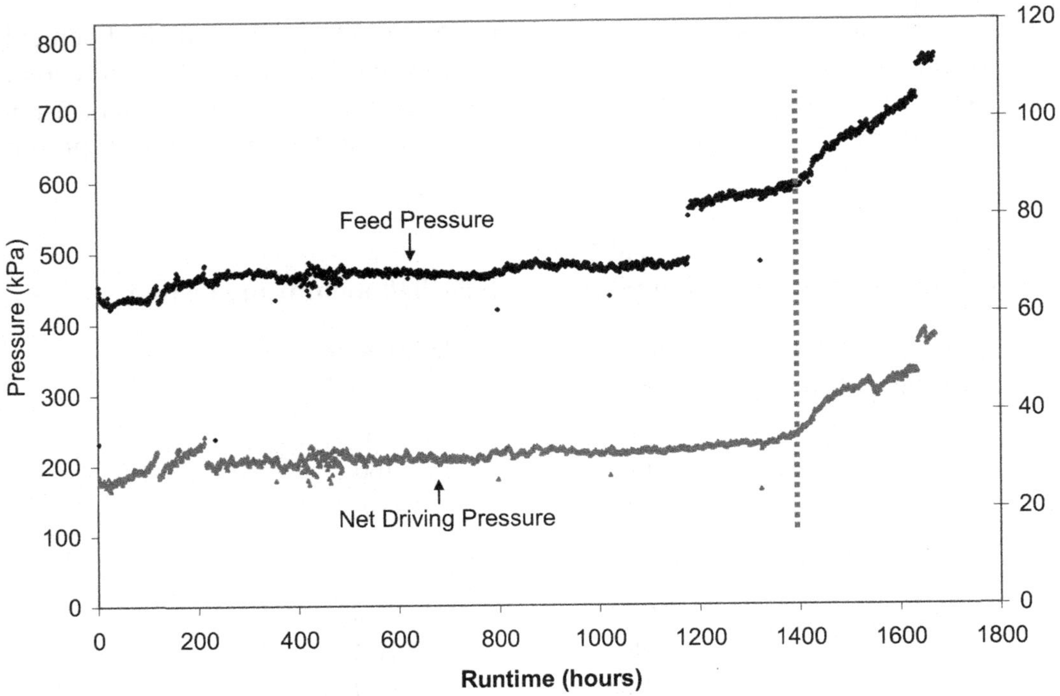

Figure 4.7 NF-4040 membrane feed pressure and net driving pressure during pilot-scale testing at the SWC

The TCSF of the NF-4040 membrane is presented in Figure 4.8. Initially, the TCSF of the NF-4040 membrane was approximately 0.1 L/m^2-hr-kPa (0.4 gfd/psi) and stabilized after two weeks at approximately 0.08 L/m^2-hr-kPa (0.33 gfd/psi). A TCSF of 0.08 L/m^2-hr-kPa for the NF-4040 membrane is approximately three times greater than the stabilized TCSF of the TMG10 and NF-90 membranes (Figure 4.2), and would lead to a considerable reduction in energy costs. After permeate backpressure was applied to the pilot-scale unit, the TCSF slowly decreased to approximately 0.075 L/m^2-hr-kPa (0.3 gfd/psi) over the next 200 hours until around 1,400 hours of runtime the TCSF dropped fairly quickly to 0.05 L/m^2-hr-kPa (0.25 gfd/psi). The considerable drop in TCSF over 250 hours indicated a severe fouling incident that was not typical of the first 1,400 hours of NF-4040 membrane operation.

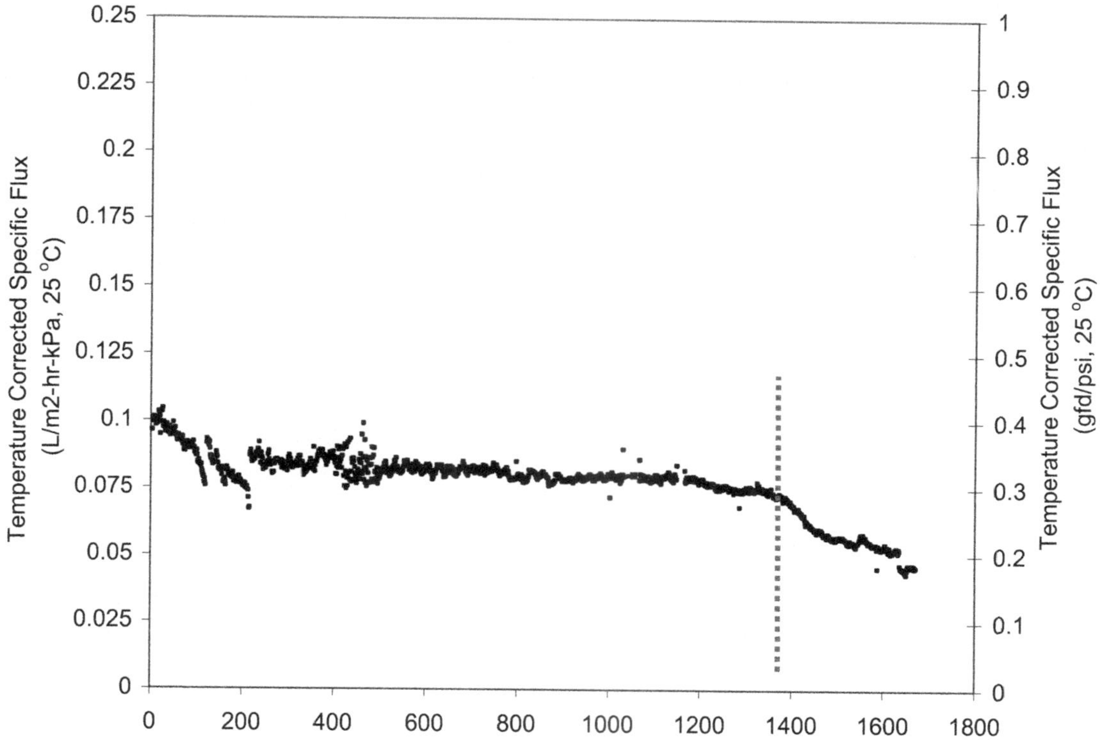

Figure 4.8 Temperature corrected specific flux of the NF-4040 membrane during pilot-scale testing at the SWC

An autopsy was performed on the lead and tail end NF-4040 elements and is presented in detail in Chapter 6. Based upon the findings of the NF-4040 membrane autopsy, it appears that the main cause of the permeability decline was biofouling, which may have been exacerbated with the application of permeate back-pressure. Subsequent monochloramine measurements performed at the end period of SWC TMG10 membrane testing revealed the absence of a monochloramine residual. WBWRP feed water, being non-nitrified secondary effluent with approximately 40 mg-N/L, carries a monochloramine residual of approximately 5 mg-Cl/L into the RO trains which has been found to effectively control biofouling. It appears that either the absence of free ammonia or insufficient chlorine dosage in the SWC hinders the formation of chloramines or makes it difficult to form a sufficient amount to carry a chloramine residual through the MF units and into the RO treatment trains.

Up until approximately 1,400 hours of operation when a severe fouling incident was observed, the NF-4040 membrane was exceeding operational performance expectations based on feed pressure and specific flux values. Although it is unclear whether the observed fouling would have eventually occurred if back-pressure had not been applied to the stage-1 permeate, fouling issues were observed throughout TMG10 membrane testing at the SWC which, was conducted after NF-4040 membrane testing (see TMG10 membrane TCSF values, Figure 4.9). During TMG10 membrane testing at the SWC, a stabilized TCSF was never sustained and operational performance fell below the value required to achieve cost savings based on historical TFC-HR and ESPA2 operation (0.022 L/m^2-hr-kPa, Figure 4.9). Because severe biofouling was identified through the TMG10 membrane autopsy (Appendix E) and TMG10 membrane operation was stable at the WBWRP, it is hypothesized that operation at the SWC, especially the lack of maintaining a monochloramine residual, was likely responsible for the poor performance of the TMG10 membrane throughout testing and the NF-4040 membrane during the final 200 hours of operation. The fouling trend of the NF-4040 at the SWC, however, would be detrimental to long term performance and would point to the need of more frequent clean-in-place (CIP) events for this membrane. Additionally, operation of Train 16 (TFC-HR membranes) at the SWC during the final 200 hours of NF4040 membrane operation displayed a similar 50 percent reduction in specific flux as a result of fouling and increasing feed pressure to maintain the flux set point (Figure 4.10).

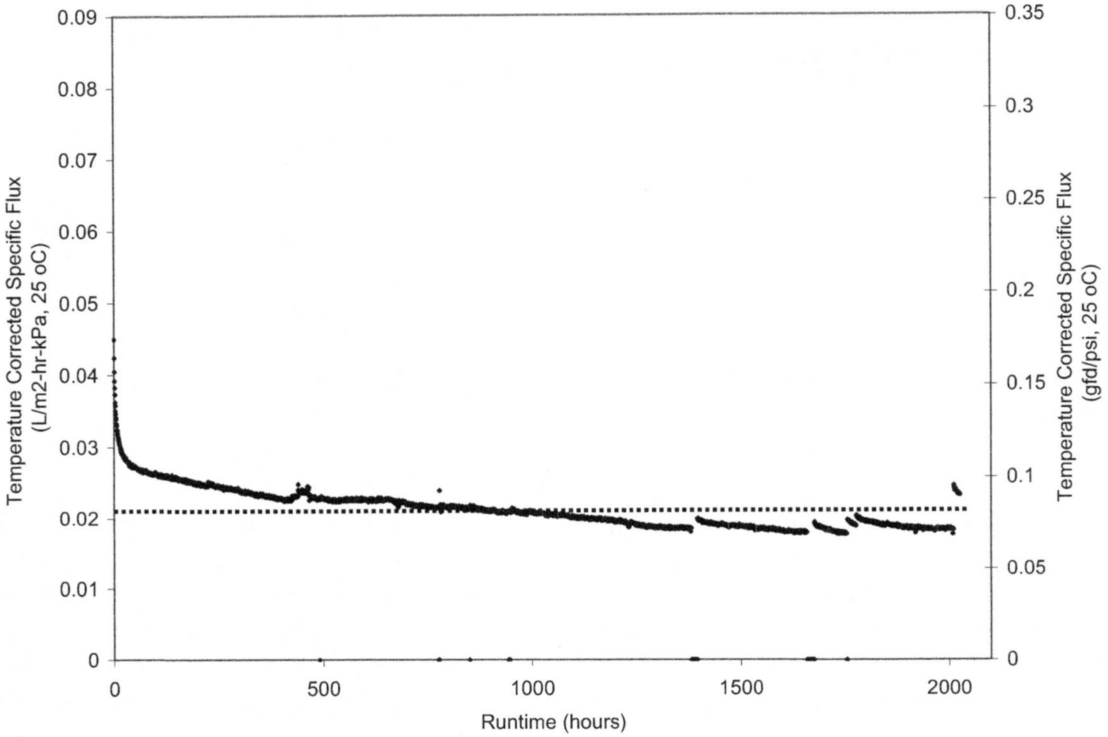

Figure 4.9 Temperature corrected specific flux of TMG10 membrane during pilot-scale testing at the SWC
(dashed line represents acceptable TCSF based on historic SWC full-scale data employing TFC-HR membranes, Appendix C, Figure C.2)

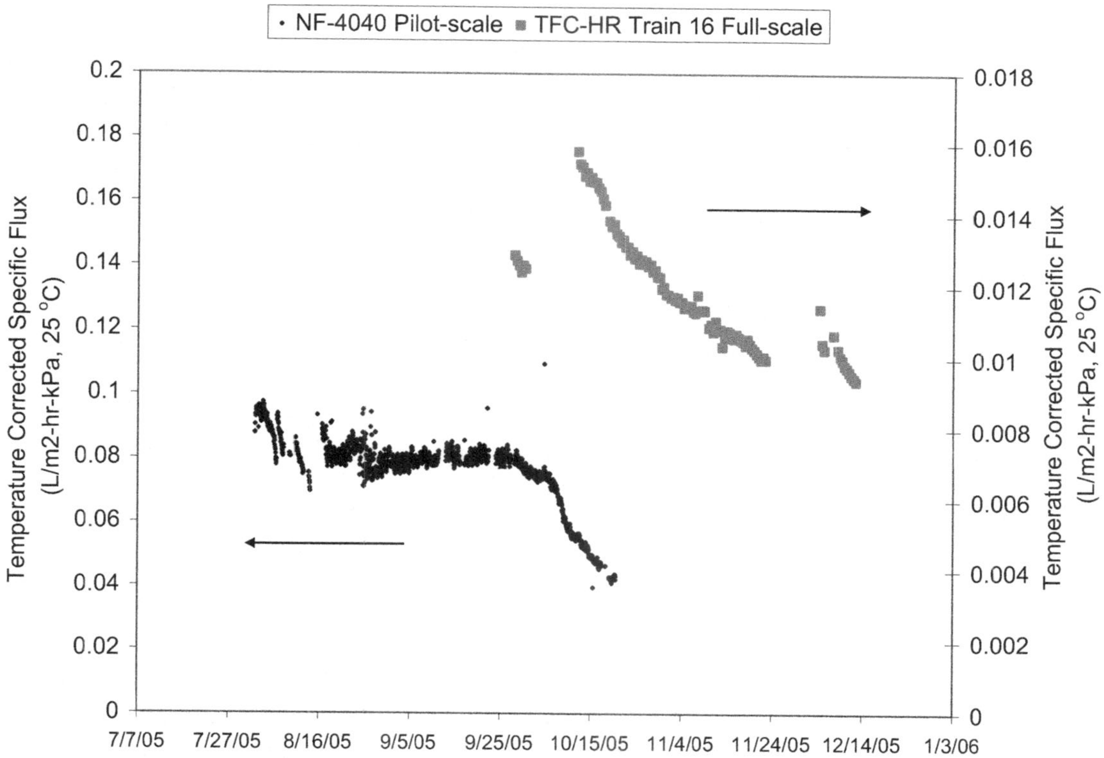

Figure 4.10 Temperature corrected specific flux values for the NF-4040 membrane on pilot-scale and the TFC-HR on Train 16 at the SWC

Rejection of TOC, Inorganic ions, and Nitrogen

Total organic carbon (TOC) concentrations and conductivity values for feed and permeate samples taken during testing for both the NF-4040 (pilot-scale) and the TFC-HR (full-scale) membranes are presented in Table 4.7. During the testing period, the TOC concentrations in the feed water were consistently reduced by approximately 94 percent resulting in permeate concentrations of 0.3 mg-C/L on average. For comparison, the TFC-HR membrane employed on full-scale achieved TOC rejection values of 96 percent resulting in permeate TOC concentrations of approximately 0.23 mg-C/L. The NF-4040 therefore, was found to adequately reject TOC and permeate concentrations were consistently lower than the CDPH draft regulation requirement of 0.5 mg-C/L during the testing period.

Table 4.7

Summarized TOC and conductivity feed and permeate concentrations and subsequent rejection values observed on pilot-scale and full-scale during NF-4040 pilot-scale testing

	Unit	Pilot-scale NF-4040 Average (Stdev) 9 samples	Full-scale TFC-HR Average 1/04 - 5/05
TOC			
Feed TOC	mg-C/L	5.5 (+/- 0.3)	5.74
Permeate TOC	mg-C/L	0.3 (+/- 0.08)	0.23
TOC Rejection	percent	94.3 (+/- 1.2)	96
Conductivity			
Feed Conductivity	μS/cm	1 696 (+/- 62)	1 650
Permeate Conductivity	μS/cm	875 (+/- 117)	26
Conductivity Rejection	percent	48.4 (+/- 7)	98.4

Not unexpectedly, the NF-4040 membrane was found to only marginally reject total dissolved solids, on average reducing feed conductivity by 48 percent (Table 4.7). ICP analysis performed on permeate samples indicated that monovalent ions such as sodium and potassium were rejected by approximately 40 percent while the removal of divalent cations such as calcium and magnesium were rejected in the 70 to 80 percent range (Figure 4.11). While anions such as sulfate were almost completely removed by the NF-4040 membrane, chloride was only marginally rejected (25 percent, Figure 4.11) and nitrate concentrations were equal to or greater than feed concentrations, often resulting in negative rejection values (Figure 4.12).

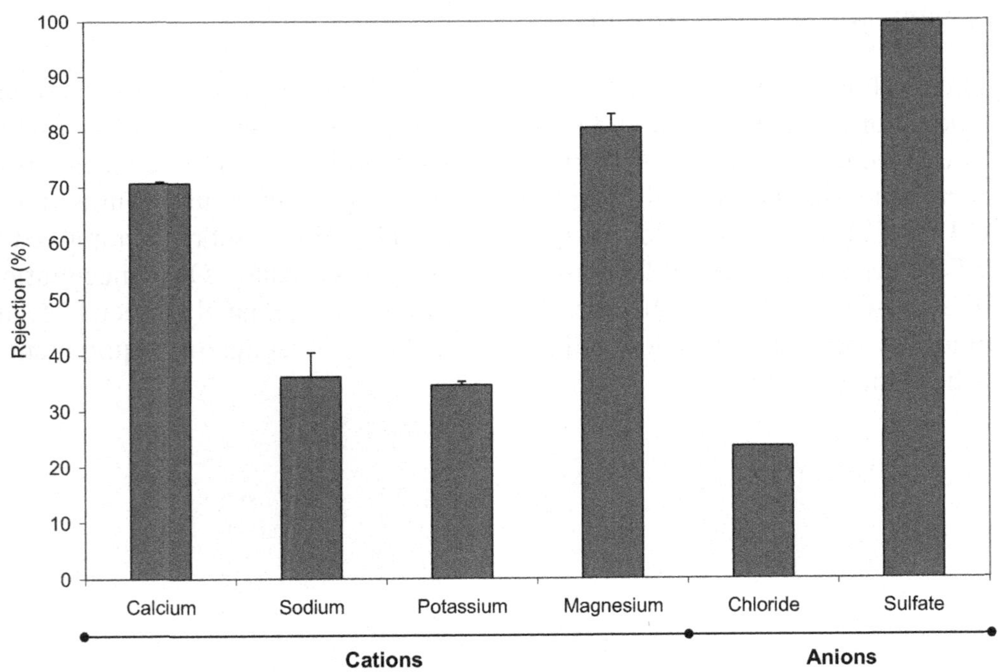

Figure 4.11. Average rejection of select ions by the NF-4040 membrane (over 1,600 hours)

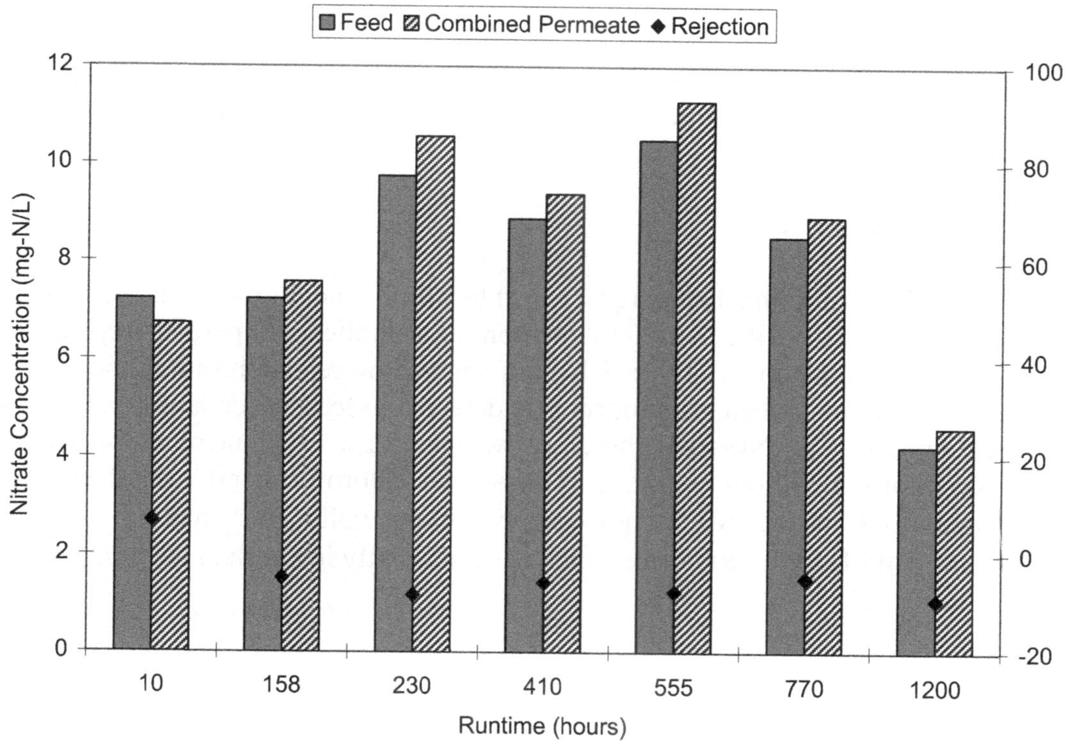

Figure 4.12 Feed, permeate concentrations and subsequent rejection values for nitrate during pilot-scale testing of the NF-4040 membrane

Poor nitrate rejection may be related to ion mobility and pore and permeate electroneutrality during operation. Researchers have shown that during NF and to a lesser extent RO experiments with ternary and quaternary ion mixtures, negative rejection of certain ions is often observed (Tsuru et al. 1991, Hagmeyer and Gimbel 1998, Schaep et al. 1999, Bowen et al. 2002). This can occur for example when a mobile co-ion (to the membrane charge) is introduced at much lower concentration than a mobile counter-ion in solution (i.e. 20 mM NaCl versus 1 mM $NaNO_3$). In this case, a small percentage of counter-ions (i.e. Na^+) in the feed water permeate (since no membrane rejects 100 percent of ions) and as a consequence forces a large percentage of the mobile less concentrated co-ions (NO_3^-) to permeate in order to maintain permeate electroneutrality. It is hypothesized that during testing, the highly mobile nitrate ion freely permeates the membrane to balance the charge of less mobile but more concentrated cations (Tsuru et al. 1991b, Bowen et al. 2002).

Regardless of the mechanism, the largest disadvantage identified for employing a low pressure NF to treat recycled water was the rejection of nitrate, Thus, NF cannot serve as the sole barrier for nitrate to meet federal primary drinking water standards or the more stringent CDPH draft groundwater recharge nitrogen regulation (CDPH 2007). However, properly designed and operated nitrification/denitrification facilities can commonly meet total nitrogen concentrations of less than 10 mg-N/L in secondary and tertiary treated effluents (Metcalf and Eddy 2003).

Although the rejection of nitrate and other monovalent inorganic ions was poor during testing, brine stream total dissolved solids concentrations of the NF-4040 membrane during testing were 45 percent less on average than the RO membrane, TFC-HR, operating at the SWC (brine conductivities of 2,970 μS/cm for the NF-4040 versus 6,580 μS/cm for the TFC-HR). In addition, the sodium adsorption ratio (SAR) of the NF-4040 membrane brine stream (SAR of 10.7) was 24 percent less than the brine stream of the TFC-HR membrane (SAR of 14). In areas where brine disposal is a major issue, such as landlocked application, reducing both TDS concentration and SAR of a brine stream increases the disposal or reuse options for high-pressure membrane application.

Rejection of Regulated Organics

The results of five sampling campaigns for chloroform and bromoform analysis are presented in Table 4.8. Initially, the NF-4040 membrane exhibited 50 percent rejection of chloroform (see sampling campaign #1, Table 4.16) but showed no rejection throughout the rest of the testing period. In addition, when bromoform was detected in feed water samples, permeate concentrations were similar. The NF-4040 therefore was the least efficient membrane of the membranes tested on pilot-scale in removing the THM solutes chloroform and bromoform. Feed water THM concentrations at the SWC, however, were generally less than 20 μg/L for chloroform and around 1 μg/L for bromoform, which is significantly lower than the current MCL for THMs of 80 μg/L.

Table 4.8
Summary of chloroform and bromoform feed and NF-4040 membrane permeate concentrations during testing at the SWC

	Chloroform		Bromoform	
	Feed	Permeate	Feed	Permeate
NF-4040 Pilot-scale	μg/L	μg/L	μg/L	μg/L
Campaign 1	15.5	8.5	n.d.	n.d.
Campaign 2	20.1	19.5	NA	0.8
Campaign 3	19.3	18.8	1.7	1.7
Campaign 4	9.5	10	0.7	0.8
Campaign 5	10.1	9.6	1.1	1.2

N/A – not analyzed

Trace Organic Contaminant Removal

Feed and permeate samples were collected at various stages of operation for trace organic contaminant analysis by the GC/MS method described in the method section (compounds listed in Table 2.1). Results of trace organic contaminant occurrence and rejection by the NF-4040 membrane for samples collected 24 hours after start-up are summarized in Figure 4.13. Of the compounds that can be identified in environmental samples at the ppt level with the GC/MS method, only 10 out of 19 compounds were detected and/or quantified in the MF treated feed water. With concentrations between 500 and 2,000 ng/L, the chlorinated flame retardants TCPP, TCEP, and TDCPP were quantified at the highest concentrations of the detected compounds. Other compounds identified in the feed stream include the pharmaceutically active residues salicylic acid, naproxen, gemfibrozil, ketoprofen, diclofenac, primidone, and carbamazepine. At start-up (after 24 hours), seven of the ten compounds detected in the feed stream were also detected in the permeate stream of the NF-4040 membrane. With the exception of ketoprofen, however, the rejection of the trace organic contaminants identified in the feed water was greater than 80 percent which led to concentrations of less than 80 nanograms per liter for all compounds quantified in permeate samples. For comparison, permeate samples taken from the TFC-HR membrane operating on full-scale had no detectable concentrations of any trace organic contaminants found in MF treated feed water.

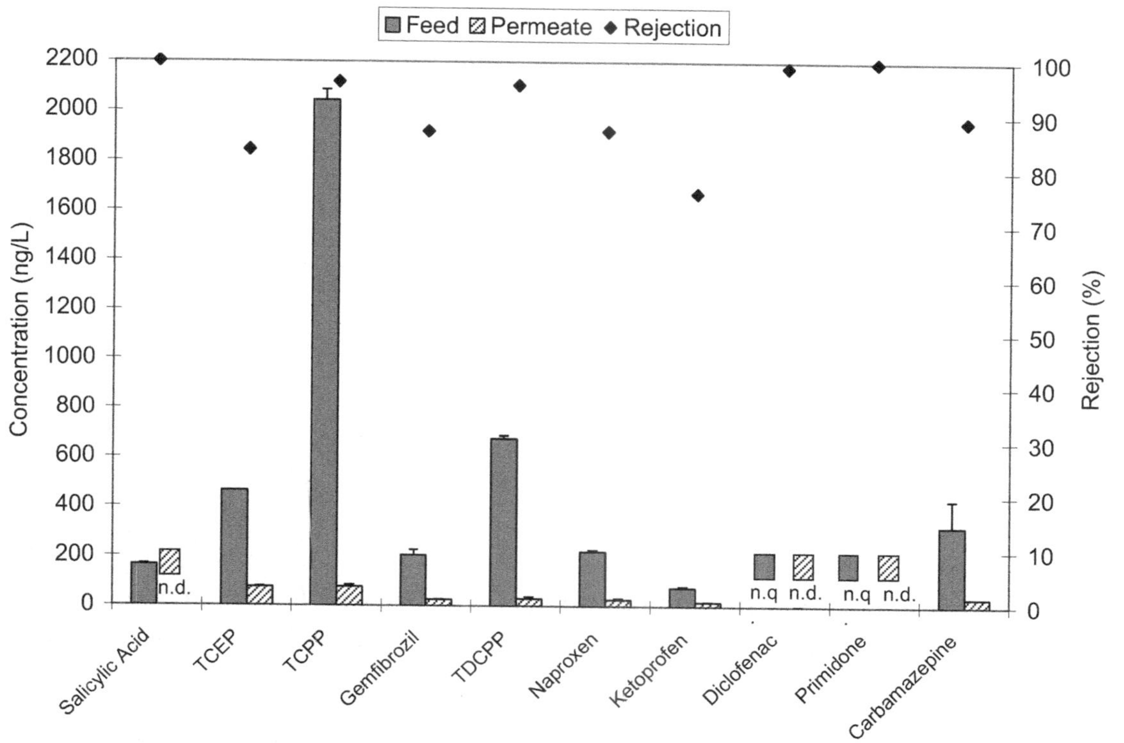

Figure 4.13 Un-spiked feed, permeate concentrations and subsequent rejection values of select trace organic contaminants during pilot-scale testing of the NF-4040 membrane (results from 6 feed and 6 permeate samples collected 24 hours after start-up. The two acronyms n.d. and n.q. are not detected (signal to noise ratio < 3:1) and not quantified (3:1 < signal to noise ratio < 11:1), respectively)

Although the rejection of the trace organic contaminants identified in feed water samples varied between 80 and 100 percent at 24 hours after start-up, rejection was observed to increase as operation continued. The results of four comprehensive sampling campaigns spaced throughout NF-4040 membrane operation in terms of rejection are presented in Figure 4.14. Rejection was observed to increase with operation and with the exception of TCEP and on one occasion naproxen, compounds present in feed water were rejected greater than 95 percent after 800 hours of operation. It is hypothesized that organic fouling, which established during the initial 100 to 200 hours of operation (Figure 4.8), and to a lesser extent membrane compaction after a period of operation could lead to a higher degree of rejection for certain constituents by NF membranes, especially organic compounds occurring at the ppt level.

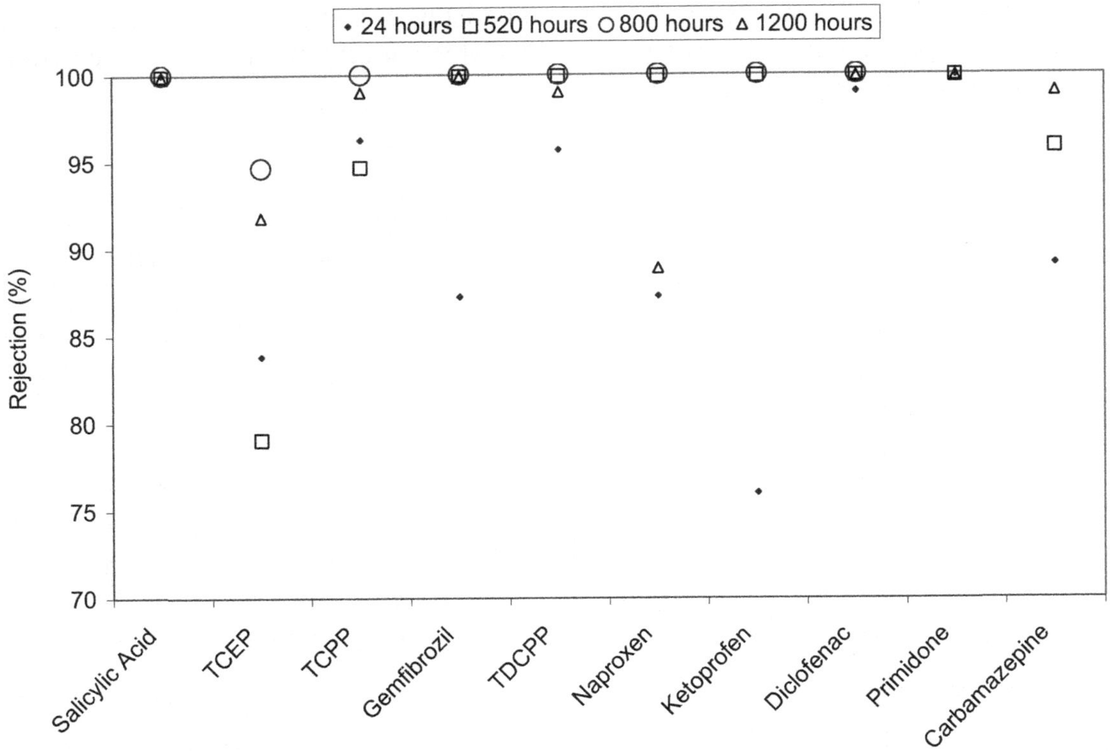

Figure 4.14 Rejection of trace organic contaminants identified in NF-4040 membrane feed water during four sampling campaigns spanning the range of testing

The effect of membrane fouling on the rejection of trace organic contaminants, however, has only recently been addressed and conclusions from these studies have been somewhat contradictory. Ng and Elimelech (2004) found that colloidal silica fouling reduced the back diffusion of solutes present in feed water and significantly decreased the rejection of hydrophilic and hydrophobic organic solutes. Xu et al. (2006a), however, determined that organic fouling resulted in an increased membrane surface charge as well as increased adsorption capacity which, resulted in an increased rejection of negatively charged organic solutes and hydrophobic organic solutes. Similarly to the findings of Ng and Elimelech (2004), Xu et al. (2006a) observed a decrease in the rejection of hydrophilic nonionic organic solutes as membrane fouling increased, but found that the rejection of the majority of solutes investigated was enhanced by effluent organic matter fouling.

To further evaluate trace organic contaminant rejection and because feed water concentrations of trace organics contaminants were often low or not detectable or quantifiable, a spiking event was conducted where a variety of compounds spanning a broad range of physical and chemical characteristics were continually added to the NF-4040 feed stream at a concentration of approximately 700 nanogram per liter at 1,200 hours of operation (testosterone and 17β-estradiol were spiked at approximately 300 nanogram per liter). Summarized results of the NF-4040 spiking experiment are presented in Figures 4.15 and 4.16. Nonionic (uncharged) trace organic contaminant rejection was found to depend mostly upon contaminant size, with the smaller molecular weight compounds acetaminophen, phenacetine and caffeine being the compounds least rejected by the NF-4040 membrane. In general, however, nonionic trace organic contaminant rejection during the spiking experiment exceeded for most compounds levels greater than 90 percent. In contrast, the rejection of the ionic negatively charged trace organics employed in the study was independent of solute size and was in excess of 97 percent resulting in permeate concentrations of less than 20 ng/L (Figure 4.16).

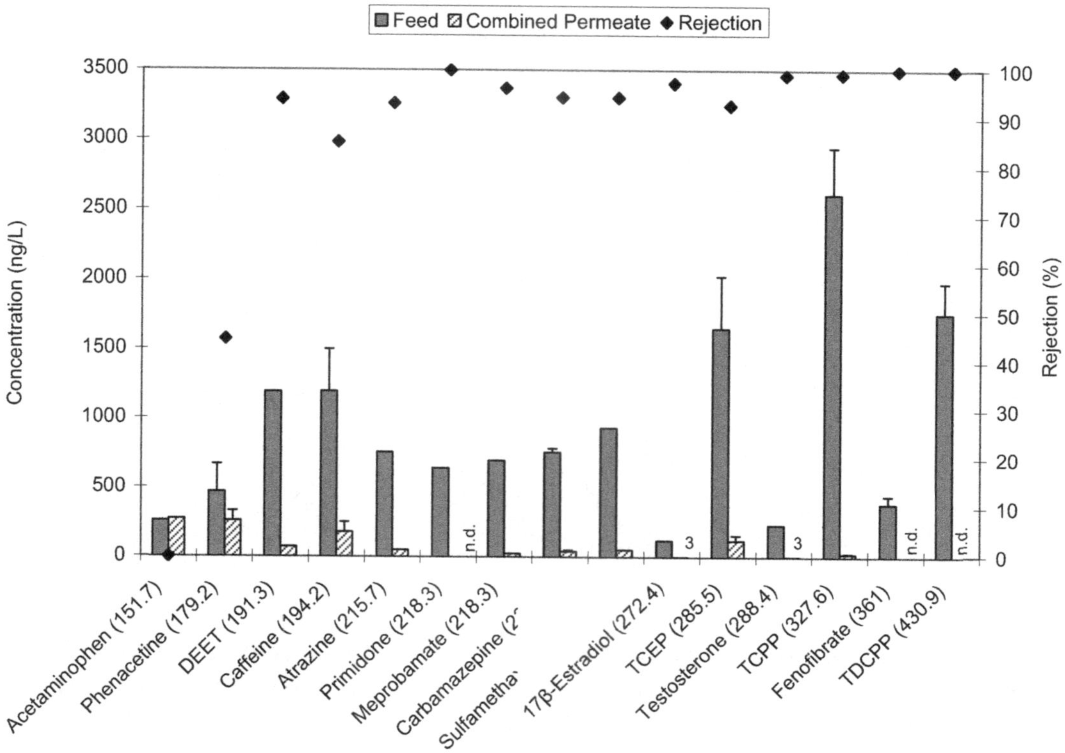

Figure 4.15. Spiked feed and permeate concentrations and subsequent rejection values of non-ionic trace organic contaminants (molecular weights in parenthesis) during pilot-scale testing of the NF-4040 membrane
(spiking and sampling occurred at end of testing period)

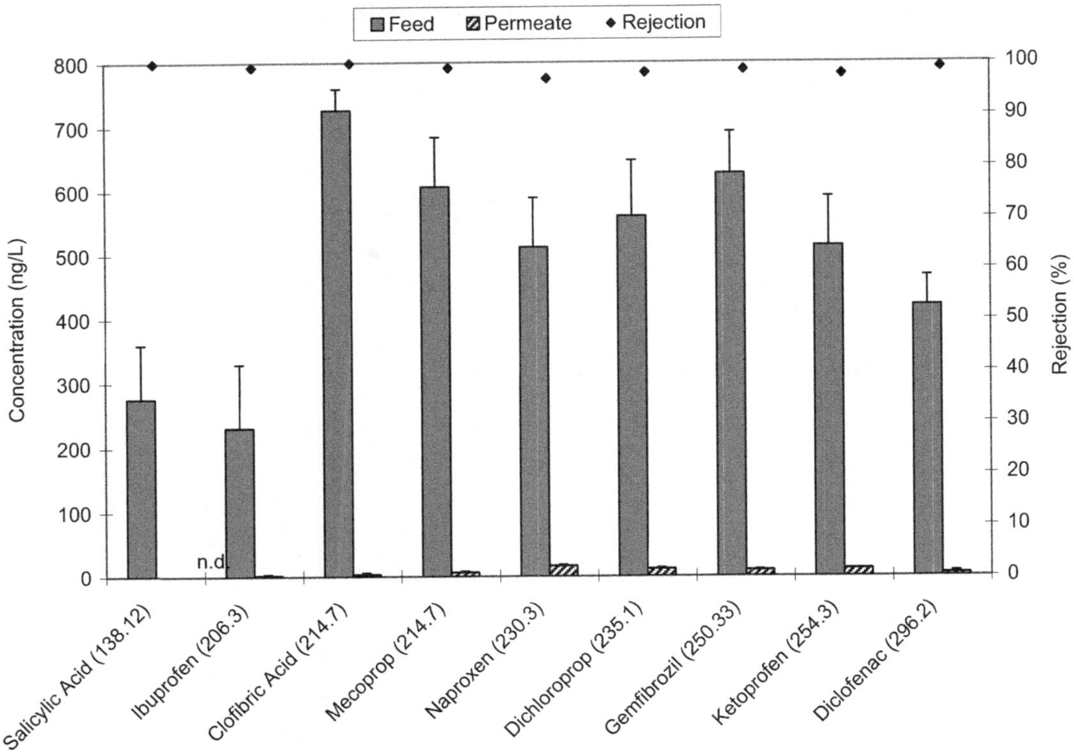

Figure 4.16 Spiked feed and permeate concentrations and subsequent rejection values of ionic trace organic contaminants (negatively charged, molecular weights in parenthesis) during pilot-scale testing of the NF-4040 membrane
(spiking and sampling occurred at end of testing period)

The rejection of nonionic trace organic compounds by NF and RO membranes has been found to depend mainly upon solute size and to a lesser extent solute properties such as hydrophobicity and hydrophilicity, and dipole moment and polarity (Kimura et al. 2003, Kimura et al. 2004, Bellona et al. 2004, Nghiem et al. 2005). Previous research conducted on the rejection of hydrophobic steroid hormones on bench-scale have determined that adsorption and partitioning of these compounds to and through a membrane inhibits rejection by NF and RO membranes (Kimura et al. 2003, Schäfer et al. 2003, Ng and Elimelech 2004, Nghiem et al. 2005, Braeken et al. 2006). Because steroid hormones can interfere with the endocrine system at very low concentrations, the complete rejection of these compounds is desired during indirect potable reuse applications. Results from the NF-4040 spiking experiment indicate that the three most hydrophobic compounds testosterone, 17β-estradiol, and fenofibrate (log K_{ow} = 3.5, 4.1, and 4.8, respectively) were rejected in excess of 95 percent by the NF-4040 membrane. Based on the molecular weights of the nonionic solutes spiked into the feed water, it can be interpolated from results present in Figure 4.14 that the theoretical molecular weight cut-off (defined as molecular weight where 90 percent rejection is observed) of the NF-4040 is approximately 200 g/mol. These findings support that size exclusion was the dominant rejection mechanism of nonionic trace organic solutes and there was no noticeable breakthrough of hydrophobic trace organic contaminants as has been reported in recent literature.

The rejection of ionic trace organic contaminants has generally been observed to be greater than nonionic solutes of the same size due to electrostatic rejection phenomena, such as electrostatic repulsion and Donnan exclusion (Bellona and Drewes 2005, Braeken et al. 2006, Sorin et al. 2006). Due to sizes greater than most inorganic ions in the feed stream, it is probable that the ionic solutes are the least mobile of the negatively charged electrolytes and therefore are excluded from the membrane and therefore the permeate stream.

Performance of TMG10 Membrane

TMG10 membrane operational set points during testing at the SWC were identical to the operational set points at the WBWRP and are presented in Table 4.1. The feed pressure of the TMG10 membrane during testing at the SWC is presented in Figure 4.17. The start-up feed pressure of the TMG10 membrane was approximately 830 kPa (120 psi) which is approximately 80 kPa (12 psi) less than at the WBWRP as a result of the feed water temperature being 6°C greater at the SWC (Figures 4.1 and 4.17).

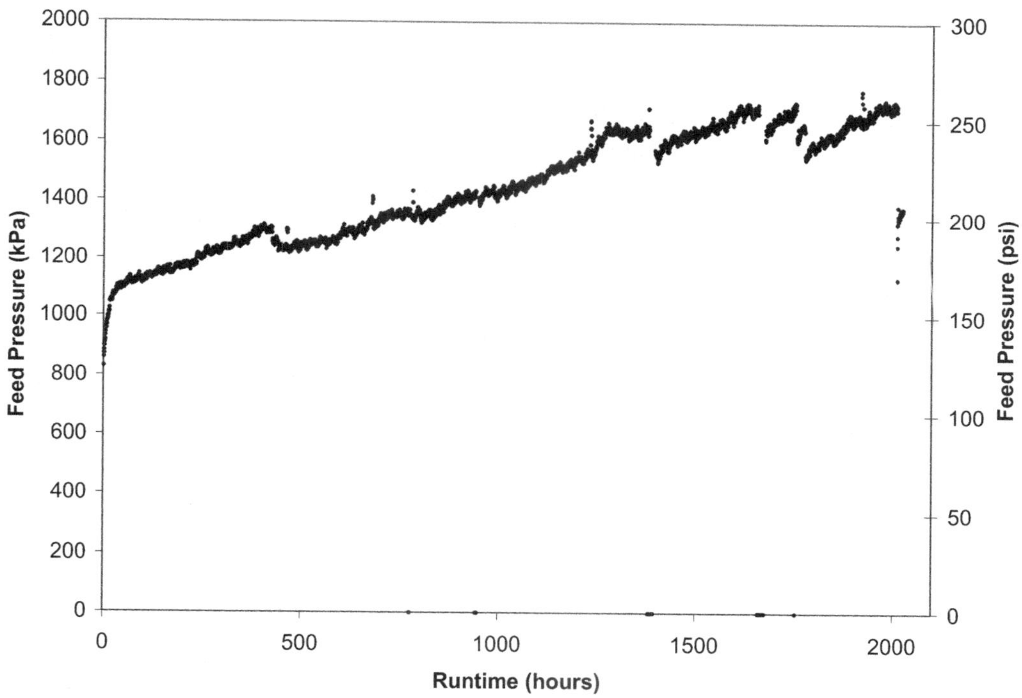

Figure 4.17 Feed pressure of the TMG10 membrane during pilot-scale testing at the SWC

Similarly to TMG10 membrane testing with non-nitrified feed water (WBWRP), TMG10 testing with nitrified/denitrified feed water resulted in a rapid increase in the feed pressure required to meet the 20 L/m^2-hr (11.8 gfd) permeate flux set point as a result of membrane compaction and initial fouling. As was discussed previously regarding NF-4040 membrane performance, the TMG10 membrane feed pressure during testing with nitrified/denitrified, however, never stabilized and was approximately 1,700 kPa (250 psi) when a clean-in-place (CIP) was initiated (after approximately 2,000 hours of runtime).

The TCSF data for the TMG10 membrane during pilot-scale testing with nitrified/denitrified wastewater effluent at the SWC was summarized previously in Figure 4.9. The initial TCSF of the TMG10 membrane at the SWC was greater than observed at the WBWRP (Figure 4.2), but declined sharply and never reached a stable level of operation. Based on historic full-scale TFC-HR membrane TCSF data from the SWC (Appendix C, Figure C.2), the TMG10 membrane TCSF value fell below what was considered an acceptable TCSF level (0.02 L/m^2-hr-kPa (0.08 gfd/psi)) in order to achieve any economic benefits from reduced operating pressures. Subsequent membrane autopsies on TMG10 membranes recovered from the pilot-scale testing unit showed severe biofouling on the membrane surface (Chapter 6). Measurements of feed water monochloramine concentrations at a runtime of 2,000 hours showed non-detectable levels of monochloramine, indicating that the failure of the TMG10 membrane at the SWC was a result of unfavorable operating conditions and not caused by the membrane itself. Since the TMG10 membrane was considered a viable membrane at the WBWRP (based on operational data), it appears that the absence of chloramines in SWC feed water resulted in biological growth on the membrane surface which in turn resulted in the observed increase in feed pressure and decrease in TCSF.

Due to significantly different operational characteristics of the TMG10 membrane operated at the WBWRP and TMG10 membrane operated at the SWC, a CIP was performed at approximately 2,000 hours of runtime. The TMG10 membranes were cleaned with a high pH solution of 1 percent sodium tripolyphosphate and 1 percent trisodium phosphate in recirculation mode. Figure 4.18 presents TMG10 membrane feed pressure prior to and after the CIP. TMG10 membrane feed pressure was initially reduced to approximately 1,100 kPa (160 psi) after CIP, but quickly stabilized at approximately 1,400 kPa (200 psi) for the remainder of the TMG10 runtime (14 hours). TMG10 membrane TCSF values before and after the CIP are presented in Figure 4.19. The CIP allowed for the recovery of approximately 20 percent of the TCSF that was lost during the 2,000 hours of operation, although the membrane was only operated for 14 hours after cleaning. The result was a TCSF after cleaning of approximately 0.022 L/m^2-hr-kPa (0.09 gfd/psi), which was higher than the TCSF reference point.

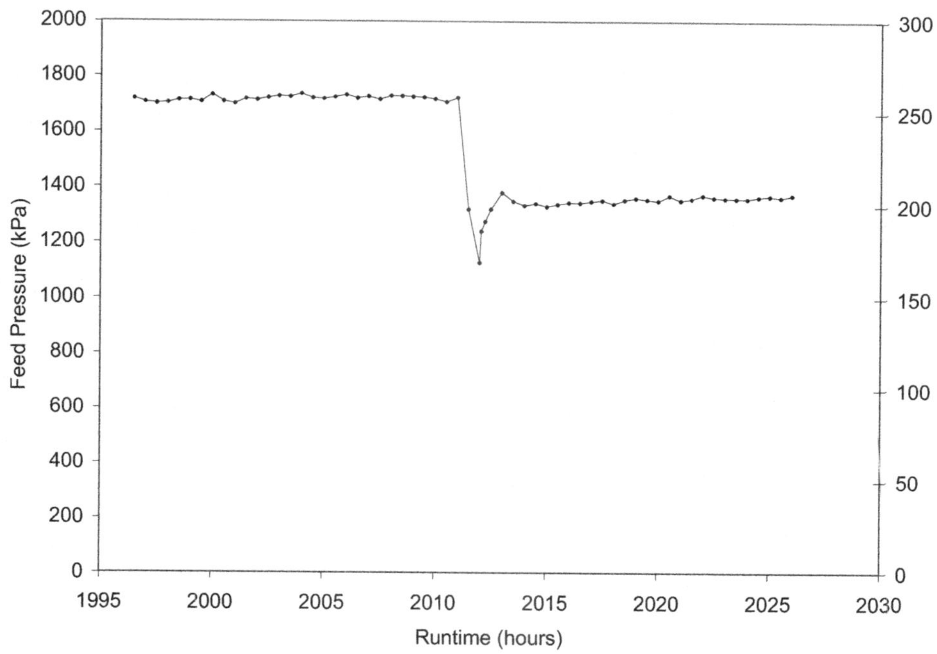

Figure 4.18 Feed pressure of the TMG10 membrane (pilot-scale) at the SWC during the clean-in-place

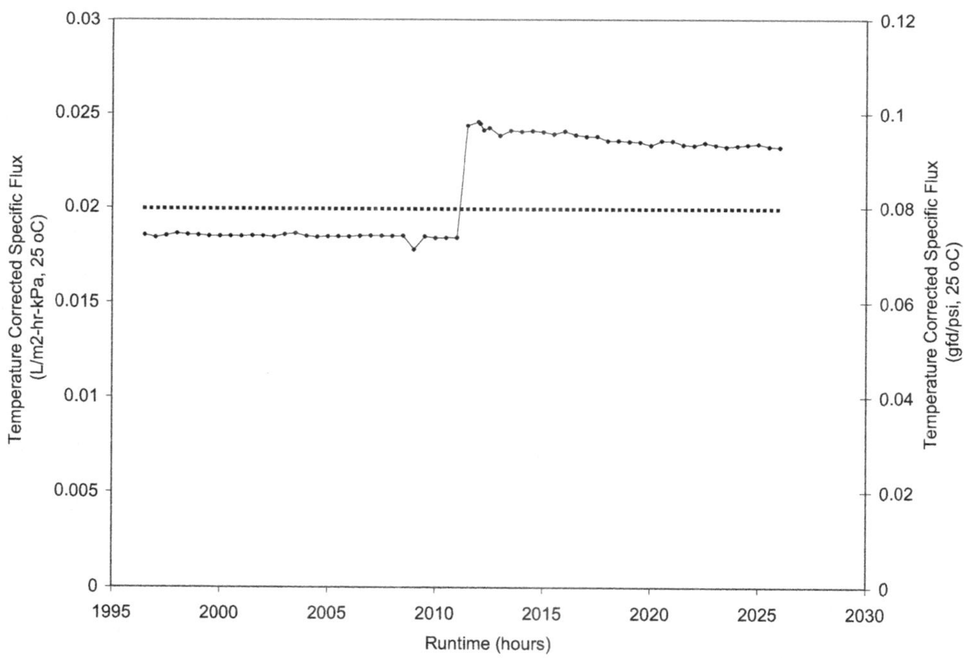

Figure 4.19 Feed pressure of the TMG10 membrane (pilot-scale) at the SWC during the clean in place
(dashed line represents acceptable TCF based on historic SWC full-scale data, Appendix C, Figure C.2)

Rejection Performance

The performance of the TMG10 membrane in rejecting constituents present in SWC feed water was similar to what was observed for the TMG10 membrane operating at the WBWRP. During the 2,000 hours of operational runtime at the SWC, the TMG10 membrane was found to significantly reduce (>97 percent) TOC and nitrate present in SWC feed water. The removal of DBPs such as chloroform, bromoform and trichloroacetic acid was observed to be similar to rejection performance observed at the WBWRP where the degree of rejection followed the trend: TCAA > BF > CF. In addition, similar to TMG10 membrane testing at the WBWRP, of the unregulated organic contaminants identified in the SWC feed water, there was only one detection (naproxen) during permeate monitoring. A more detailed summary of the rejection performance of the TMG10 membrane during testing at the SWC is presented in Appendix D.

CHAPTER 5
FULL-SCALE MONITORING AND TESTING

PERFORMANCE OF A SELECTED CANDIDATE MEMBRANE DURING FULL-SCALE TESTING

Testing Conditions

This portion of the study included full-scale verification of the most promising ULPRO membrane by operation in the 9.46 ML per day (2.5 mgd) RO Train 3 at WBWRP. One pressure vessel in the first stage of the train was equipped with seven TMG20-430 membranes. TMG20-430 is the model number for the 8-inch equivalent of the 4-inch TMG10 membrane used in previous portions of this study. Operation began on April 26, 2005 and continued through November 27, 2005 for approximately 5,000 hours of testing. Train 3 operates on microfiltered effluent with a typical feed water quality as listed in Table 2.3.

Two Train 3 pressure vessels had been previously modified to include independent instrumentation, allow sampling, and allow independent control of permeate and concentrate flows for data collection. One of the two modified test vessels contained seven TMG20-430 membranes. The second modified vessel contained seven ESPA2 membranes, a thin film composite membrane which has performed effectively in several full-scale systems operating on microfiltered secondary effluent at West Basin as well as other facilities. The ESPA2 membranes acted as the control during this testing. The control using the ESPA2 membrane provided confirmation that the feed water quality was not unusual and verified the expected performance of a known membrane. A process and instrumentation diagram showing the instrumentation and controls on each test vessel is shown in Appendix E (Figure E.1).

WBWRP had abundant operational experience with standard thin-film composite reverse osmosis membranes such as Hydranautics ESPA2 and Koch TFC-HR on pretreated secondary wastewater. In these applications, the membranes were typically run at an average flux rate of 20.4 L/m^2-hr (12 gfd). Analysis of this data demonstrated an initial period of rapid specific flux decrease (feed pressure increase) after startup. This permeability decline continued for approximately 6 to 8 weeks until the membrane performance stabilized. Subsequently, the membranes continued to lose permeability but at a much slower and acceptable pace.

Under this study, seven new TMG20-430 membranes were tested and compared with the control membranes. Each was initially tested at a flux rate of 20.4 L/m^2-hr (12 gfd). The 20.4 L/m^2-hr (12 gfd) test continued for 15 weeks to analyze the membrane performance. The specific flux was anticipated to decline over the first 8 week period and then stabilize. The post "break in period" permeability decline was observed during the remaining 7 weeks. As part of a parallel study performed by West Basin MWD, after 15 weeks, the membranes were operated at fluxes of 17.0, 18.7, and 22.1 L/m^2-hr (10, 11, and 13 gfd) for periods of 1 week each. Samples for water quality were collected at 24 and 72 hours after startup, and then after week 9 (20.4 L/m^2-hr, 12 gfd), week 16 (17.0 L/m^2-hr, 10 gfd), week 17 (18.7 L/m^2-hr, 11 gfd), week 18 (22.1 L/m^2-hr, 13 gfd), and week 25 (20.4 L/m^2-hr, 12 gfd). Samples for unregulated organics were collected at 2 and 72 hours after start-up, after week 16 (17.0 L/m^2-hr, 10 gfd), after 18 weeks (22.1 L/m^2-hr, 13 gfd) and after 25 weeks (20.4 L/m^2-hr, 12 gfd, after CIP). After week 18, the membranes were cleaned and the 20.4 L/m^2-hr (12 gfd) test was repeated. The final 12 week test brought the total operating time to 5,000 hours and demonstrated the effectiveness of the cleaning.

The operating parameters for this test were based upon the active area of the membranes tested. The permeate flow was adjusted to attain a flux rate of 20.4 L/m^2-hr (12 gfd) initially for each of the two membranes types. For example, the Hydranautics ESPA2 membrane contained 37.2 m^2 (400 square feet) of membrane area, and the pressure vessels in WBWRP Train #3 contained 7 membranes (Figure 5.1). Therefore, the permeate flow for the 20.4 L/m^2-hr (12 gfd) operation on ESPA2 was:

$$\left(\frac{20.4\text{ Liters}}{\text{m}^2\text{ - hour}}\right) \text{x } (7\text{ membranes}) \text{ x} \left(\frac{37.2\text{ m}^2}{\text{membrane}}\right) \text{x} \left(\frac{1\text{ hour}}{60\text{ minutes}}\right) = 88.2\text{ L/min} \qquad (5.1)$$

$$\left(\frac{12\text{ gallons}}{\text{ft}^2\text{ - day}}\right) \text{x } (7\text{ membranes}) \text{ x} \left(\frac{400\text{ft}^2}{\text{membrane}}\right) \text{x} \left(\frac{1\text{ day}}{1440\text{ minutes}}\right) = 23.3\text{ gpm} \qquad (5.2)$$

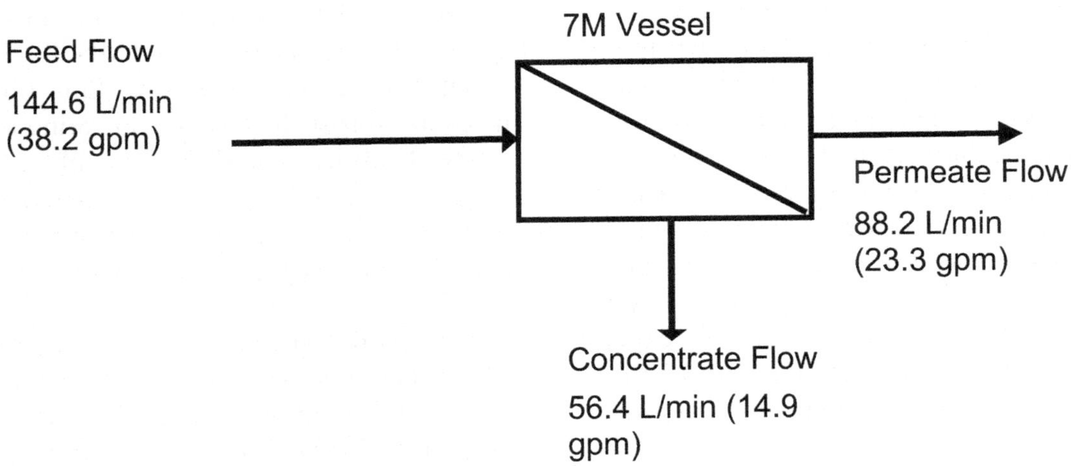

Figure 5.1 Hydranautics ESPA2 membrane operating parameters at 20.4 L/m^2-hr (12 gfd)

Because the full-scale RO Train #3 production varies day to day, the flow of the test vessels required daily adjustment. The permeate flow was adjusted by throttling a valve on the feed piping of the test vessels while concentrate flows were adjusted by throttling a valve on the concentrate discharge piping from the vessels. The valve adjustments allowed for uniform permeate and concentrate flows to achieve desired flux and recovery set points. The vessel recovery was 61 percent to approximate the first stage recovery of a two stage array operating at 85 percent recovery. A summary of the run parameters for the various test periods and test elements are listed in Table 5.1.

Table 5.1
Test Operating Parameters

Membrane Model	Units	Hydranautics ESPA2				Toray Membrane America TMG20-430			
Element	m^2	37.2				39.9			
Membrane Area	ft^2	400				430			
Vessel Membrane	m^2	260.1				279.6			
Area	ft^2	2800				3010			
Recovery	%	61				61			
Permeate Flux	L/m^2-hr	20.4	17.0	18.7	22.1	20.4	17.0	18.7	22.1
	gfd	12	10	11	13	12	10	11	13
Permeate Flow	L/min	88.2	73.4	81.0	95.8	95.0	79.1	87.1	103
	gpm	23.3	19.4	21.4	25.3	25.1	20.9	23.0	27.2
Concentrate Flow	L/min	56.4	46.9	51.9	61.3	60.6	50.7	55.6	65.9
	gpm	14.9	12.4	13.7	16.2	16.0	13.4	14.7	17.4
Feed Flow	L/min	144	120	132	157	155	129	142	168
	gpm	38.2	31.8	35.1	41.5	41.1	34.3	37.7	44.6

The same cleaning procedures were applied to each membrane. First, the membranes were cleaned with 2 percent citric acid solution, maintained between pH 2.0 to 2.5, in order to remove any inorganic fouling that may have occurred. Next, the membranes were cleaned with a high pH solution of 1 percent sodium tripolyphosphate, 1 percent trisodium phosphate, and 1 percent tetrasodium EDTA, maintained between pH 11 to 11.5, to remove any organic fouling present. New batches of each cleaning solution were used for each membrane. In all cleanings, the solutions were heated to 35 to 38°C and circulated through the individual test vessels for one hour at 151.4 L/min (40 gpm).

Performance of TMG20-430 ULPRO Membrane

The feed pressure of the TMG 20-430 and ESPA2 membranes are presented in Figure 5.2. Immediately following startup of the test, the feed pressures for the TMG20-430 and ESPA 2 membranes were 786 kPa (114 psi) and 855 kPa (124 psi), respectively. Both membranes fouled to similar conditions within the first 150 hours of operation. Over the course of the first 2,500 hours of operation, the two membranes maintained similar feed pressures ranging between 896 kPa (130 psi) and 965 kPa (140 psi). After 2,500 hours, the membrane was operated at various flux conditions as part of West Basin MWD's parallel study. The feed pressures were observed to increase and decrease proportionally to the permeate flux value. Following the period of flux variation, the membranes were returned to 20.4 L/m^2-hr (12 gfd) and cleaned after approximately 3,000 hours of operation. The membrane cleaning temporarily reduced the feed pressure of the TMG20-430 to 807 kPa (117 psi), approximately 3 percent higher than the startup value. The TMG20-430 feed pressure gradually increased to 1,110 kPa (161 psi) by the end of the 5,000 hour test. The cleaning of the ESPA2 reduced the feed pressure to 745 kPa (108 psi), approximately 13 percent lower than the startup pressure. By the end of the 5,000 hour test, the ESPA2 membrane had reached a feed pressure of 979 kPa (142 psi).

The specific flux, or water permeability, was calculated from the operating data collected during testing. The specific flux for the two test membranes is shown in Figure 5.3. The initial specific flux values for TMG20-430 and ESPA2 membranes were 0.041 L/m^2-hr-kPa (0.167 gfd/psi) and 0.037 L/m^2-hr-kPa (0.149 gfd/psi), respectively. As noted for the feed pressures, the membranes fouled within the first 150 hours to attain similar specific flux values. During the first 2,500 hours, the ESPA2 membrane stabilized at a specific flux value of approximately 0.030 L/m^2-hr-kPa (0.120 gfd/psi). The TMG20-430 membrane was observed to have a slightly lower permeability of approximately 0.028 L/m^2-hr-kPa (0.115 gfd/psi), but remained within 5 percent of the ESPA2. No significant change in the specific flux was observed during operation at 17.0, 18.7, or 22.1 L/m^2-hr (10, 11, or 13 gfd). Membrane cleaning temporarily returned the TMG20-430 to a specific flux of 0.035 L/m^2-hr-kPa (0.144 gfd/psi), approximately 85 percent of the startup value. The TMG20-430 completed the 5,000 hour test with a specific flux of 0.024 L/m^2-hr-kPa (0.099 gfd/psi). Membrane cleaning temporarily elevated the specific flux of the ESPA2 membrane to 0.039 L/m^2-hr-kPa (0.160 gfd/psi), approximately 107 percent of its startup value. This single post-cleaning data point may reflect a temporary effect on membrane performance caused by the cleaning pH. The ESPA2 completed testing with a specific flux of 0.029 L/m^2-hr-kPa (0.116 gfd/psi).

The performances of the two test membranes are considered similar during the first 3,000 hours of testing. However, there was a notable difference in the membrane performances observed during the last 2,000 hours of operation. The discrepancy in the performances is considered to be a consequence of the cleaning effectiveness of each membrane. The TMG20-430 membrane cleaning was not as effective as the cleaning of the ESPA2 membrane. This is evidenced by the inability of the TMG20-430 membrane to return to the startup feed pressure and specific flux following cleaning. The data supports that the TMG20-430 membrane did not respond as well as the ESPA2 membrane to the cleaning regime utilized in this study. Further investigation into this cause of the less effective cleaning was not included in the scope of this study, but the difference in cleaning effectiveness may have been related to differences in the application of each cleaning procedure, individual membrane chemistries, membrane surface characteristics, or membrane element construction.

The startup performance of the pilot-scale TMG10 membrane was slightly better than the full-scale TMG20-430 membrane. The initial specific flux of the TMG10 membrane was 0.042 L/m^2-hr-kPa (0.170 gfd/psi), while the startup value for the TMG20-430 membrane was 0.041 L/m^2-hr-kPa (0.167 gfd/psi). In order to directly compare the long-term performances without considering the effect of membrane cleaning, it was necessary to compare the performance after approximately 2,500 hours of operation. At approximately 2,500 hours, the specific flux of the TMG20-430 membrane was 0.028 L/m^2-hr-kPa (0.114 gfd/psi). After a similar period of operation, the TMG10 membrane specific flux was 0.028 L/m^2-hr-kPa (0.115 gfd/psi). The similar specific flux values confirm similar performance on this application between the full-scale and pilot-scale operations.

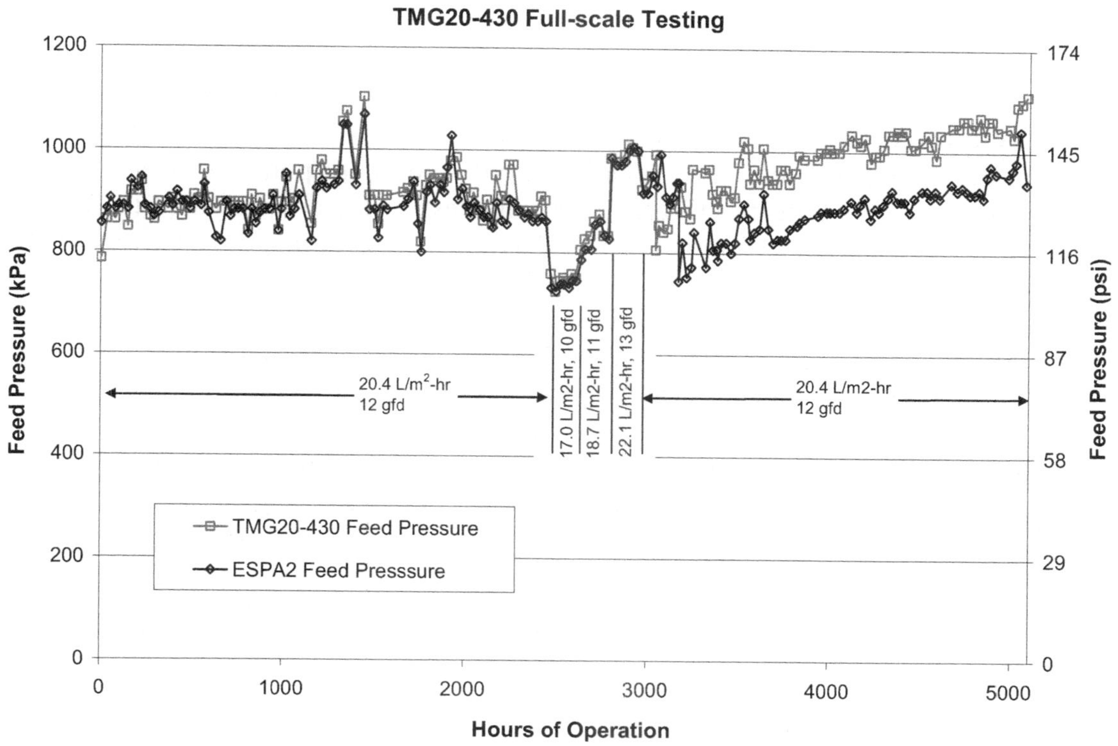

Figure 5.2 Feed Pressure of 8-inch TMG20-430 and ESPA2 membranes

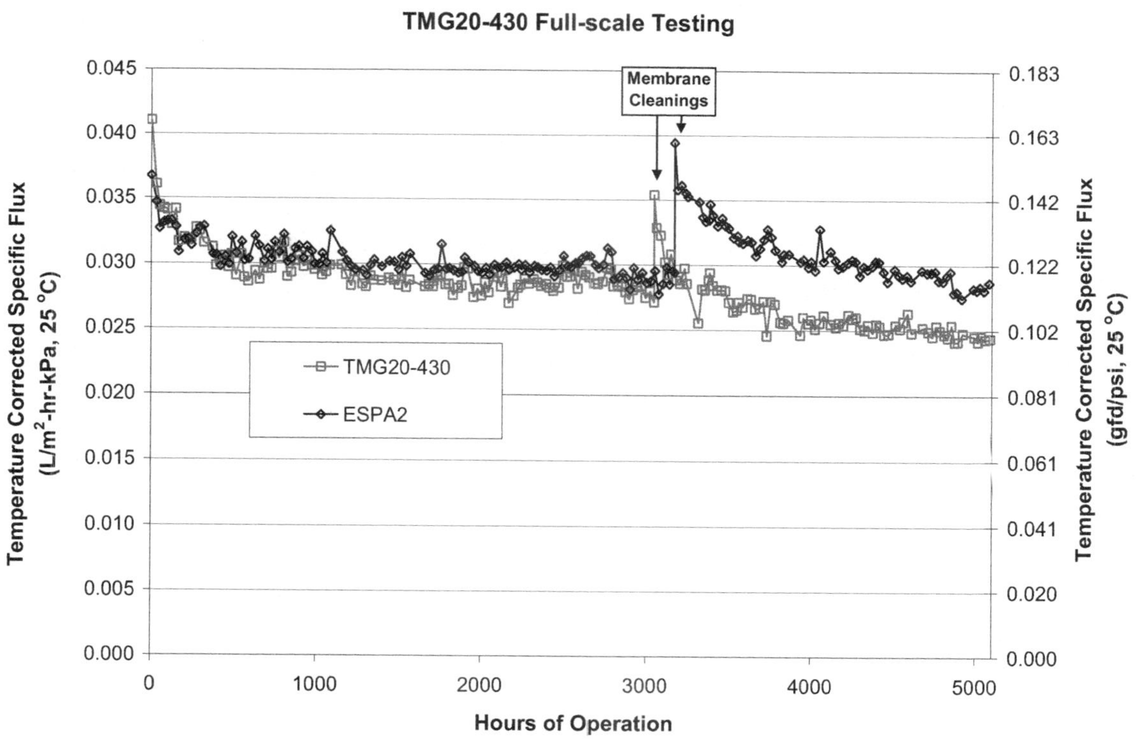

Figure 5.3 Specific Flux of 8-inch TMG20-430 and ESPA2 membranes

Rejection of TOC and Nitrogen

During TMG20-430 membrane full-scale testing, samples for total organic carbon (TOC) and ammonia were taken and analyzed by United Water staff at the WBWRP. Summarized results for the feed and permeate TOC concentrations and corresponding rejection values reported for operation at 20.4 L/m^2-hr (12 gfd) during the test period are presented in Table 5.2. The average permeate TOC concentration was 0.20 mg-C/L, resulting in an average TOC rejection value of 98.3 percent during the TMG20-430 membrane testing period. The average permeate TOC concentration for the ESPA2 membrane was 0.21 mg-C/L. The average TOC rejection value for the ESPA2 membrane was 98.2 percent. These rejection values are based on feed concentrations (eq. 5.3) and operation at 61 percent recovery.

Table 5.2 Full-scale feed and permeate TOC concentrations and TOC rejection

Sample Date	4/26/2005	4/28/2000	6/13/2005	9/8/2005	10/13/2005	Average
Feed (mg-C/L)	12	12	12	11	12	12
TMG20-430 Permeate (mg-C/L)	0.22	0.21	0.14	0.34	0.10	0.20
TMG20-430 Rejection (%)	98.2	98.2	98.8	96.9	99.2	98.3
ESPA2 Permeate (mg-C/L)	0.30	0.31	0.14	N/A	0.10	0.21
ESPA2 Rejection (%)	97.6	97.4	98.8	N/A	99.2	98.2

N/A – not available

It was useful to compare ammonia rejections based both on feed concentrations and a feed/brine average concentration. The ammonia rejection calculated based on the feed concentration (Equation 2.2) represents the removal of the ammonia from the feed stream. However, in order to compare systems not operated at the same conditions, ammonia rejection based on a feed/brine average ammonia concentration (Equation 2.3) is provided. This value takes into account the increased concentrations of feed ammonia in the downstream membranes in a system setting, and offers a more useful comparison for two systems operated at different recoveries. It is not useful to compare TOC values with this method because TOC membrane passage appears to be more closely related to a sieve mechanism and is not necessarily dependent upon system recovery.

Neither nitrate nor nitrite was detected in any samples collected during the test period, so ammonia represented all of the inorganic nitrogen in the system. Summarized results for feed and permeate ammonia samples and corresponding rejection values are presented in Table 5.3. The average permeate ammonia concentration for the TMG20-430 membrane was 1.6 mg-N/L, corresponding to an average ammonia rejection (based on feed ammonia) of 95.1 percent. The average permeate ammonia concentration for the ESPA2 was 1.3 mg-N/L, corresponding to an average ammonia rejection (based on feed ammonia) of 96.1 percent. As expected, the respective rejections based on the feed/brine average ammonia concentrations were proportionally higher because the operating conditions were similar between the two systems.

Table 5.3
Full-scale feed and permeate ammonia concentrations and ammonia rejection

Sample Date	4/26/2005	4/28/2000	6/13/2005	9/8/2005	10/13/2005	Average
Recovery (%)	61	61	61	61	61	61
Feed (mg-N/L)	35	32	30	35	36	34
TMG2-430						
Permeate Ammonia (mg-N/L)	1.5	1.6	1.7	1.5	1.9	1.6
Ammonia Rejection (%) – Feed	95.7	95.0	94.3	95.7	94.7	95.1
Ammonia Rejection (%) – F/B Average	97.2	96.8	96.3	97.2	96.6	96.8
ESPA2						
Permeate Ammonia (mg-N/L)	1.0	1.2	1.4	N/A	1.5	1.3
Ammonia Rejection (%) – Feed	97.1	96.3	95.3	N/A	95.8	96.1
Ammonia Rejection (%) – F/B Average	98.1	97.6	97.0	N/A	97.3	97.5

N/A – not available

Average Tier 1 analysis data for the TMG20-430 membrane samples collected at 20.4 L/m^2-hr (12 gfd) is presented in Table 5.4. Average Tier 1 analysis data for the pilot-scale TMG10 membrane is also presented in Table 5.4. TOC permeate concentrations for the TMG20-430 membrane were similar to combined permeate samples taken during pilot-scale testing and were approximately 0.2 mg-C/L. Average TOC rejections for the two systems were also similar at 98.3 percent.

Average permeate ammonia concentrations were slightly higher from the full-scale vessel than from samples taken during pilot-skid monitoring. The average ammonia rejection for the TMG20-430 membrane based on the feed/brine average method was slightly lower than the TMG10 at pilot-scale.

Table 5.4
Comparison of full-scale and pilot-scale TOC and ammonia data

	Full-scale TMG20-430 Average	Pilot-scale TMG10 Average
Recovery (%)	61	82
TOC		
Feed TOC (mg-C/L)	12	12
Permeate TOC (mg-C/L)	0.20	0.20
TOC Rejection (%) – Feed	98.3	98.3
Ammonia		
Feed (mg-N/L)	34	32
Permeate Ammonia (mg-N/L)	1.6	1.2
Ammonia Rejection (%) – Feed	95.1	96.3
Ammonia Rejection (%) – F/B Average	96.8	98.2

Rejection of Regulated Organics

Sampling campaigns were performed on the 8 inch research vessel for the quantification of Tier 2 DBP compounds including chloroform, bromoform, dichloroacetic acid, and trichloroacetic acid. Summarized results from chloroform analysis conducted on feed and permeate samples from the TMG20-430 and ESPA2 membranes are presented in Figure 5.4. The TMG20-430 and ESPA2 membranes rejected chloroform similarly, displaying between 50 and 60 percent rejection during the sampling campaigns. With the exception of the second sampling campaign, feed concentrations of chloroform were below 20 μg/L and permeate samples were below 10 μg/L. Chloroform results generated during sampling of the TMG10 membrane on pilot-scale indicate that the rejection of chloroform during the 8 inch research vessel test was higher than on pilot-scale (Figure 4.3). In addition, chloroform data generated through sampling full-scale treatment Train 3 (TFC-HR, Figure 4.3) and 4 (ESPA2, Figure 4.3) resulted in chloroform rejection between 20 and 40 percent.

Summarized results of TMG20-430 and ESPA2 8 inch research vessel feed and permeate concentrations of bromoform, DCAA, and TCAA are presented in Table 5.5. Bromoform was quantified at the low μg/L range in around half of the feed samples analyzed and was not detected in permeate samples taken from either membrane (Table 5.5). The rejection of DCAA and TCAA by both the TMG20-430 and ESPA2 membrane was greater than 96 percent and quantifiable permeate concentrations were below 2 μg/L for both compounds (Table 5.5).

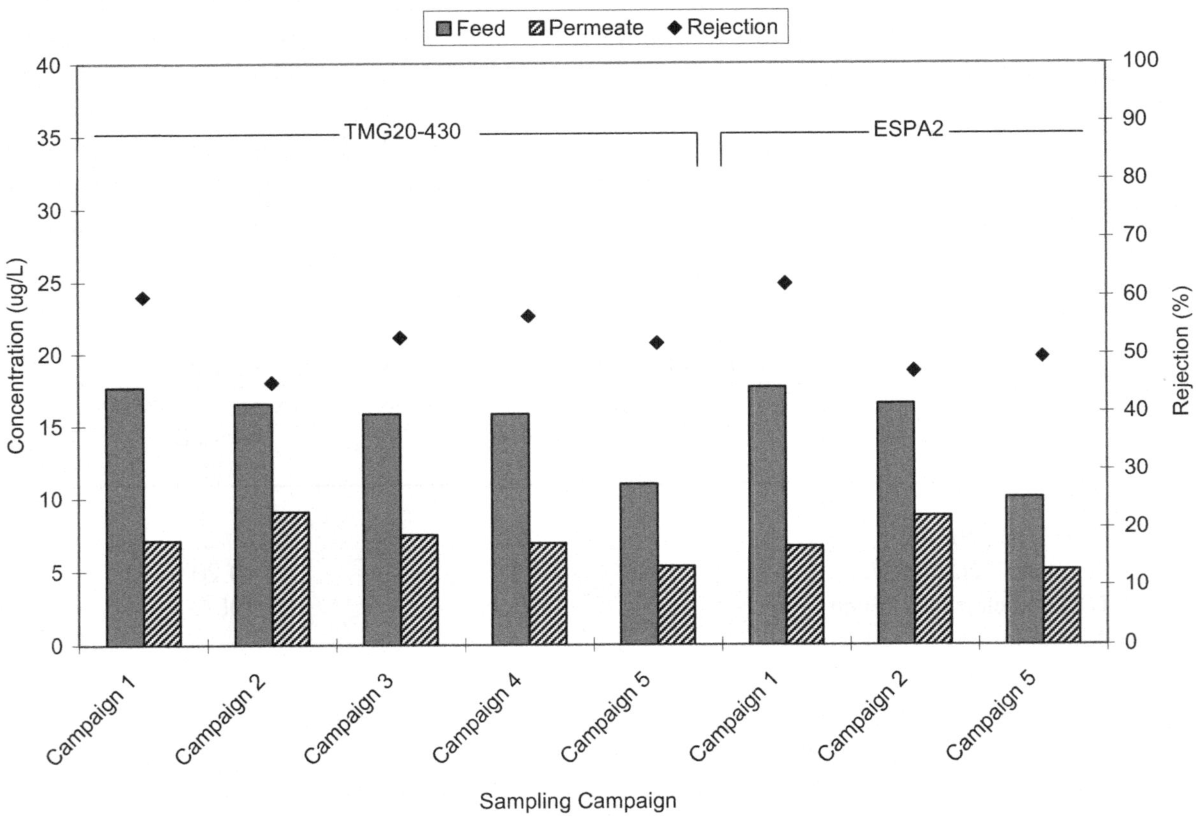

Figure 5.4 Feed and permeate chloroform concentrations and rejection values for the research vessel 8 inch TMG20-430 and ESPA2 membranes

Table 5.5

Summary of bromoform, dichloroacetic acid, and trichloroacetic acid feed and permeate concentrations (μg/L) for the research vessel 8 inch TMG20-430 and ESPA2 membranes

	TMG20-430					ESPA2		
Sampling Campaign	Campaign 1	Campaign 2	Campaign 3	Campaign 4	Campaign 5	Campaign 1	Campaign 2	Campaign 5
Bromoform								
Feed	0.8	3.1	1.5	n.d.	n.d.	0.8	3.1	n.d.
Permeate	n.d.	n.d.	n.d.	n.d.	n.d.	n.d.	n.d.	n.d.
Dichloroacetic acid								
Feed	44	27	22	34.5	N/A	44	27	NA
Permeate	1	n.d.	n.d.	n.d.	N/A	1.2	n.d.	NA
Rej.	97.7	N/A	N/A	N/A	N/A	97.2	N/A	NA
Trichloroacetic acid								
Feed	13.2	19	12.5	13.4	N/A	13.2	19	NA
Permeate	0.43	n.d.	n.d.	n.d.	N/A	0.5	n.d.	NA
Rej.	96.7	N/A	N/A	N/A	N/A	96.2	N/A	NA

N/A – not available
n.d. – not detected

One sampling campaign was performed on the 8 inch TMG20-430 membrane for NDMA analysis. Table 5.6 presents feed and permeate NDMA concentrations for the TMG20-430 during 8 inch research vessel testing and as a comparison, results from ESPA2 membrane during full-scale operation on Train 4. The rejection of NDMA by the TMG20-430 membrane was very similar to NDMA rejection observed during pilot-scale TMG10 testing (Figure 4.5), which was between 50 and 60 percent for both tests. ESPA2 membrane NDMA rejection was slightly lower at 34 percent, but neither membrane met the CDPH drinking water NDMA reporting limit of 10 ng/L.

Table 5.6

NDMA feed and permeate concentrations for the TMG20-430 membrane 8 inch research vessel compared to the ESPA2 membrane on full-scale

Membrane	ESPA2	TMG20-430
Date Sampled	3/22/2005	4/28/2005
Location	Train 4 full-scale	Research Vessel
Feed (ng/L)	32	27
Permeate (ng/L)	21	12
Rejection (%)	34	56

Rejection of Unregulated Organics

The occurrence of the unregulated Tier 2 compounds (Table 2.1) in WBWRP feed water is summarized in the box and whisker plot presented in Figure 4.6. Five sampling campaigns were performed to determine the rejection of unregulated Tier 2 compounds in the 8 inch research vessel employing the TMG20-430 and ESPA2 membranes. Permeate samples were collected within two hours of start-up, at 72 hours of runtime, at approximately 2,500 hours of runtime, and at approximately 4,300 hours of runtime. The sampling frequency was designed to study the influence of membrane fouling on the rejection of trace organic contaminants while quantifying the removal efficiencies of the individual membranes. Samples were collected in duplicate from TMG20-430 and ESPA2 membrane permeate streams during sampling events. Sample analysis for unregulated Tier 2 compounds yielded non-detects for all compounds in all permeate samples taken during TMG20-430 and ESPA2 membrane testing.

Economic Assessment of Select Candidate Membranes as compared to Bench-mark RO Membranes

As discussed above, the TMG membrane was operated in its 8-inch diameter configuration (TMG20-430) in parallel with a control membrane whose full-scale performance is well documented and has been considered a benchmark for current state-of-the-art best performance in this application. In the case of this testing the ESPA2 membrane serve the role as the control membrane. An economic assessment comparing the total cost of water production with TMG versus the ESPA2 membrane should consider purchase cost (capital), expected life and operating expenses.

Based on recent bids received by municipal agencies for replacement membrane, the TMG membrane has been priced competitive with the ESPA2 membrane. These products compete directly and it is expected they will continue to be offered at similar cost. Therefore purchase price is not considered a factor in making an economic comparison of the TMG and ESPA2 membranes. Expected lives of the membranes are also considered to be equal.

With regard to operating costs, operating pressure is the only TMG operating parameter considered to deviate from the benchmark ESPA2 membrane. Pretreatment requirements and recovery rate are the same. Electrical consumption will be directly proportional to the required operating pressure. From Figure 5.2 we see that the operating pressure of the TMG and ESPA2 membranes tracked very closely during operation at 20.4 L/m^2-hr (12 gfd). The operating pressure at 2,500 hours was approximately 896 kPa (130 psi) for both. For a full-scale installation at 85 percent recovery this relates to an electrical consumption of 370 kWh/ML (1.40 kWh/kgal) of permeate produced. The two membranes exhibited different response to the chemical cleanings which were performed at approximately 3,000 hours. While the cleaning procedures were identical, the TMG membrane did not recover as much permeability as the ESPA2 membrane. Subsequent permeability trends tracked similar slopes, resulting in higher operating pressure for the TMG membrane during the final 2,000 hours of testing. Estimates of full-scale electrical consumption based on the performance at 5,000 hours are 479 kWh/ML (1.81 kWh/kgal) for the TMG membrane and 412 kWh/ML (1.56 kWh/kgal) for the ESPA2 membrane, a difference of 67 kWh/ML (0.25 kWh/kgal). Based on an example electrical cost of $0.12/kWh, this difference would amount to higher cost for the TMG membrane of $8.04/ML ($0.03/kgal) of permeate.

This cost differential was due entirely from the difference in effectiveness of the cleaning, not the rate of fouling during operation. The possibility exists that a modified or different cleaning procedure may offer better results for the TMG membrane. The scope of this work did not include optimization of the cleaning procedure for the TMG membrane. If an equally effective cleaning were accomplished with the TMG membrane, the operating costs would be expected to be the same as for the ESPA2 membrane.

CHAPTER 6
ASSESSMENT OF THE FOULING POTENTIAL OF NF/ULPRO MEMBRANES

INTRODUCTION

Membrane fouling is considered one of the main impediments in the development and operation of membrane processes. The common constituents usually found in various water sources, such as dissolved organic substances, colloidal and particulate matter, sparingly salts, and microorganisms, can adsorb, accumulate, or precipitate within the membrane polymer matrix and on membrane surfaces leading to membrane fouling. Adsorption within the matrix of the membrane is generally considered irreversible fouling while cake-layer fouling due to accumulation and precipitation of material on the membrane surface is often considered reversible subject to chemical cleaning. The presence of chemical reactants, such as chlorine or ozone, can also cause damage to polyamide membranes. Membrane fouling causes gradual decline in productivity of membrane filtration, shortened membrane life, and increased operation and maintenance costs.

Pretreatment using microfiltration (MF) or ultrafiltration (UF) to remove particles, colloids and bacteria can significantly reduce the fouling potential of RO feed water over conventional pretreatment. Integrated membrane systems (IMS) consisting of MF/UF followed by NF/RO have shown to be capable of treating secondary effluent and maintain stable performance of product flow and operating pressure (USEPA 2004). Nevertheless, extensive field pilot tests are still required to select appropriate membranes to meet operational performance and water quality criteria.

A fundamental understanding of the mechanisms and formation of fouling is of great importance to make this technology more technically feasible and economically competitive. A number of controlled laboratory experiments have been performed to understand factors contributing to membrane fouling, including feed water composition, membrane properties, and operating conditions. As compared to other constituents of natural organic matter (NOM), colloidal and non-colloidal hydrophilic constituents were shown as major organic foulants to negatively charged membranes due to their high neutral character (Kwon et al., 2005). Divalent calcium ions could cause severe NOM fouling by complexation and subsequent formation of intermolecular bridge among organic foulant molecules, which also affected significantly the interaction between chemical cleaning agent and the fouling layer (Li and Elimelech 2004, Seidel and Elimelech 2002). For a surface water exhibiting a high calcium concentration and moderate dissolved organic carbon (DOC) after pretreatment, the extent of membrane fouling was significant dependent upon seasonal variation of water quality and temperature; high flux decline was observed at high DOC, high divalent cations, high alkalinity, and low temperature (Her et al. 2000). NOM fouling occurred at high temperature while scale fouling was observed at high and low temperatures (Her et al. 2000). Membrane fouling could be alleviated by reducing permeate flux and increasing crossflow velocity (Seidel and Elimelech 2002, Beverly et al. 2000, Koyuncu et al. 2004). Physicochemical properties of membranes also affect membrane fouling significantly. Colloidal fouling of RO and NF membranes was found nearly perfectly correlated with membrane surface roughness, regardless of physical and chemical operating conditions (Vrijenhoek et al. 2001).

Membrane biofouling is very difficult to control even with addition of biocides and MF/UF to diminish the number of microorganisms in feed water. The small amount of viable bacteria left in the feed water can still adhere and multiply on membrane surfaces resulting in biofilm growth. Ridgway et al. (1984) investigated the kinetics of adhesion of a *Mycobacteriun* sp. to cellulose diacetate RO membrane. They found that adhesion of mycobacteria to membrane surfaces occurred within 1 to 2 hours at 30°C and exhibited saturation-type kinetics which conformed closely to the Langmuir adsorption isotherm. This suggested the cellulose diacetate membrane surface might possess a finite number of available binding sites to which the bacteria could adhere. Ivnitsky et al. (2005) used synthetic effluents (glucose-based, and starch-based) and prefiltered secondary effluent to study NF biofouling. After 16 days of operation, fully developed biofilm was found on the membranes and the fouling layers were dominated by bacteria and extracellular polymers. Regardless of the feed water quality applied, the size and composition of biofilm population seemed to be convergent to finite numbers. Since organic substances are nutrients for bacteria growth, effective removal of organic matter can alleviate membrane biofouling.

Previous studies have contributed greatly in understanding various types of membrane fouling during surface water treatment and wastewater reclamation. The experiments conducted under well controlled laboratory conditions, however, can not completely represent full-scale treatment processes where colloidal fouling, inorganic scaling, biofouling and organic fouling occur simultaneously. In addition, the time period of operation, hydrodynamic conditions, variation of water quality, effectiveness of pre-treatment, and other operating issues can also affect extrapolating laboratory results to full-scale applications.

In this chapter, we will first present autopsy results to discuss the effect of water quality and operation on membrane fouling during the pilot skid operation. In addition, we will examine the effect of chemical cleaning on removal of scalants and foulants based on autopsy analyses. Lastly, we will evaluate the suitability of the bench-scale fouling assessment protocol by comparing the fouling potential predicted from the bench-scale experiments with the operational performance observed at pilot-scale.

EFFECT OF WATER QUALITY AND OPERATION ON MEMBRANE FOULING

Membranes selected for Autopsy

The ULPRO membrane TMG10, which was tested both with not nitrified and nitrified/denitrified feed water and the NF membranes, NF-90 and NF-4040, which were tested with not nitrified and nitrified/denitrified feedwater, respectively, were selected for membrane autopsies (Table 6.1). Upon completion of each pilot-scale investigation, the lead element of the first stage and the tail-end element of the last stage were removed from the skid and soaked in 1 percent sodium metabisulfite solution for preservation. After shipping to the CSM laboratory, the membrane elements were prepared for autopsy analyses. The membranes were unrolled, cut into small pieces, and rinsed carefully with deionized (type II) water to remove any residual sodium metabisulfite solution.

Table 6.1
Membranes selected for autopsy

Membrane	Feedwater Quality	Time of Operation	First stage Lead Element	Last stage Tail Element
TMG10	microfiltered not nitrified	12/10/03-4/5/05 2,965 hrs	X	X
TMG10	microfiltered nitrified/denitrified	10/23/05-2/10/06 2,000 hrs	N/A	X
TMG10 after CIP	microfiltered nitrified/denitrified	2/10/06-2/12/06 25 hrs	N/A	X
NF-90	microfiltered not nitrified	4/7/05-6/3/05 1,300 hrs	X	X
NF-4040	microfiltered nitrified/denitrified	8/2/05-10/21/05 1,600 hrs	X	X

N/A – not analyzed

Visible Inspection

For all membranes selected for autopsy, no visible flaws, cracks and patches of scales, and biological growth were observed on membrane leaves, feed spacers and permeate carriers of any of the lead membrane elements. For the membranes taken from the tail-end elements, however, regular small brown patches were observed on the spots contacted with spacers, pointing to potentially more severe fouling than the lead elements.

Membrane Surface Morphology and Elemental Composition

The membrane surface structure and morphology was characterized using environmental scanning electron microscopy (ESEM). The elemental composition of virgin and fouled membrane specimens was quantified by energy dispersive spectroscopy (EDS) mounted in the ESEM. The morphology of each of the fouled membranes was compared to its virgin stage. As an example, the ESEM micrographs of the TMG10 virgin membrane specimen are shown in Figure 6.1. Micrographs of the NF-90 and NF-4040 membranes are presented in Appendix F. The TMG10 virgin membrane contained a network of valleys and ridges, which was more apparent at magnification of x20,000 (Figure 6.1(b)). The EDS scan revealed mainly oxygen, carbon, and sulfur, which are representing the elemental composition in the virgin membrane (Figure 6.2). Since the electron beam generating the emission of X-rays that can penetrate the sample by 1 micron, the EDS analysis provides an elemental identification of both the thin active layer (about 250 nm) as well as the polysulfone microporous support layer as indicated by the detection of a significant amount of sulfur (Figure 6.2). The virgin NF-90 membrane exhibited a very similar morphology and an elemental composition predominantly consisting of oxygen, carbon and sulfur (Figures F.1 and F.2, Appendix F) as observed for the TMG10 membrane, whereas the NF-4040 virgin membrane exhibited a similar elemental composition but a rather smooth active layer (Figure F.6, Appendix F).

After several hundreds of hours of operation, fouling layers formed on membrane surfaces were dependent upon the feed flow direction and element stage. For the TMG10 membrane, which was tested for 3,000 hours on not nitrified feed water, the ESEM micrographs indicated heterogeneous fouling layer on the membrane leaves near the inlet of the lead element (herein feed-end of lead-element). Most of the membrane leaf was not covered by foulants

(Figure 6.3(a)), which was more apparent at higher magnifications (Figure 6.3(b)). As compared to virgin membrane specimen (Figure 6.1), the fouled membrane showed a more porous structure with ridge-valley network, and the ESEM micrograph at 20,000 magnification revealed that no foulants were observed on the surface (Figure 6.3(b)). However, a small fraction of the membrane leaf was covered by a porous fouling layer (Figure 6.3(c)) and the formation of a gel-like layer is more clearly displayed in Figure 6.3(d). The slimy and coherent film and the bacteria embedded in the extracellular matrix suggest that a biofilm began to form on the membrane surface (Figure 6.3(d)).

In general, biofouling is initiated by the irreversible attachment of viable bacteria to the membrane surface. The attachment is then strengthened by the synthesis and extrusion of extracellular polymeric substances (EPS), which stabilizes cell attachment and protects biofilm organisms from the effects of chlorine and other biocides. Cell attachment is mediated by molecular interactions between EPS and the membrane polymers. Using nutrients present in the feedwater, such membrane-bound bacteria are able to grow and multiply, often rapidly, to produce a confluent biofilm within days to weeks of placing new membranes in service. The ESEM observation of the embedded bacteria in EPS polymers confirms that EPS served as a defense barrier that retards convective flow and hindered the penetration of biocides into the biofilm. The formation of EPS layer made the bacteria resistant to the chloramines added to the not nitrified feed water at the WBWRP and more difficult to eliminate.

The EDS analysis showed a very small amount of sodium and chlorine being present in the fouled membrane (Figure 6.4). Sodium was likely introduced by the preservation solution sodium metabisulfite, which was probably not completely removed by DI water rinsing prior to the characterization. Chloride was likely attributed to the addition of chloramines in the feedwater to control biogrowth. No other additional elements were detected by the EDS analysis indicating that the membrane operated on the not nitrified feed water did not experience any inorganic scaling.

In contrast to the feed-end of the lead-element membrane, scanning on the brine-end of the TMG10 lead-element membrane (near outlet section) did not reveal the formation of a fouling layer; the membrane surface remained as a porous ridge-valley structure (Figure 6.5). The ESEM micrograph results suggest that membrane biofouling developed along the feed water flow direction from the inlet to outlet. Similar to the feed-end of the lead-element membrane, traces of sodium and chlorine were also observed on the brine-end of the lead-element membrane.

The ESEM micrographs showed that the surfaces of the membrane specimens taken from the TMG10 tail-end element were covered completely by the gel-like biofouling layer; the ridge-valley structure on virgin membrane surface was no longer observable (Figures 6.6 and 6.7). Bacteria were completely embedded in the EPS polymers. The feed-end tail-element membrane, however, displayed less porous morphological features than the brine-end membrane and the fouling layer seemed more compact and condensed. The elemental composition of the feed-end of TMG10 tail-element membrane specimens was similar to the lead element with trace amounts of sodium and chlorine introduced during operation in addition to oxygen, carbon and sulfur detected in the virgin membrane. Sodium was not observed in the EDS spectrum of the brine-end TMG10 tail-element membrane specimen likely due to a thorough rinsing prior to characterization.

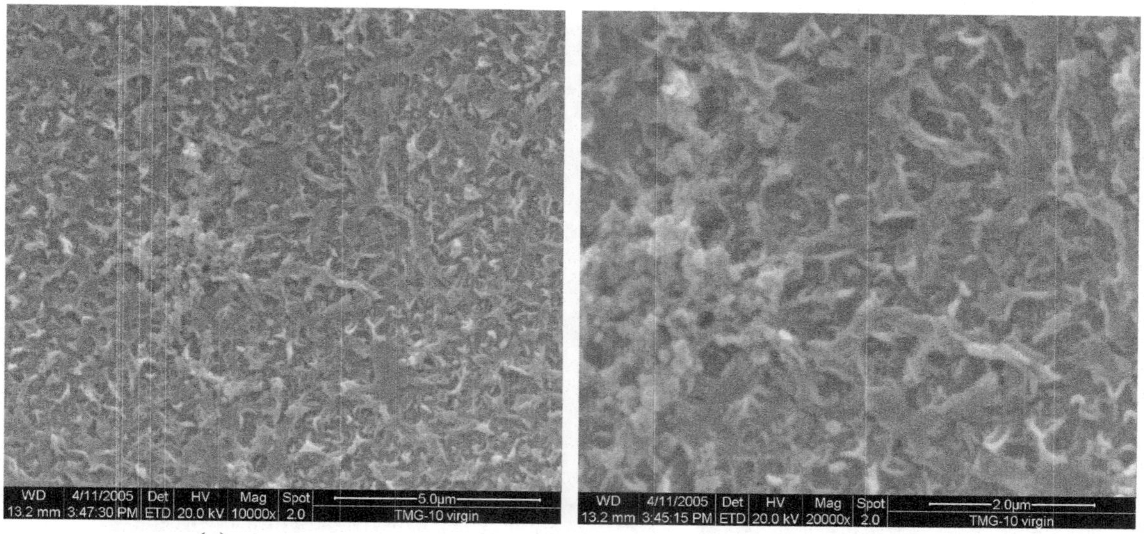

(a) (b)

Figure 6.1 ESEM micrographs of TMG10 virgin membrane (a) x10,000 magnification; (b) x20,000 magnification

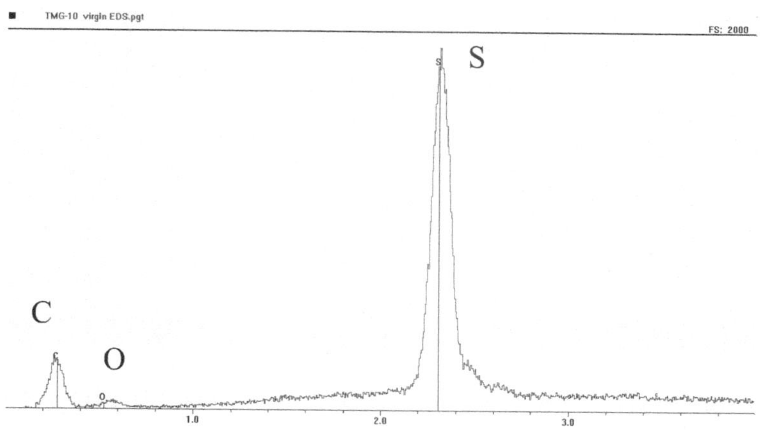

Figure 6.2 EDS spectrum of TMG10 virgin membrane

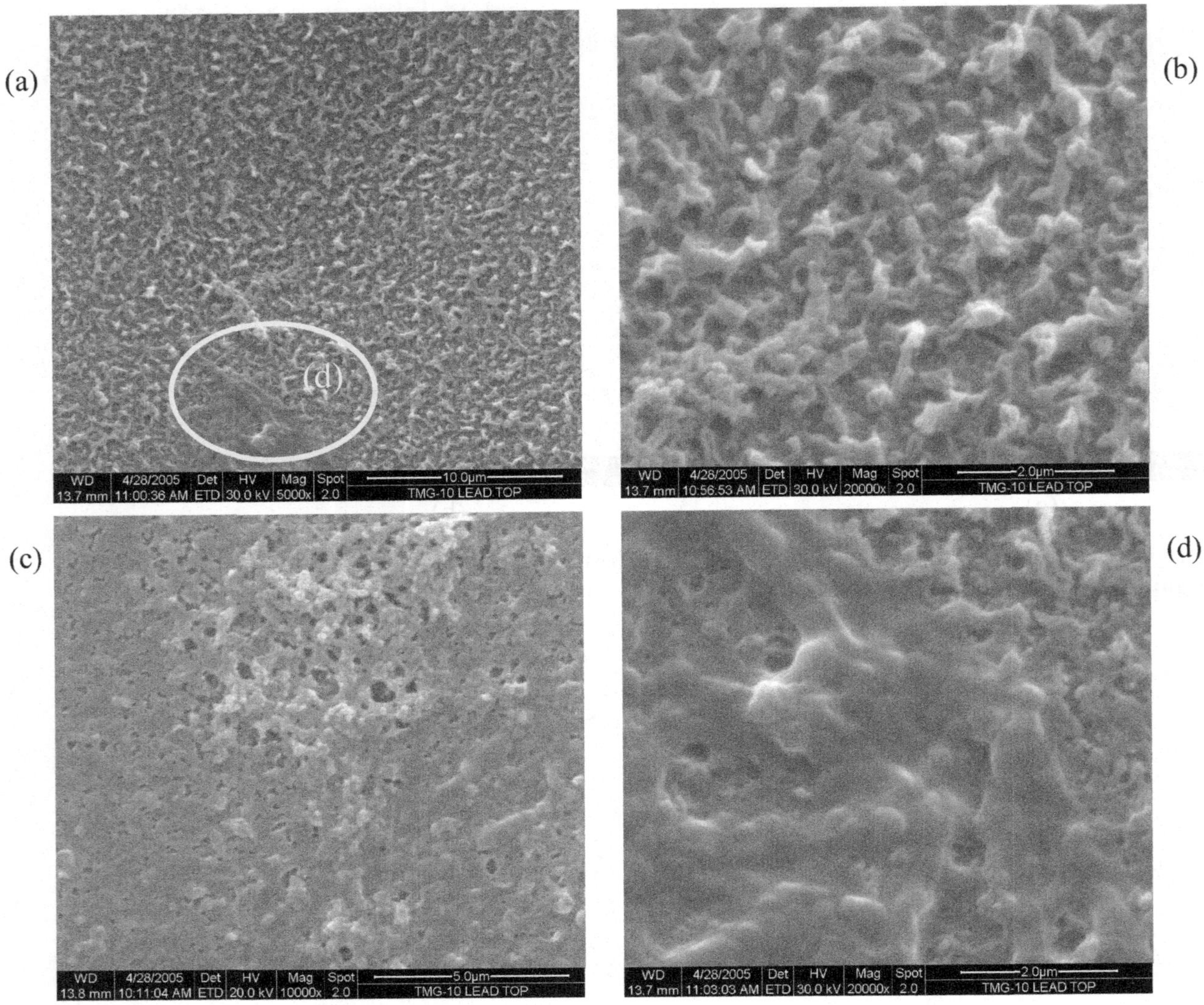

Figure 6.3 ESEM micrographs of the feed-end lead-element membrane specimen at different magnifications (a) x5,000; (b) x20,000; (c) 10,000; (d) x20,000

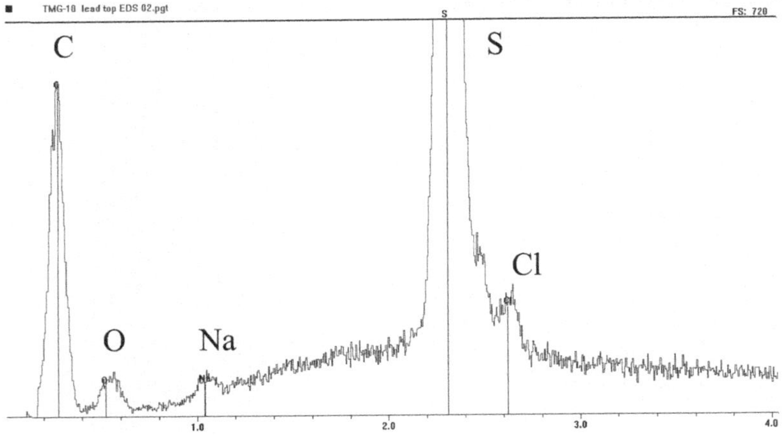

Figure 6.4 EDS spectrum of the feed-end lead-element TMG10 membrane specimen

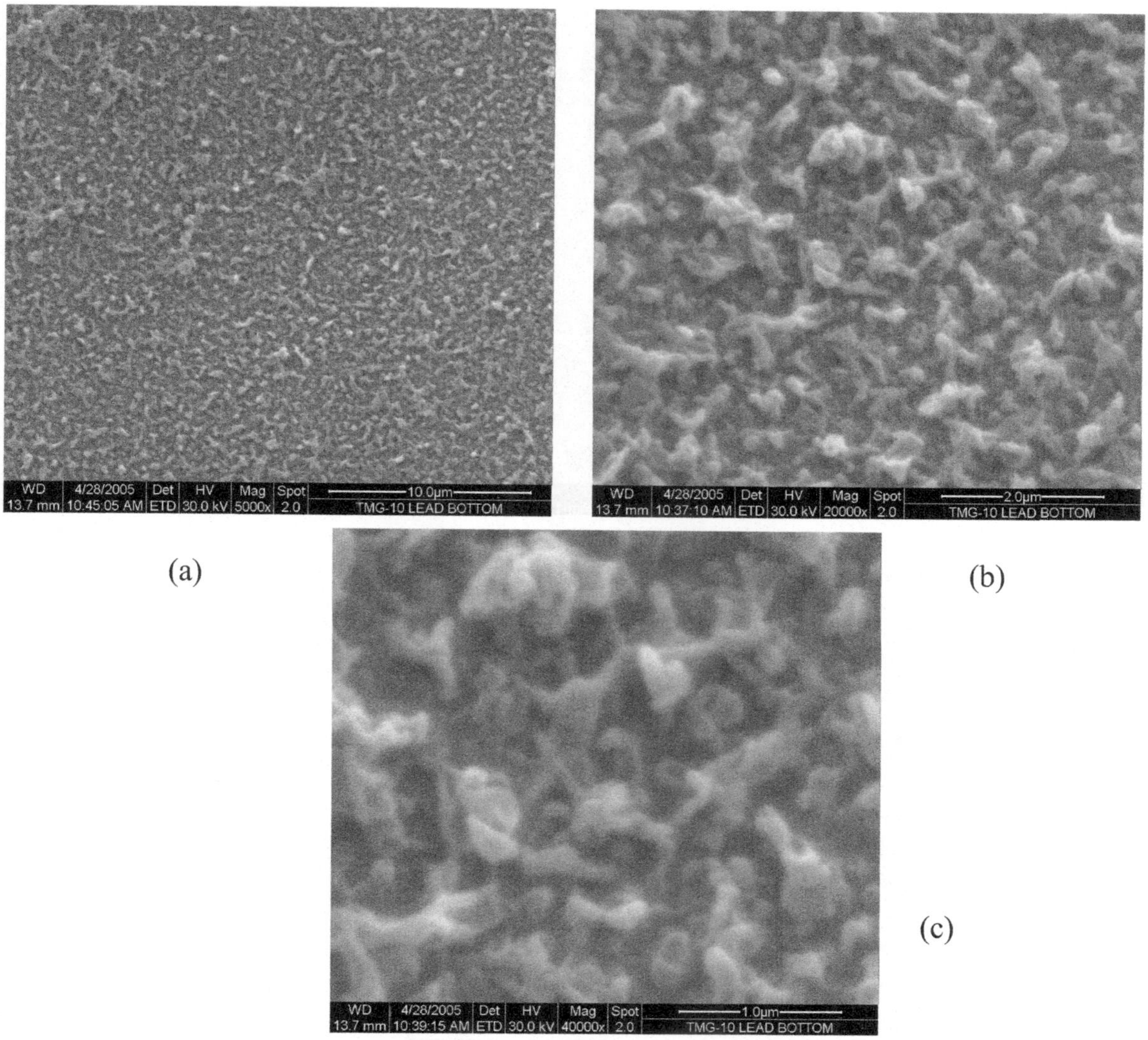

Figure 6.5 ESEM micrographs of the brine-end lead-element TMG10 membrane specimen at different magnifications (a) x5,000; (b) x20,000; (c) x40,000

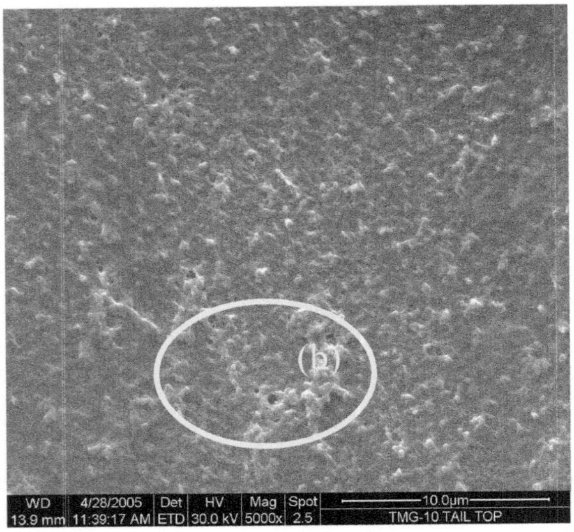

(a)

(b)

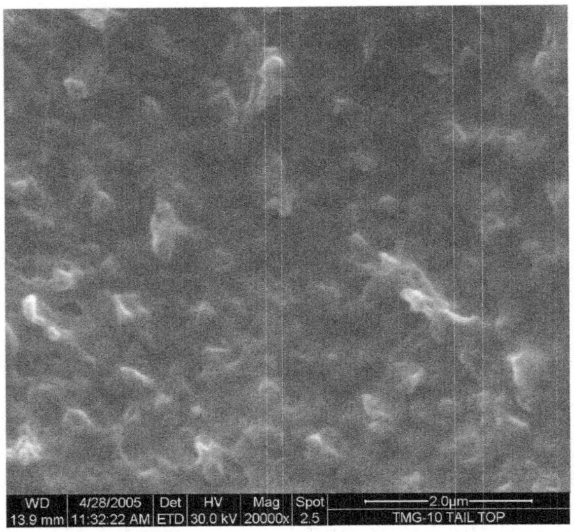

(c)

Figure 6.6 ESEM micrographs of the feed-end tail-element TMG10 membrane specimen at different magnifications (a) x5,000; (b) x10,000; (c) x20,000

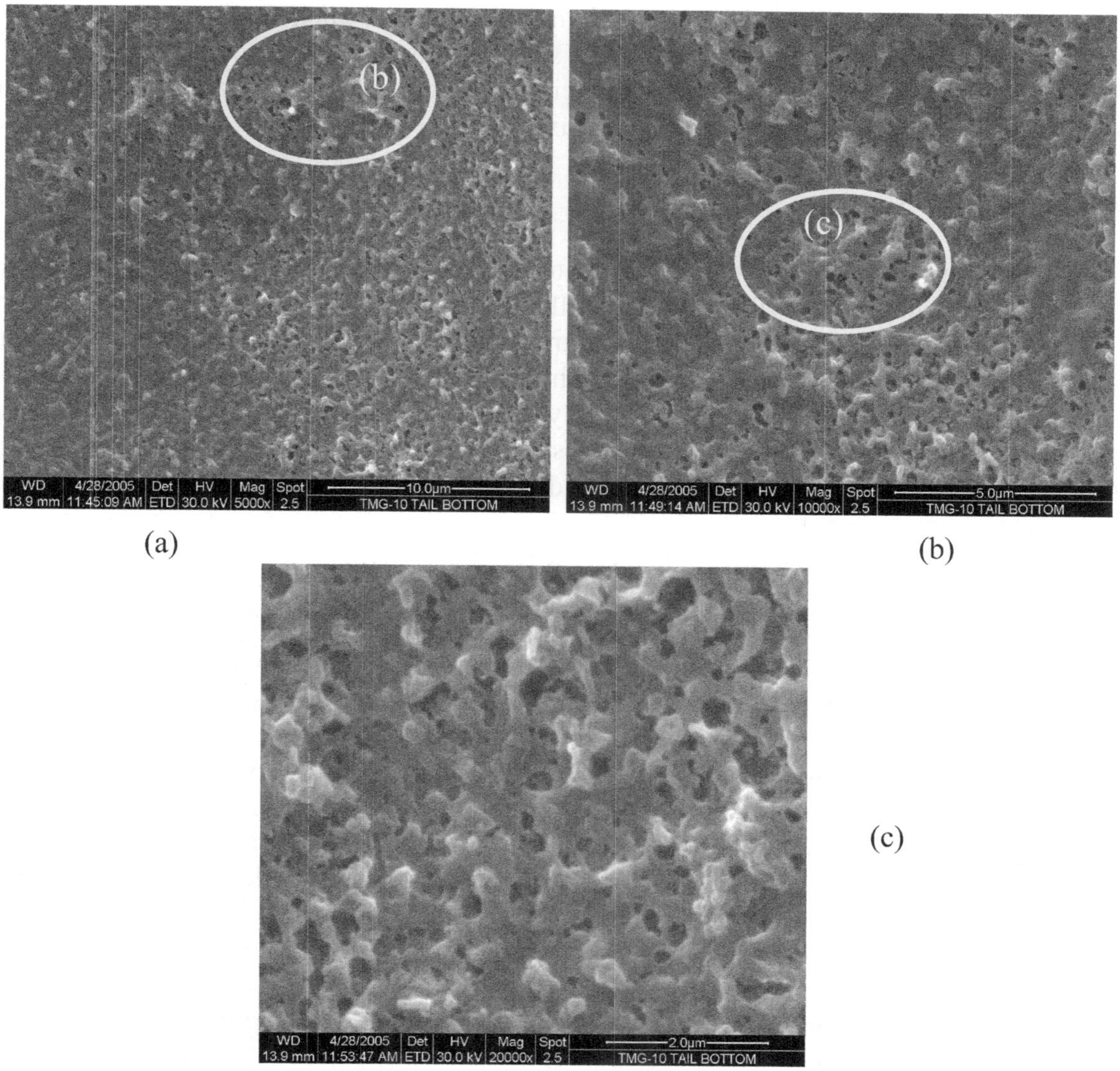

Figure 6.7 ESEM micrographs of the brine-end tail-element TMG10 membrane specimen at different magnifications (a) x5,000; (b) x10,000; (c) x20,000

After 1,300 hours of operation, no obvious foulants were observed by ESEM analysis for both lead and tail-end NF-90 elements (Figure F.3, Appendix F). Similar to the TMG10 membrane operated at WBWRP, the EDS analysis exhibited very small amounts of sodium and chlorine present in the fouled membrane (Figure F.4, Appendix F).

In contrast to the TMG10 and NF-90 operated at WBWRP, the ESEM micrographs of the lead NF-4040 element showed microbial colonies developed on the feed-end membrane surface after 1,600 hours of operation at the SWC (Figure F.7, Appendix F). Some precipitate grains varying in size between 0.1-10 µm were observed on the membrane surface. By performing EDS spot scanning on the grains, the major constituents of the grains were found as barium (Figure F.8(a), Appendix F). The brine-end of the lead element displayed similar fouling layer as the feed-end with combination of microbial colonies and precipitate grains (Figure F.9, Appendix F). The EDS spot scanning showed high amount of aluminum and calcium in the inorganic precipitates with trace amount of barium (Figure F.10, Appendix F). The feed-end and brine-end membrane specimens of the tail-end NF-4040 element exhibited a thick fouling layer consisting of a combination of a biomass layer and inorganic scalants (Figures F.11 and F.12, Appendix F). The EDS scanning detected large amount of calcium, aluminum, and phosphorus elements as inorganic scalants (Figures F.13 and F.14, Appendix F).

Phospholipids Analyses

Phospholipids extraction has been used to estimate the amount of viable microbial biomass in soil (Findlay et al. 1989, Rauch-Williams and Drewes 2006) and on membrane surfaces (Speth et al. 1998). Phospholipids are components of biological cell membranes (about 6 percent of dry cell mass) and are degraded within minutes to hours of cell death by cellular enzymes (Vestal and White 1989). In this study phospholipid analysis was used to reveal biomass in the fouling layers. Virgin membrane specimens were used as blank control for all membranes. The results of biomass (phospholipids) measurements for the various membrane elements are summarized in Table 6.2.

For the TMG10 membrane operated with not nitrified feed water (WBWRP), phospholipids were quantified as 0.21±0.23 and 0.38±0.20 nmol PO_4/cm^2 for the specimens of the brine-end membranes from the lead element and tail-end elements, respectively. Consistent with the ESEM micrographs, the phospholipids results supported the previous observation that biofouling in the tail-end element was more severe and the foulant biomass was almost twofold higher than in the lead element. The biofilm mass quantified in the lead element exhibited a high relative standard deviation of up to 110 percent (percent deviation) as compared to 54 percent for the tail-end element. This further confirms the results from ESEM characterization that the distribution and formation of a biological active layer in the lead element is clustered and rather heterogeneous, which likely explains the high relative standard deviation of foulants measurement.

Table 6.2
Biomass (Phospholipids) measured for fouled membrane specimens

Membrane	Lead Feed-end Mean and Stdev (nmol PO_4/cm^2) RSD (%)	Lead Brine-end Mean and Stdev (nmol PO_4/cm^2) RSD (%)	Tail Feed-end Mean and Stdev (nmol PO_4/cm^2) RSD (%)	Tail Brine-end Mean and Stdev (nmol PO_4/cm^2) RSD (%)
TMG10 not nitrified feed (after 2,965 hrs)	N/A	0.21±0.23 110	N/A	0.38±0.2 54
NF-90 not nitrified feed (after 1,300 hrs)	0.06±0.02 38.8	n.d.	0.36±0.33 94.1	0.47±0.42 88.3
NF-4040 nitrified/denitrified feed (after 1,600 hrs)	1.64±0.44 27	1.05±0.29 27.6	1.6±0.73 45.6	1.45±0.55 38

Stdev – Standard deviation
RSD – relative standard deviation
n.d. – not detected
N/A -not analyzed

The NF-90 membrane, which was also operated with not nitrified feed water (WBWRP), exhibited biomass accumulations in the lead element that varied between undetected for the brine-end to 0.06±0.02 nmol PO_4/cm^2 for the feed-end. The amount of biomass in the NF-90 tail element was higher than the lead element between 0.36±0.33 and 0.47±0.42 nmol PO_4/cm^2 for the specimens of the feed-end and brine-end, respectively. This findings confirms the pattern observed for the TMG10 (at WBWRP) that biofouling tends to be more established in the tail rather the lead membrane elements for the pilot skid.

Phospholipid analyses of the NF-4040 membrane, which was employed using nitrified/denitrified feed water (SWC), revealed severe biofouling in all of the examined membrane elements. The amount of phospholipids in the fouling layer exceeded 1.1 nmol PO_4/cm^2 for both the lead and tail-end elements. The relative standard deviation of the biomass measurements was low further supporting the presence of an evenly distributed biomass layer across the membrane surfaces.

ATR-FTIR Measurement

The functional groups of the virgin and fouled membranes were characterized by ATR-FTIR. All the virgin and fouled membranes exhibited similar distinct and sharp absorption peaks. As an example, the ATR-FTIR spectra of virgin and brine-end tail-element membrane specimens are shown in Figure 6.8 for the TMG10 (operated at WBWRP). The broad absorption peak at 3,300 to 3,400 cm^{-1} indicated polysaccharides-like substances as foulants. Similar to the EDS analysis, infrared light can penetrate through the thin active layer into the polysulfone support layer of the membrane. Therefore, the virgin membrane specimens exhibited similar ATR-FTIR spectra with indicative peaks at 1,615 cm^{-1}, 1,545 cm^{-1} and 1,670 cm^{-1} (amide groups), 1,592 cm^{-1} and 1,110 cm^{-1} (aromatic double bonded carbon), 1,016 cm^{-1} (ester groups), 1,250 cm^{-1} (carboxylates groups), and at 1,151 cm^{-1} and 694 cm^{-1} (sulfone groups).

Although trace amounts of chlorine were detected in all fouled membrane specimens by the EDS analysis, any potential structural changes within the polymer in response to chloramine exposure were not revealed by the ATR-FTIR measurement. The major disadvantage of polyamide membranes is the sensitivity to free chlorine. A few studies suggested that membrane failure resulted from chlorine attack on amide nitrogen and aromatic rings (Glater et al. 1994). Amide nitrogen is vulnerable to chlorine because of electron withdrawing effects of the carbonyl group, resulting in formation of N-chloro derivatives. Due to the greater electron affinity of chlorine (Lide 2004-2005), aromatic rings and sulfonyl groups are susceptive to electrophilic substitution by chlorine. ATR-FTIR scans of chlorine attacked membranes showed the sulfony peak at 1,000 to 1,050 cm^{-1} missing, and an additional chlorine stretching peak occurring at 1,575 to 1,595 cm^{-1} (Beverly et al. 2000). The resulting substitution products may cause deformation in the polymer chain and cleavage at amide linkages. Since no additional functional groups were detected in the fouled TMG10 membrane specimens, findings of this study imply that the polymer structure of the TMG10 elements was not significantly altered during the 3,000 hours of operation at WBWRP.

The virgin and fouled NF-90 membranes exhibited similar distinct and sharp absorption peaks, except the fouled membranes displayed higher absorption peak at 3,300 cm^{-1} indicating the presence of polysaccharides as foulants. The ATR-FTIR spectra of virgin and brine-end tail-element membrane specimens are shown in Figure F.5 (Appendix F). Since no additional functional groups were detected in the fouled membrane specimens, it implies that the polymer structure and functionality of the NF-90 elements were not significantly changed during 1,600 hours of operation.

The ATR-FTIR spectra revealed that the major fractions of the foulants on the NF-4040 membrane surfaces were polysaccharides-like substances (Figure F.15, Appendix F). As compared to the intensity of the absorption peaks of the virgin membrane, sulfonic acids, sugars, proteins, aliphatic and aromatic acids, fulvic and humic acid were also present in the fouling layer in the lead element. Since the fouling layer of the tail-end membranes was likely too thick, the infrared light could not penetrate through the fouling layer. The ATR-FTIR spectra of the tail-end element membranes showed broad and overlapped absorption peaks, which increased the difficulty in identification of specific functional groups. It is assumed that polysaccharides and proteins were the major foulants of the tail-end element.

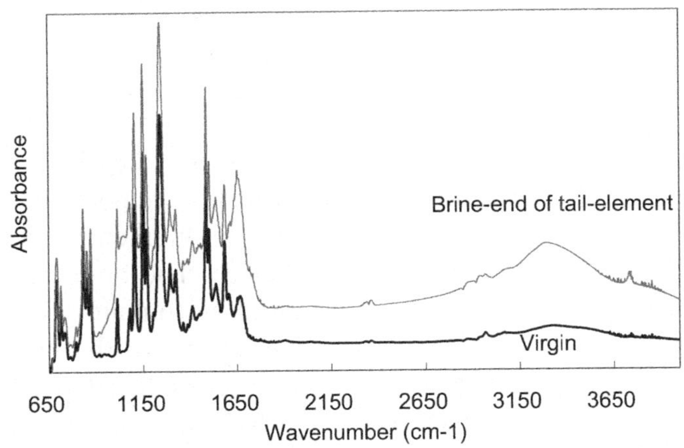

Figure 6.8 ATR-FTIR spectra of virgin and brine-end tail-element TMG10 membrane specimens operated at WBWRP

Contact Angle Measurement

Contact angle measurement is a standard method to determine the hydrophobicity of a surface. A larger contact angle corresponds to a more hydrophobic surface. As compared to the virgin membrane specimens, the fouled membrane specimens of all membranes exhibited more hydrophobic features after termination of the pilot tests (Table 6.3).

Table 6.3
Contact angle measured for the virgin and fouled TMG10 membrane specimens

Membrane	Virgin Contact angle (degree)	Lead Feed-end Contact angle (degree)	Lead Brine-end Contact angle (degree)	Tail Feed-end Contact angle (degree)	Tail Brine-end Contact angle (degree)
TMG10 not nitrified feed	54±4	67±6	68±5	61±3	60±5
NF-90 not nitrified feed	63±1	71±4	72±3	80±3	75±2
NF-4040 nitrified/denitrified feed	30±2	60±3	59±3	32±2	31±5

N/A – Not analyzed

The TMG10 and NF-90 virgin membranes were characterized as moderate hydrophobic and the hydrophobicity was significantly increased as a result of adsorption of hydrophobic substances from wastewater onto membrane surfaces or into the polymer matrix. Roudman and

DiGiano et al. (2000) also observed a significant increase of hydrophobicity for nanofiltration membranes after adsorption of natural organic matter. The adsorption of hydrophobic compounds was found to cause the decline of permeate flux resulting in membrane fouling (Van der Bruggen et al. 2002).

The lead TMG10 element membranes, however, were found more hydrophobic than the tail-end element with no obvious difference between the inlet or outlet sections. The decrease of hydrophobicity in the tail-end element was likely caused by the accumulation of hydrophilic polysaccharides substances on membrane surfaces. Microorganisms produce hydrophilic organic compounds such as polysaccharides, urea, etc., which represent the extracellular enzyme of microorganisms. During biological treatment of wastewater, large amounts of soluble microbial materials are produced from substrate metabolism and biomass decay. These soluble microbial products (SMP) can result in severe organic fouling by precipitate onto membrane surfaces (Park et al. 2005, Shon et al. 2004, Ivnitsky et al. 2005, Jarusutthirak et al. 2002). Contact angle of the fouled membranes was found much lower than that of virgin membranes (Shon et al. 2004). Better removal of organic substances from biologically treated wastewater resulted in significantly lower permeate flux decline and higher contact angle (Shon et al. 2004).

In this study, the decrease in contact angle observed on TMG10 membranes was probably the combination result of biomass growth on membrane surfaces as measured by phospholipids analyses, and the precipitation of soluble microbial products present substantially in the microfiltered secondary effluent. Contact angle measurement was proven to be a sensitive and complementary tool to other characterization methods, such as phospholipids analyses, ATR-FTIR and ESEM.

Unlike the TMG10 operated at the SWC, the fouled NF-4040 membrane specimens exhibited very different hydrophobicity features for the lead and tail-end elements. As compared to the hydrophilic virgin membrane, the lead NF-4040 element exhibited a very hydrophobic character due to the adsorption of hydrophobic substances from wastewater onto membrane surfaces (Table 6.3). The tail-end NF-4040 element exhibited very hydrophilic features as a result of precipitation of polysaccharides substances and the formation of hydrophilic EPS polymers during biofouling. Although the NF-4040 lead-element exhibited a similar amount of biomass (in terms of phospholipids) in the fouling layer as the tail-end element, the fouling layer on the tail-end element membrane was much thicker than on the lead element membrane (see ESEM micrographs in Figures F.6, F.7, F11, and F.12, Appendix F). The large amount of polysaccharides and EPS polymers in the tail-end element overwhelmed the hydrophobic features occurred during the earlier stage of fouling.

REMOVAL OF FOULANTS AFTER CLEAN-IN-PLACE

TMG10 Autopsy – Nitrified/Denitrified Wastewater Effluent

The ULPRO membrane TMG10 was tested in the pilot-scale membrane skid over a course of 2,000 hours from October 23, 2005 to February 12, 2006, at the Scottsdale Water Campus (SWC). Upon completion of the pilot-scale investigation, a tail-end element of the last stage was removed from the skid and shipped back to the CSM laboratory for immediate autopsy. The TMG10 membranes employed in the pilot skid were cleaned in place with 380 L (100 gallons) of chemical solution (2.5 percent sodium dodecylbenzene sulfonate and 2 percent sodium tripolyphosphate with pH 10 to 11 and temperature 35 to 38°C). Prior to applying the

CIP solution, the pilot skid was first flushed by pumping RO permeate water and drained for several minutes. Subsequently, the chemical solution was circulated through each vessel for one hour at a flow rate of 38 L/min (10 gpm) for stage 2 and 57 to 76 L/min (15 to 20 gpm) for stage 1, maintaining 35 to 38°C. After chemical cleaning, each stage was rinsed individually until inlet and outlet conductivities were similar. A cleaned tail-end element was then removed from the skid for immediate autopsy. Both elements selected for autopsy were unrolled, cut into small pieces, and rinsed carefully with DI water (type II water) prior to autopsy analysis.

Visible Inspection

No visible flaws and cracks were observed in membrane elements. Brown patches of scales were found on both cleaned and uncleaned membrane leaves of active layer. Membrane permeate sides and carriers were free of any visible foulants.

Membrane Surface Morphology and Elemental Composition

The ESEM micrographs of the TMG10 membrane specimen prior to CIP revealed that the feed-end membrane surface was partially covered by a gel-like biofilm (Figure 6.9(a)). The ESEM micrograph at high magnification revealed that bacteria were embedded in EPS polymers. Numerous precipitate grains varying in size between 1-10 μm were observed on the membrane surface. By performing EDS spot scanning on the white grains, the major constituents of the grains were identified as calcium with trace amount of phosphorus (Figure 6.10 (a)). EDS screen scanning across the entire specimen sample did not detect inorganic scalants (Figure 6.10 (b)). The presence of carbon, sulfur, and oxygen in the foulants and scalants was not distinguishable from the membrane support layer. The biofilm initiated at discrete locations in the feed-end developed into a strong, coherent fouling layer at the brine-end (Figure 6.11 (a) and (b)). The EDS analysis revealed calcium and phosphorus as the major components of foulants with small amount of magnesium, aluminum, and iron present (Figure 6.12).

The ESEM/EDS analysis exhibited more severe fouling/scaling in treating the nitrified/denitrified wastewater than the non-nitrified wastewater although the TMG10 membrane was operated for a shorter time (2,000 hours versus 3,000 hours) and at the same recovery of 82 percent. The water quality analysis showed that the nitrified/denitrified wastewater at SWC exhibited higher hardness but lower silica concentrations than the non-nitrified wastewater (calcium concentration 75 *versus* 51 mg/L, and silica concentration 12 *versus* 24 mg/L, repectively). Other major elements that can cause scaling were similar Al (0.03 mg/L), Fe (0.05 mg/L), Mg (30 mg/L), P (2 mg/L), Zn (0.02 mg/L) in the nitrified/denitrified wastewater, and Al (0.01 mg/L), Fe (0.10 mg/L), Mg (21 mg/L), P (7 mg/L), Zn (0.02 mg/L) in the non-nitrified wastewater, respectively. The differences in inorganic scaling might be caused by the different antiscalants used, *Pretreat Plus 0100* (King Lee Technologies, San Diego, CA) at the WBWRP and *Hypersperse MDC700* (GE Betz, Trevose, PA) at the SWC. According to the manufacturer, *Hypersperse MDC700* controls deposits of particulate matter and scales including calcium carbonate, calcium sulfate, barium sulfate, and strontium sulfate. Containing an organic dispersants, *MDC700* is used to reduce colloidal iron fouling and inhibit scale formation when metal fouling is a primary concern. *Pretreat Plus 0100* is used to control scales including calcium carbonate, calcium sulfate, barium sulfate, strontium sulfate, calcium fluoride, iron, aluminum, and can control silicates to some extent.

After chemical cleaning, the feed-end TMG10 membrane specimen from the tail-element was still partially covered by a biofilm with incorporated precipitate grains, which is similar to the uncleaned feed-end membrane surface of the tail-element (Figure 6.13 (a) and (b)). Unlike the uncleaned membrane, individual bacteria were observed on the cleaned membrane surfaces (Figure 6.13 (c) and (d)). It is inevitable that the biofilm deposits survive from chloramines disinfection protected by EPS and they are difficult to remove by chemical cleaning. Portions of the biofilm were likely released, detached during chemical cleaning and spread through the membrane surfaces. Some of the viable cells could re-attached to the membrane and started multiplying (Figure 6.13 (c) and (d)). The EDS analysis revealed silica and calcium as the major elemental constituents of the scalants (Figure 6.14 (a)). The EDS spot scanning revealed that the grains were silica precipitates bonded with calcium and aluminum (Figure 6.14 (b)).

After chemical cleaning, the thick and compact fouling layer observed in the brine-end of the uncleaned membrane element was sloughed. Membrane surface was partially covered by foulants and the EPS biofilm still dominated the fouling layer (Figure 6.15). EDS screen scanning did not detect any scalants on membrane surfaces although inorganic grains were still present. EDS spot scanning detected the inorganic grains were mainly composed of aluminum, silica, and potassium.

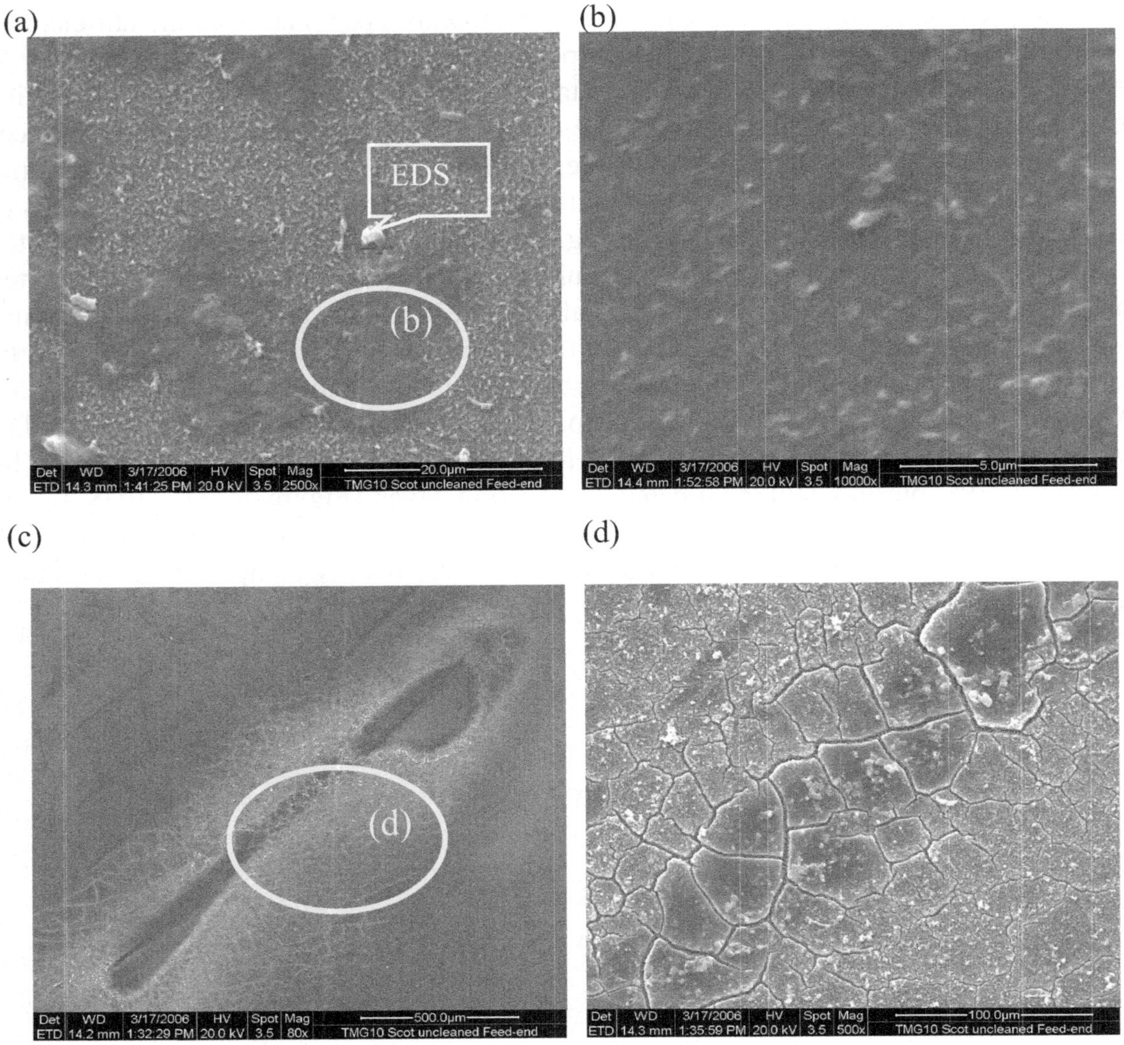

Figure 6.9 ESEM micrographs of the uncleaned feed-end tail-element TMG10 membrane specimen at different magnifications (a) x2,500 inorganic scaling (white precipitates) formed with biofouling; (b) x10,000 bacteria embedded in EPS polymers; (c) x80 feed spacer imprint; (c) x500 foulants formed around the feed spacer imprint

(a)

(b)

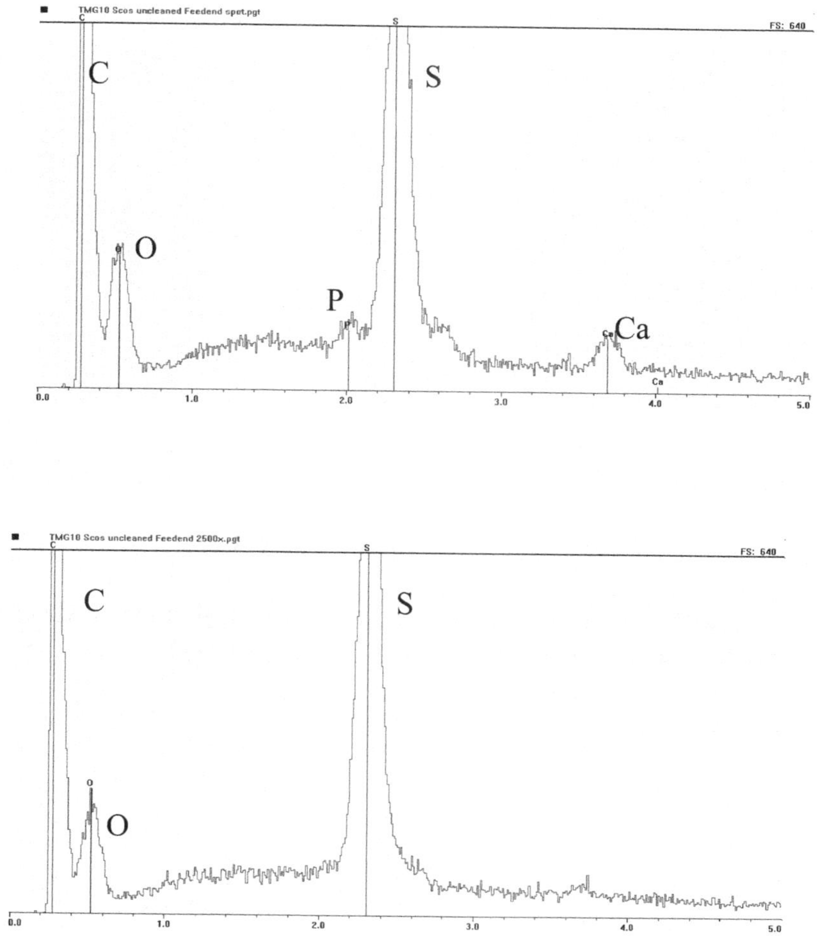

Figure 6.10 EDS spectra of the uncleaned feed-end tail-element TMG10 membrane specimen (a) spot scanning on the inorganic grain (Figure 6.9(a)); (b) screen scanning

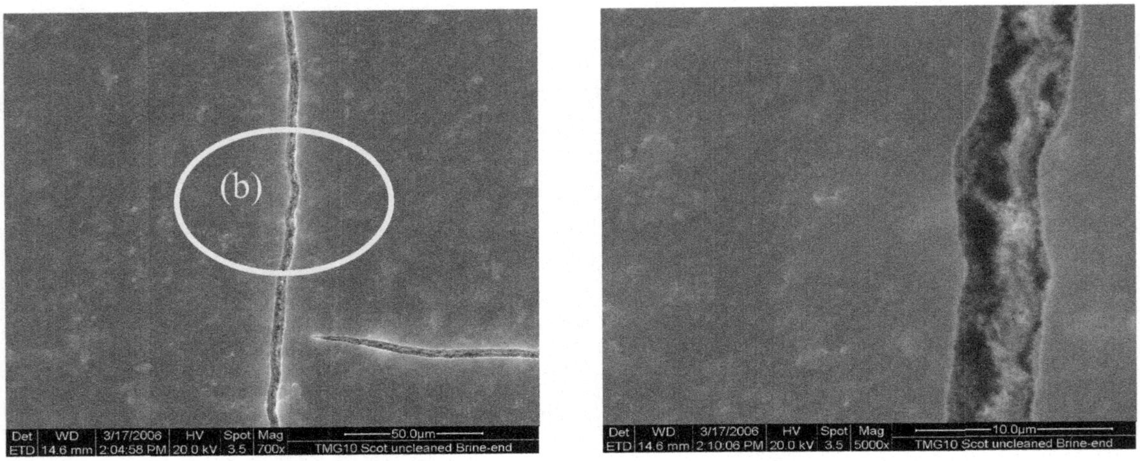

(a) (b)

Figure 6.11 ESEM micrographs of the uncleaned brine-end tail-element TMG10 membrane specimen at different magnifications (a) x700; (b) x5,000

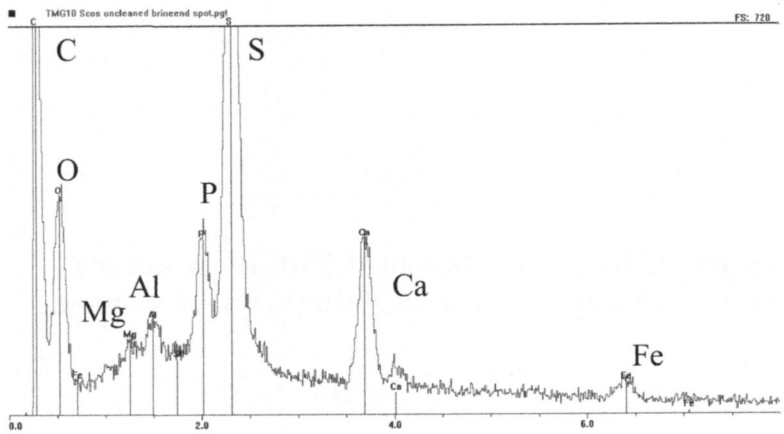

Figure 6.12 EDS spectrum of the uncleaned brine-end tail-element TMG10 membrane specimen

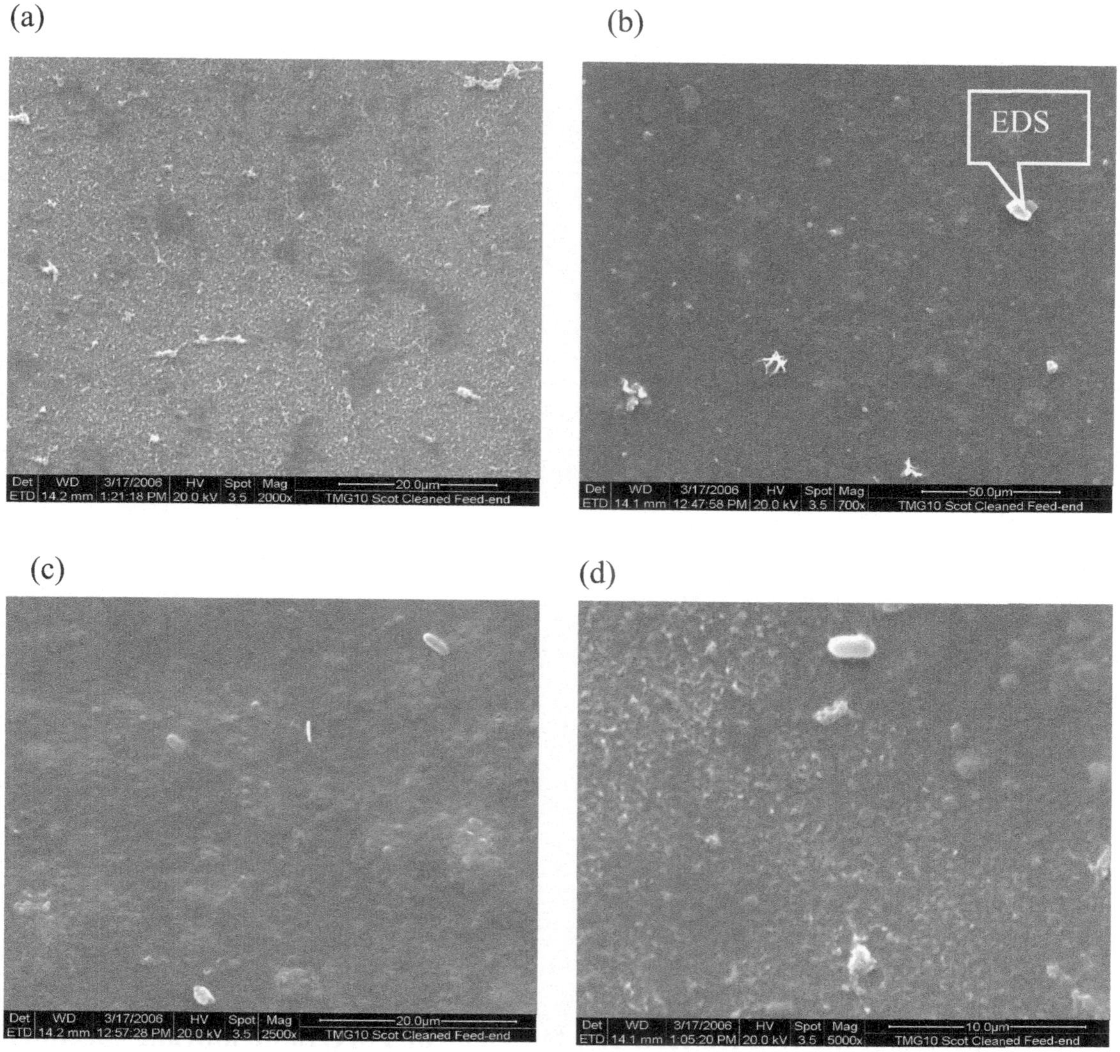

Figure 6.13 ESEM micrographs of the chemical-cleaned feed-end tail-element TMG10 membrane specimen at different magnifications (a) x2,000 partially fouled surface; (b) x700 inorganic scaling in combination with biofouling; (c) x2,500 biofouling with individual bacteria; (c) x5,000 biofouling with individual bacteria

(a)

(b)

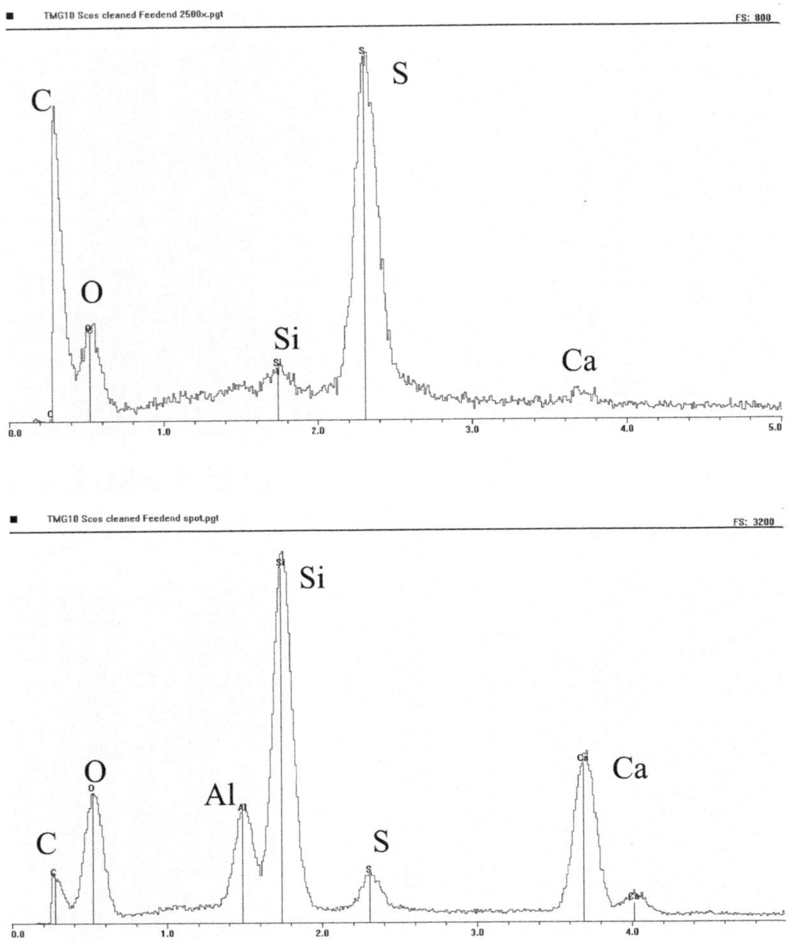

Figure 6.14 EDS spectra of the cleaned feed-end tail-element TMG10 membrane specimen (a) screen scanning; (b) spot scanning on the inorganic grains

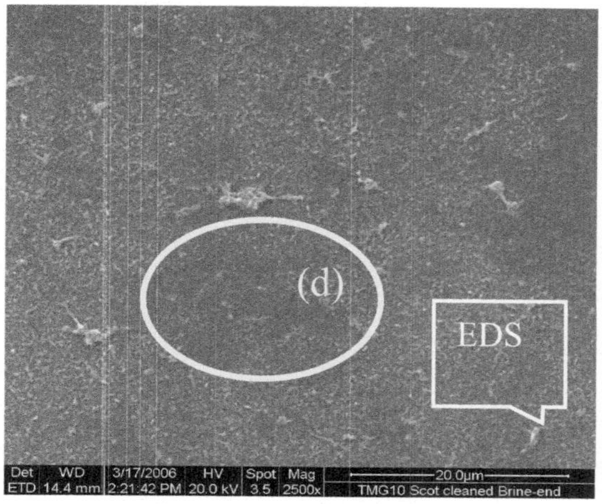

(a)

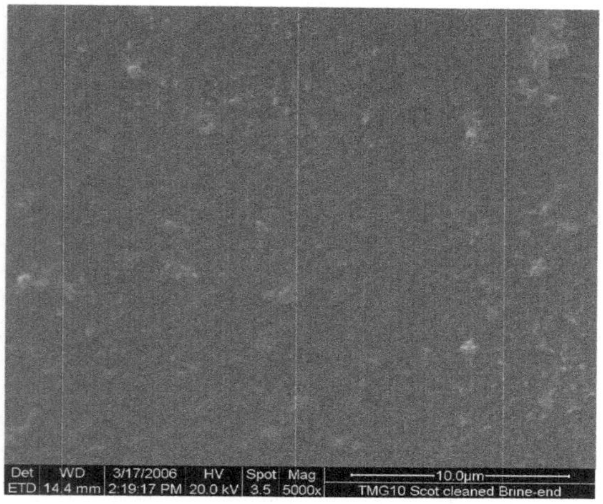

(b)

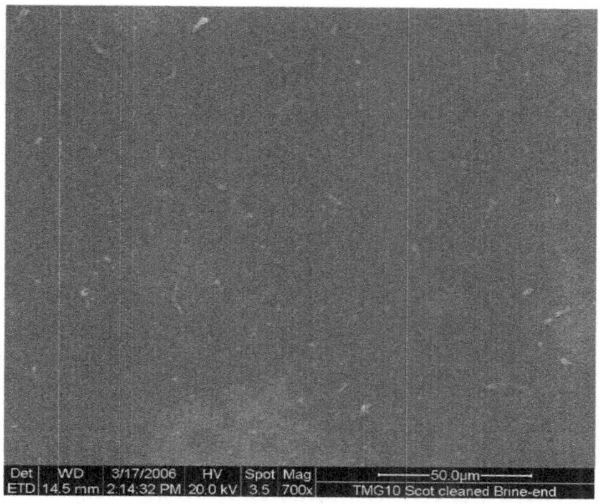

(c)

Figure 6.15 ESEM micrographs of the chemical-cleaned brine-end tail-element TMG10 membrane specimen at different magnifications (a) x2,500 partially fouled surface; (b) x5,000 biofilm; (c) x700 inorganic scaling in combination with biofouling

(a)

(b)

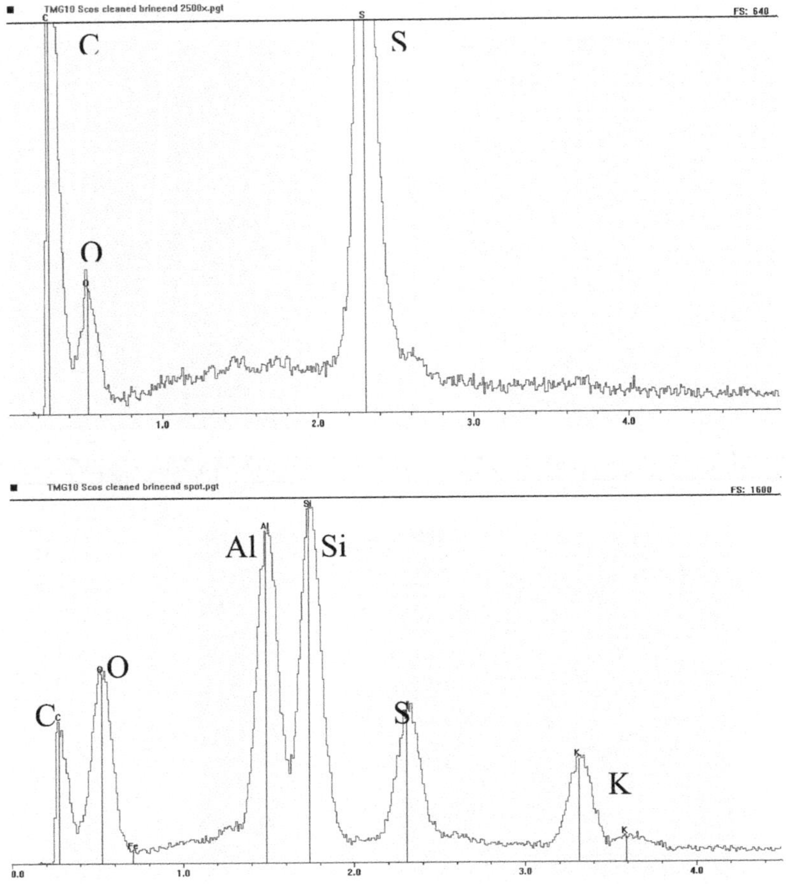

Figure 6.16 EDS spectra of the cleaned brine-end tail-element TMG10 membrane specimen (a) screen scanning; (b) spot scanning on the inorganic grains

Phospholipids Analyses

Biomass (phospholipids) measurements of the TMG10 membrane specimens after operating at the SWC prior to and after CIP are summarized in Table 6.4. The TMG10 virgin membrane was used as blank control for the phospholipids analysis. Consistent with the ESEM micrographs, the amount of phospholipids in the feed-end was lower than in the brine-end for both uncleaned and chemical-cleaned membranes. In general, the biomass quantified for the uncleaned membranes was much higher than in the chemical-cleaned membranes. Thus, chemical cleaning can partially restore permeate flux by removing portions of the biofilm.

Table 6.4
Biomass (Phospholipids) measured for membrane specimens prior to and after CIP

Membrane	Lead Feed-end Mean and Stdev (nmol PO_4/cm^2) RSD (%)	Lead Brine-end Mean and Stdev (nmol PO_4/cm^2) RSD (%)	Tail Feed-end Mean and Stdev (nmol PO_4/cm^2) RSD (%)	Tail Brine-end Mean and Stdev (nmol PO_4/cm^2) RSD (%)
TMG10 nitrified/denitrified feed (after 2,000 hrs)	N/A	N/A	0.32±0.12 36.9	0.58±0.11 18.1
TMG10 nitrified/denitrified feed after CIP (after 24 hrs)	N/A	N/A	0.07±0.09 126.4	0.35±0.21 59.3

Stdev – Standard deviation
RSD – relative standard deviation
n.d. – not detected
N/A -not analyzed

ATR-FTIR Measurement

The ATR-FTIR spectra revealed that the major fractions of the foulants on the TMG10 membrane surfaces were polysaccharides-like substances, which exhibited large absorption peaks around 3,300 to 3,400 cm^{-1} (Figure 6.17). As compared to the absorption peaks of virgin membranes at 1,615 cm^{-1}, 1,545 and 1,670 cm^{-1} (amide groups), 1,592 cm^{-1} and 1,110 cm^{-1} (aromatic double bonded carbon), 1,016 cm^{-1} (ester groups), 1,250 cm^{-1} (carboxylates groups), and at 1,151 cm^{-1} and 694 cm^{-1} (sulfone groups), the high absorption intensities of the uncleaned membrane specimens indicated that sulfonic acids, sugars, proteins, aliphatic and aromatic acids, fulvic and humic acid, and silicates were major components of the foulant layer. The decline of absorption intensity at the above wave numbers for cleaned membranes provides evidence that chemical cleaning removed a significant portion of polysaccharides, sulfonic acids, sugars and proteins from membrane surfaces.

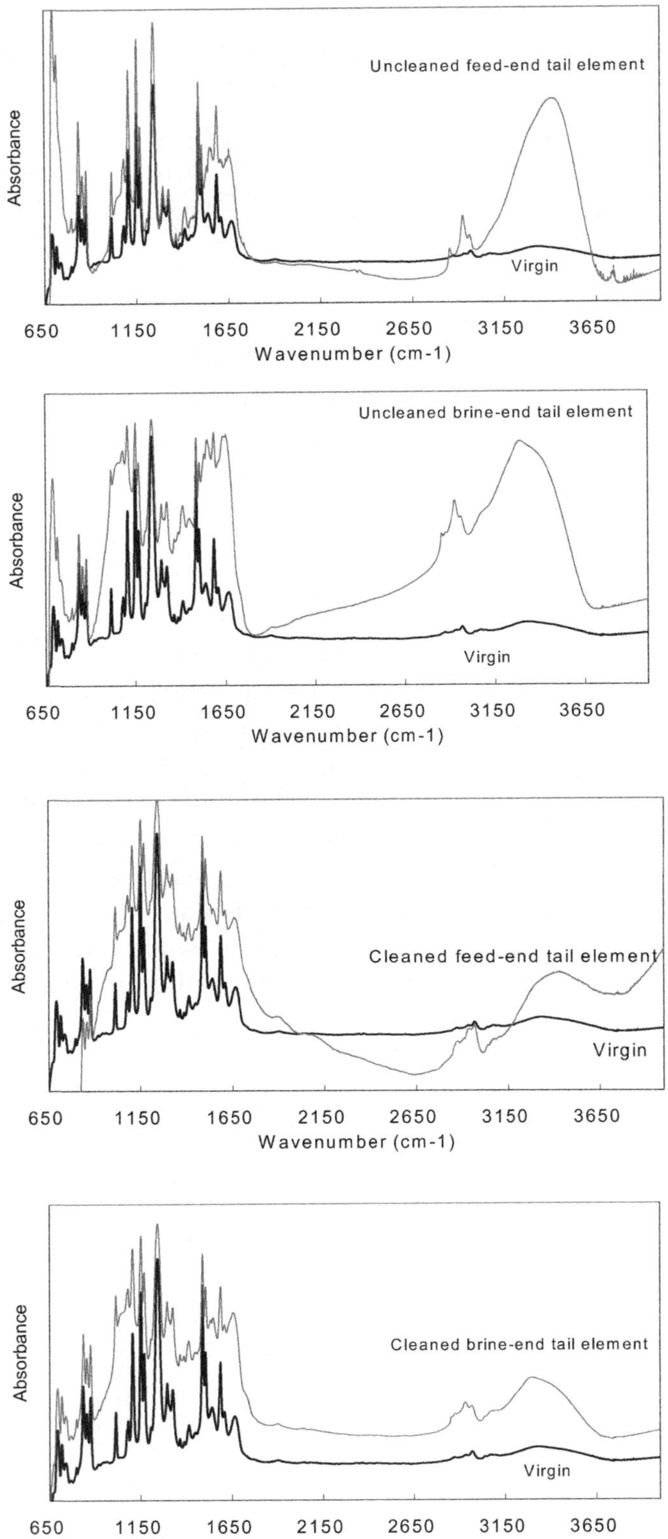

Figure 6.17 ATR-FTIR spectra of virgin, cleaned, and uncleaned TMG10 membrane specimens from the tail-end elements

Contact Angle Measurement

As compared to the virgin TMG10 membrane, the fouled membrane specimens after 2,000 hours of operation at SWC exhibited more hydrophilic features (Table 6.5), which was likely due to the large amount of hydrophilic extracellular polysaccharides substances associated with biogrowth. These polymers overwhelmed the hydrophobic features of the membrane.

Table 6.5 Contact angle measured for the virgin, uncleaned and cleaned TMG10 membrane specimens from the tail-end elements after operating at the SWC

Membrane	Virgin Contact angle (degree)	Lead Feed-end Contact angle (degree)	Lead Brine-end Contact angle (degree)	Tail Feed-end Contact angle (degree)	Tail Brine-end Contact angle (degree)
TMG10 nitrified/denitrified feed	54±4	N/A	N/A	48±3	44±3
TMG10 nitrified/denitrified feed after CIP	N/A	N/A	N/A	38±5	39±6

N/A – Not analyzed

EVALUATION OF BENCH-SCALE FOULING ASSESSMENT PROTOCOL

The extent of membrane fouling is an important element that needs to be considered in membrane selection. There are several methods to test membrane fouling, of which two are currently recommended by the Information Collection Rule of the USEPA. These methods are the rapid bench-scale membrane test and the single-element membrane test. Both methods use a continuous-flow configuration and require at least several hundreds of liters of water to assess membrane fouling and the effectiveness of cleaning for a typical five-day test. These methods provide a realistic membrane configuration (e.g., spiral-wound) but the test cannot be done conveniently at an off-site laboratory. In this study, we employed an alternative batch method that uses a smaller sample volume, smaller membrane surface and a bench-scale unit operated in recycling mode. This design has been previously used by researchers studying membrane fouling in laboratory studies (Speth et al. 1998, DiGiano et al. 2000, Li and Elimelech 2004, Hoek et al. 2003).

The fouling potential estimated from the bench-scale laboratory experiments are compared in this section to the flux decline observed during pilot-scale field testing. The initial membrane compaction (i.e. flux decline within 48 hours in the pilot studies) was not considered in the assessment. The membranes used in the laboratory test were, however, pre-compacted with DI water.

TMG10 Treating Microfiltered Non-nitrified Wastewater Effluent

During the TMG10 membrane pilot-scale study at WBWRP, the temperature corrected specific flux decreased from approximately 0.04 L/m^2-hr-kPa (0.17 gfd/psi) to approximately 0.032 L/m^2-hr-kPa (0.128 gfd/psi) during the first 200 hours, and gradually declined to 0.027 to 0.03 L/m^2-hr-kPa (0.11 to 0.12 gfd/psi) within 700 hours of operation, where it leveled off throughout the remainder of the testing period (Figure 4.2). The specific flux of the TMG10 membrane decreased by approximately 8 percent over the first 200 hours of operation and stabilized at 15 percent decline throughout the remainder of the testing, which was comparable to the flux decline of 13 percent obtained from the bench-scale experiments (Figure 3.1).

NF-90 Treating Microfiltered Non-nitrified Wastewater Effluent

During the NF-90 membrane pilot study at the WBWRP, the temperature corrected specific flux decreased from approximately 0.09 L/m^2-hr-kPa (0.35 gfd/psi) over the first 200 hours to 0.03 L/m^2-hr-kPa (0.12 gfd/psi) after 600 hours of operation (Figure 4.2). The specific flux of the NF-90 decreased by 15 percent over the first 200 hours of operation and decreased to 23 percent after 1,300 hours of field testing, which was lower than the flux decline of 34 percent observed during the bench-scale experiments (Figure 3.1).

NF-4040 Treating Microfiltered Nitrified/denitrified Wastewater Effluent

The NF-4040 membrane exhibited a rather minor initial flux decline in treating the nitrified/denitrified wastewater. During the pilot testing at the SWC, the temperature corrected specific flux of the NF-4040 declined initially from 0.1 L/m^2-hr-kPa (0.4 gfd/psi) and stabilized after 2 weeks at approximately 0.08 L/m^2-hr-kPa (0.33 gfd/psi) (Figure 4.8). After approximately 1,100 hours of operation, the flux decreased suddenly to 0.05 L/m^2-hr-kPa (0.25 gfd/psi) within 500 hours likely due to a lack of chloramines addition to the feed water. The NF-4040 exhibited a flux decline of 24 percent over 200 hours of operation in laboratory experiments, which is higher than the flux decline of 18 percent observed during the same period of operation at pilot-scale.

TMG10 Treating Microfiltered Nitrified/denitrified Wastewater Effluent

The TMG10 membrane exhibited a higher flux decline in treating the nitrified/denitrified wastewater than the non-nitrified wastewater. After initial membrane compaction, the temperature corrected specific flux decreased from approximately 0.04 L/m^2-hr-kPa (0.17 gfd/psi) to approximately 0.025 L/m^2-hr-kPa (0.1 gfd/psi) over the first 200 hours of operation and continuously declined to 0.018 L/m^2-hr-kPa (0.07 gfd/psi) throughout the remainder of the testing period. The specific flux of the TMG10 decreased by 9 percent over 200 hours of operation and gradually decreased by 36 percent over 2,000 hours during the pilot-scale testing (Figure 4.9). The flux decline of the TMG10 during the first 200 hours was comparable to the flux decline of 13 percent obtained from the bench-scale experiments (Figure 3.1).

The findings from this study indicate that bench-scale data could not directly predict the degree of fouling of membranes employed at full-scale due to different operating parameters and hydrodynamic conditions. Bench-scale tests of membrane fouling, however, provide a quick screening approach in comparing the performance of different membranes and in exploring the fouling propensity of the candidate membranes.

SUMMARY

In addition to operational parameters (such as flux decline, feed pressure, and recovery) to describe the fouling behavior of the tested membranes, a suite of analytical techniques was employed to characterize membrane fouling established on lead elements and tail-end elements removed from the pilot skid after operation. The nature of the foulants was analyzed by attenuated total reflection (ATR) Fourier transform infrared spectroscopy (FTIR) to determine functional groups, environmental scanning electron microscopy (ESEM) to characterize membrane surface and morphology, energy dispersive spectroscopy (EDS) to characterize elemental composition, phospholipids analysis to quantify biofouling, and contact angle measurements to characterize membrane surface hydrophobicity.

Findings of this study revealed that all membranes tested exhibited a very similar fouling tendency with a more severe biofouling and inorganic scaling occurring in the tail-end elements. The nature and stages of fouling, however, correlate closely with feed water quality and effectiveness of fouling control measures.

After treating the microfiltered non-nitrified wastewater effluent for 3,000 hours, ESEM micrographs of the TMG10 membrane specimens revealed a heterogeneous fouling layer that was formed on the lead membrane leaves. The ridge-valley network on membrane surface was partially covered by a slimy and coherent extracellular biofilm. The surfaces of the membrane specimens taken from the tail-end element were covered completely by the gel-like biofouling layer. Contact angle measurement indicated adsorption of hydrophobic organic substances occurred in the lead and tail-end elements. EDS analysis did not detect any inorganic scaling in the lead and tail-end elements. ATR-FTIR spectra indicated polysaccharides substances in the fouling layer but the functional groups of the membrane were not significantly modified due to fouling. Consistent with findings from ESEM analysis, phospholipids analyses confirmed higher biomass in the tail-end element.

The TMG10 membrane exhibited a larger flux decline in treating the nitrified/denitrified wastewater than the non-nitrified wastewater. After 2,000 hours of operation at SWC, ESEM micrographs revealed a heterogeneous fouling layer that was formed on the membrane leaves of the TMG10 tail-end element. In the feed-end the ridge-valley network on membrane surface was partially covered by biofilm incorporated with inorganic precipitate grains. The biofilm initiated at discrete locations in the feed-end developed into a strong, coherent whole fouling layer at the brine-end. EDS scanning revealed Ca, Mg, Al, P, Si and Fe as the major constituents of scales. The ATR-FTIR spectra showed that the major fractions of the foulants on the TMG10 membrane surfaces were polysaccharides-like substances, sulfonic acids, sugars, proteins, aliphatic and aromatic acids, fulvic and humic acid, and silicates.

After 2,000 hours of treating the nitrified/denitrified wastewater, the membranes were cleaned through a clean-in-place. After chemical cleaning, the thick and compact fouling layer observed in the brine-end of the uncleaned TMG10 membrane tail element was sloughed. The EDS analysis revealed calcium, aluminum, and silica as the major elemental constituents in the fouling layer. The ATR-FTIR spectra displayed that chemical cleaning removed a significant portion of polysaccharides, sulfonic acids, sugars and proteins from membrane surfaces. Phospholipids analyses indicated that chemical cleaning removed more than half of the biomass from the membrane surface. ESEM micrographs, however, showed some of the viable cells re-attached to the membrane and started multiplying after chemical cleaning. It is inevitable that the biofilm deposits, protected by EPS, survive chloramine disinfection and they are difficult to remove by chemical cleaning.

The NF-90 membrane exhibited a similar fouling behavior as the TMG10 in treating the non-nitrified wastewater at WBWRP. Membrane autopsy revealed the adsorption of hydrophobic organic substances and initiation of biofouling. No inorganic scaling was observed after 1,300 hours of operation.

The NF-4040 membrane exhibited severe biofouling and inorganic scaling in treating the nitrified/denitrified wastewater. After 1,600 hours of operation, the NF-4040 lead element exhibited highly hydrophobic characteristics due to adsorption of hydrophobic organic substances. The modification of the originally hydrophilic membrane surface made it favorable for cell attachment. The large amount of adhesion of polysaccharides substances and synthesis of EPS polymer due to biogrowth overwhelmed the hydrophobic feature of the tail-end element and changed the membrane surface to a very hydrophilic nature.

Membrane autopsy enables developing improved understanding of membrane fouling at different stages and with different water qualities. Membrane fouling is a combination of physical (hydrodynamic condition, recovery, feed flow rate, permeate flux), chemical (feed water quality, addition and types of antiscalants), and biological factors (addition and effectiveness of biocides). Membrane fouling does not occur spontaneously, but rather in stages. Initially hydrophobic organics substances adsorb and modify membrane surface making it favorable for cell adhesion. Attached viable cells produce extracellular polymeric substances (EPS) to stabilize the attachment and protect biofilm organisms from chemical cleaning and biocides. The EPS matrix can also act as nucleation sites for the precipitation of crystals of sparingly salts, chemisorb both organic and inorganic molecules, and entrap colloidal or other particles (Schneider et al. 2005). The formation of confluent biofilm incorporating with precipitation of other organic matters, particles and inorganic solutes causes the gradual decline of permeate flux, increased operating pressure, and higher frequency of chemical cleaning. The findings of this study indicate that it is critical to establishing effective methods of controlling biofouling in wastewater desalination and reclamation. The addition of chloramines controlled biofouling more effectively for the non-nitrified wastewater than nitrified/denitrified feed water.

Our results are consistent with findings reported by Schneider et al. (2005). Schneider et al. (2005) conducted membrane autopsies on specimens removed from first and second stage of a RO unit used for pre-treated surface water filtration. Membranes from all elements were covered by a thick, slimy, and compact biofouling layer. Inorganic fouling was observed only in the last element of the second stage. Speth et al. (1998), however, observed a different fouling pattern on NF membranes used for surface water treatment. The first of three elements in series exhibited the thickest fouling layer. Both acid and alkaline/detergent cleaning yielded only short-term flux recovery. All these studies showed the fouling layer had a strong biological nature.

The bench-scale testing protocol developed in the study can provide a conservative estimate on membrane fouling propensity over the first several hundred hours where flux decline is mainly associated with colloidal fouling, organic fouling and inorganic scaling. The bench-scale experiments cannot predict long-term flux decline caused by biofouling and continuous accumulation of foulants. With effective pre-treatment, appropriate antiscalants and biocides, biofouling, colloidal fouling and inorganic scaling can be controlled resulting in a steady specific flux as demonstrated during the pilot-scale study with the non-nitrified wastewater.

CHAPTER 7
MODELING FRAMEWORK

INTRODUCTION

In order to develop a qualitative and quantitative assessment of the solute mass transport, several models were reviewed regarding their suitability to model the retention of organic trace contaminants during NF and ULPRO treatment. Many models have been proposed to describe membrane separations with two major types of models: transport models and irreversible thermodynamic models. Transport models are directed to model rejection mechanisms within the membrane, whereas irreversible thermodynamic models treat the membrane as a black box that provide limited insight into the membrane:solute interactions. Four models were selected for review based on their prevalence in the literature and potential suitability for this application. These models include the extended Nernst-Planck equation, the solution-diffusion model, the surface force pore flow (SFPF) model, and the Spiegler-Kedem model.

The solution diffusion model has successfully been applied to solute transport in RO and certain NF membranes (Wijmans et al. 1995, Schutte 2003, Van der Bruggen et al. 2002). This model is based on the assumption that pores do not exist in the membrane and solute transport is exclusively attributed to diffusion. The rejection is predicted based on the solubility of the compound within the membrane and the rate at which it diffuses through the membrane (Wijmans et al. 1995). While this model is widely accepted and used for RO membranes, it is not necessarily suitable to be applied to all NF membranes because of the exclusion of convective transport in the model (Berg et al. 1997, Van der Bruggen et al. 2002).

The surface force pore flow model is a mechanistic transport model that accounts for membrane structure and membrane:solute interactions (Jain et al. 2004). It has been used for UF membranes since pore flow is the dominant mechanism for these membranes. The model uses parameters like permeation volume, permeation rate, concentration at the pore outlet and solute rejection for a single pore. These parameters are then added up for a collection of pores and a prediction for solvent and solute flux can be made (Kosutic et al. 2000, Kastelan-Kunst et al. 1997). However, due to its complexity, numerical techniques are required to solve the model and therefore, this model is difficult to apply to real water matrices (Jain et al. 2004).

The successful implementation of the extended Nernst-Planck equation to describe and/or predict the transport of ionic solutes, mostly salts, across a membrane has been detailed in numerous papers (Tsuru et al. 1991a and 1991b, Dresner 1972, Hagmeyer and Gimbel 1998, Wang et al. 1995, Bowen and Mukhtar, 1996). Historically, the extended Nernst-Planck equation has been favorably applied to membrane modeling, since it describes pore solute transport due to solvent volume flux as well as diffusive transport due to concentration and electrical potential gradients (Dresner 1972, Tsuru 1991a, Bowen and Mukhtar 1996). The extended Nernst-Planck equation coupled with the Teorell-Meyer-Sievers assumption, which includes electroneutrality equations, null-current conditions, Donnan ion partitioning equations and volumetric flux equations were used to describe the transport of salts within and through a membrane (Dresner et al. 1972, Tsuru et al. 1991b, Wang et al. 1995, Hagmeyer and Gimbel 1998, Bowen and Mukhtar 1996). Bowen and Mukhtar (1996) proposed a hybrid model, termed the Donnan steric pore model (DSPM), based on these equations, which allowed for the treatment of a membrane as porous through the use of diffusive and convective hindrance factors. The major advantages of the DSPM model are the use of structural and electrical properties of a membrane for use in describing solute transport. Many recent studies investigating the transport of salts through membranes have employed and/or built upon the DSPM, thus demonstrating the applicability of this approach to describe fundamental mass transfer processes during membrane separations (Schaep et al. 1999, Hagmeyer and Gimbel 1998, Bowen and Welfoot 2002a and 2002b, Bowen et al. 1997, Bowen and Mohammad 1998, Bandini and Vezzani 2003, Lee and Lueptow 2001). The DSPM has also been applied to describe the transport of uncharged solutes through membranes, although the use of the DSPM for this purpose has been generally limited to the characterization of a membrane in terms of pore size (Santos et al. 2006, Bowen et al. 2002, Bowen et al. 1997). The greatest limitation of the DSPM as pointed out in Bowen et al. (2002) and Bandini and Vezzani (2003) is the complex fitting procedures that require extensive iterations and experimentation to characterize the membrane. In addition, the DSPM has mainly been used to either a) describe the separation of salt mixtures by NF and RO membranes or b) optimize the separation of individual components of salt/organic dye mixtures. In the case of municipal wastewater reclamation projects where the prediction of ionic and nonionic organic solutes would be greatly beneficial, the utilization of the full DSPM is hindered by its complexity and the limited range in which it has been used.

The Spiegler-Kedem model was developed based on irreversible thermodynamics (or non-equilibrium thermodynamics (Spiegler et al. 1966). Some researchers have doubted the applicability of this model because it treats the membrane system as a black box with a minimal fundamental understanding of the transport and rejection mechanisms and because it is not able to characterize the structural and electrical characteristics of the membrane (Wijmans et al. 1995, Yoon et al. 2002). Additionally, this model does not take charge effects into consideration. For charged compounds, the predicted value will serve as a lower limit for rejection since charged compounds are better rejected than non-ionic compounds (Berg et al. 1997, Bellona et al. 2004). However, the ease of use and the availability of parameters, even for complex systems, have made this model attractive (Yoon et al. 2002, Jain et al. 2004). Therefore, this research further explored the Spiegler-Kedem model and their suitability to model the rejection of non-ionic and ionic compounds during NF and ULPRO treatment. Findings are summarized in Appendix G.

Qualitative Model of Solute Rejection

Due to the complexity of the approaches taken to model membrane performance as well as the relatively large number of possible membrane models, there is a need to develop simple and effective predictors of membrane rejection performance. Requirements for a relatively simple and effective model would be the condition of a minimum number of membrane and solute parameters in order to make a qualitative prediction of rejection. If possible, membrane and solute parameters would be relatively easy to obtain and the model would be useful to non-engineers and non-scientists.

A comprehensive literature review was performed by members of the research team to determine the factors driving the rejection of organic solutes by RO, ULPRO and NF membranes. The culmination of the literature review was a review paper (Bellona et al. 2004) outlining dominant rejection mechanisms as a function of molecular and membrane properties. From the literature reviewed, three dominant rejection mechanisms were identified for the removal of organic solutes by high-pressure membranes including: (1) size/steric exclusion (also including hindered diffusion), (2) electrostatic repulsion/exclusion, and (3) adsorption. The primary solute parameters affecting rejection were identified as: molecular weight (Dalton), molecular size (length and width), acid disassociation constant (pK_a), hydrophobicity/hydrophilicity (log K_{ow}), and diffusion coefficient. The primary membrane parameters affecting rejection were identified as: molecular weight cut-off (Dalton), pore size, surface charge (quantified as zeta potential), hydrophobicity/hydrophilicity (quantified as contact angle), and surface morphology (quantified as roughness).

From the knowledge gained during the literature review, a rejection diagram was proposed, which qualitatively allows prediction of solute rejection if certain solute and membrane properties are known. The rejection diagram with modifications from the original version published in Bellona et al. (2004) is presented in Figure 7.1. The rejection of a particular solute and membrane is performed by making decisions at different levels regarding molecular size versus membrane pore size, solution pH versus solute pK_a, solute log K_{ow}, and membrane surface charge. Solutes are sorted by determining: whether solute molecular weight is less than the molecular weight cut-off of a membrane, whether the solute is charged (negatively charged only), whether the solute is estimated as having hydrophobic interactions with a membrane and whether the molecular width of the solute is less than the pore radius. After making it through the possible decisions, a qualitative prediction of rejection is made as one of the following: low (0 to 30 percent), moderate (30 to 60 percent), high (60 to 90 percent), and very high (> 90 percent).

Table 7.1 presents rejection results from various membrane experiments that were conducted during the completion of this study and rejection predicted by employing the rejection diagram presented in Figure 7.1. The membranes chosen represent the membranes that were employed in a quantitative modeling exercise utilizing the Spiegler-Kedem model and therefore were characterized in terms of MWCO and average fictive pore size. The results of the qualitative modeling exercise indicate that the rejection diagram is an effective tool for estimating rejection. For example, phenacetine, a relatively small pharmaceutically active residue was estimated to be rejected by the NF-90 membrane between 30 and 60 percent due to steric exclusion. During bench-scale experiments phenacetine was observed to be rejected at approximately 45 percent. Solutes in which the qualitative model provided a good estimate of rejection include NDMA rejection by the TFC-HR membrane at the WBWRP (Figure 4.5), and ibuprofen rejection by the NF-90 during bench-scale (Appendix B, Figure B.19) and pilot-scale (WBWRP, Table 4.3) experiments. Phenacetine rejection was over estimated by the qualitative model as compared to rejection observed during laboratory-scale TMG10 membrane experiments (Figure 3.3). Chloroform rejection during NF-90 membrane pilot-scale testing at the WBWRP was observed to be higher than predicted (Figure 4.3) indicating that adsorption and partitioning was less important during full-scale testing.

Because the qualitative model presented in Figure 7.1 was developed through the identification of dominant rejection mechanisms and the effect of solute and membrane interactions on rejection, estimated rejection is usually within the range of observed rejection. The qualitative model, however, does not include operating conditions such as solvent flux, recovery, feed water quality, and concentration and therefore is limited to relatively simple systems and was developed only as a tool for estimating rejection. Although the diagram simplifies rejection mechanisms, the rejection diagram highlights the complexity of membrane rejection performance while considering solutes possessing different physicochemical properties. Just by observing the various paths that can be taken through the rejection diagram it becomes apparent that it may be difficult to employ a single quantitative model for predicting the rejection of common organic constituents present in municipal wastewater effluents. Quantitative modeling approaches proposed during this study, therefore, were identified by developing approaches to predict the rejection of ionic and nonionic trace organic contaminants.

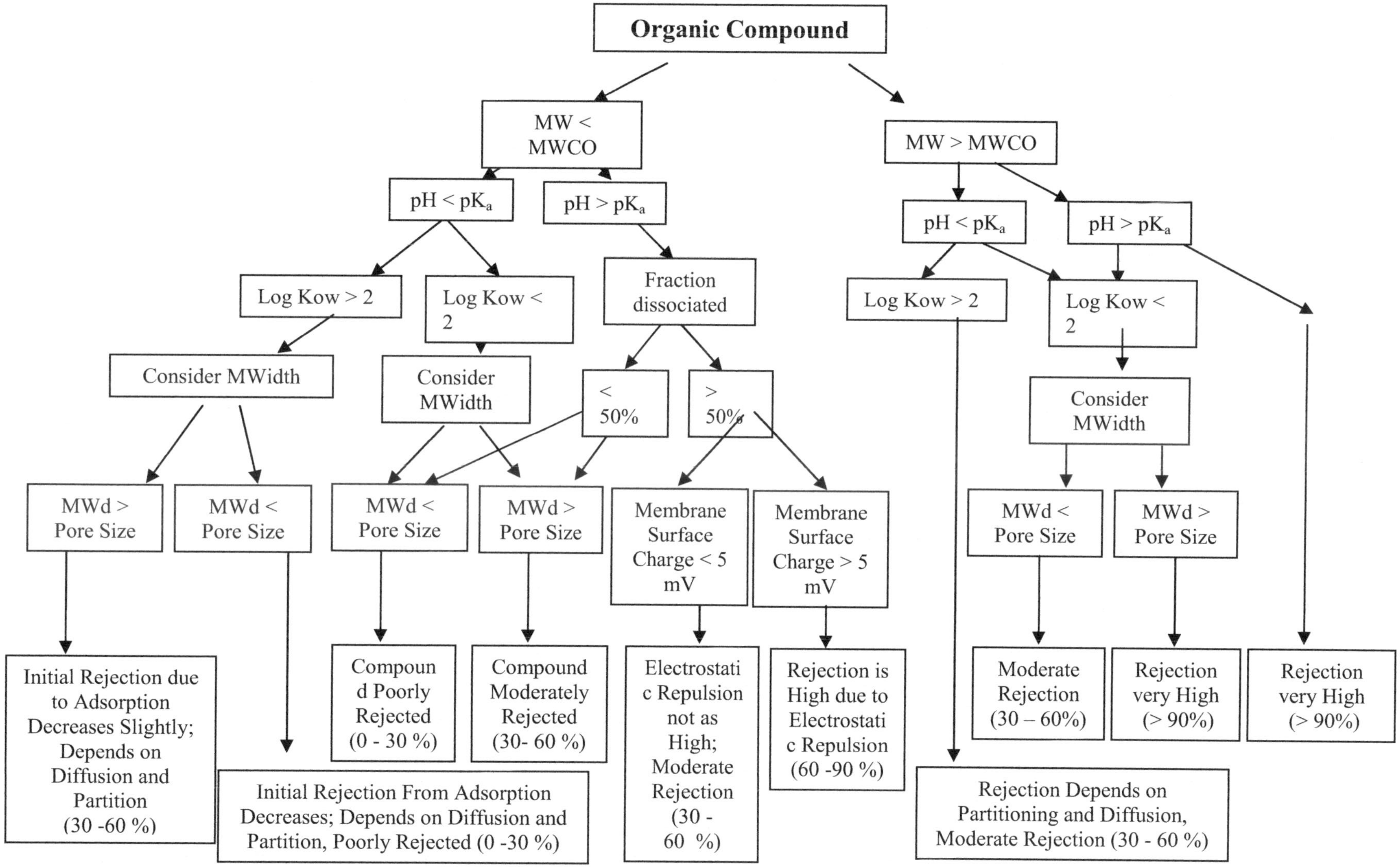

Figure 7.1. Rejection diagram for qualitatively predicting the rejection of organic solutes

Table 7.1.
Properties of solutes and membranes employed in the study and observed and estimated rejection utilizing the rejection diagram presented in Figure 7.1

Solute	Membrane	Scale	MW > MWCO†	pH < pKa	Log Kow < 2	Average Fictive Pore Size (Å)‡	MWd > Pore Size	Fraction Dissassociated at Feed Water pH of 6.5	Membrane Surface Charge at Feed Water pH of 6.5	Estimated Rejection	Observed Rejection
NDMA	TFC-HR	Full – WBWRP	No	Yes	Yes	7.7	No	0	- 10 mV	0 - 30 %	35%
Primidone	TFC-HR	Bench-	Yes	Yes	Yes	7.7	Yes	0	- 10 mV	> 90%	> 99
Ibuprofen	NF-90	Bench-	Yes	No	Yes	10	Yes	> 90%	-20 mV	> 90%	96%
Ibuprofen	NF-90	Pilot-WBWRP	Yes	No	Yes	10	Yes	> 90%	-20 mV	> 90%	> 90%
Chloroform	NF-90	Pilot-WBWRP	No	Yes	Log Kow = 2	10	No	0	-20 mV	0 - 30%	38 -55%
Phenacetine	NF-90	Bench-	No	Yes	Yes	10	Yes	0	-20 mV	30 - 60%	45%
Phenacetine	TMG10	Laboratory-	Yes	Yes	Yes	7.8	Yes	0	NA	> 90%	80%

†The MWCO (MW at 90% rejection) of TFC-HR was approximated at 100 Daltons, NF-90 was 200 Daltons, TMG10 was 100 Daltons during Spiegler-Kedem model development

‡Fictive pore radius calculated during Spiegler Kedem model development

CHAPTER 8
CONCLUSIONS AND RECOMMENDATIONS

CONCLUSIONS

The focus of this research study was to explore alternatives of an integrated membrane system involving nanofiltration (NF) and ultra-low pressure reverse osmosis (ULPRO) in place of reverse osmosis (RO), potentially representing a more cost-effective system because of anticipated lower pressure requirements and the greater selectivity for organic micropollutants. High-pressure membranes, encompassing RO, ULPRO, and NF, may provide an effective treatment barrier for regulated and unregulated organic compounds, which commonly occur in impaired water sources. These compounds represent a broad range of physicochemical properties such as hydrophilic ionic, hydrophilic non-ionic and hydrophobic non-ionic, and are associated with potential adverse effects for human health and aquatic life. The trace organic compounds being emphasized in this research include disinfection by-products (e.g., trichloroacetic acid, chloroform, bromoform, N-nitrosodimethylamine), pesticides, endocrine disrupting compounds (e.g., 17β-estradiol, testosterone, bisphenol A), pharmaceutical residues (e.g., ibuprofen, naproxen, gemfibrozil, carbamazepine, primidone), and chlorinated flame retardants. In general, the molecular weight of the target compounds selected in this study ranged from about 60 to 300 g/mol, the hydrophobicity expressed as the log K_{ow} value ranged from 4.5 to –3.2, and the acid dissociation constants, pK_a, of the ionizable compounds ranged from 0.3 to 4.9 indicating that they were negatively charged in the operating range of common membrane installations (pH 6–8). It is assumed that compounds not selected in this study but with physicochemical properties represented by these groups, would exhibit a similar rejection behavior.

A key element of this investigation was to explore whether NF and ULPRO membranes can consistently meet potable water quality requirements with respect to TOC, total nitrogen, and both regulated and unregulated trace organic compounds. While meeting these water quality goals, the operating characteristics of NF and ULPRO membranes (such as flux, fouling/scaling, and cleaning frequencies) should be comparable to thin-film composite RO membranes currently employed at full-scale facilities with operating feed pressure requirements significantly lower than the pressure range currently employed at facilities using conventional RO. In addition, findings from rejection studies at the laboratory-, pilot- and full-scale should be developed into a model framework providing the basis to reliably predict – a priori – rejection of organic micropollutants by RO, NF and ULPRO membranes.

Since there is little guidance in the water industry in identifying and selecting potentially viable NF and ULPRO membranes to treat water of impaired quality, the work presented here involved the development and validation of a laboratory-scale membrane testing protocol to select viable membranes for pilot- and full-scale operation. This selection protocol balanced operational characteristics (such as pressure, flux and potential for fouling) with product water quality and allowed for a pre-selection of potentially viable candidate membranes. Membranes considered for selection were characterized as thin-film composite (TFC) polyamide membranes and included commercially available ULPRO (TFC-ULP, Koch Membrane Systems; XLE, Dow/Filmtec; TMG10, Toray America; RE-BLR, Saehan) and NF membrane products (NF-90, NF-270, NF-200 and NF-4040, Dow/Filmtec; TFC-S, Koch Membrane Systems; ESNA1-LF, Hydranautics; NE-90, Saehan). Three candidate membranes were selected and tested using a 70

L/min (18 gpm) membrane pilot skid that was designed, constructed and operated during this study. The three selected candidate membranes were each tested for at least 1,300 hours on microfiltered feed water at two full-scale facilities. The feed waters represented non-nitrified microfiltered effluent provided by the West Basin Water Recycling Plant (WBWRP) in El Segundo, CA and nitrified/denitrified microfiltered effluent provided by the Scottsdale Water Campus (SWC), AZ. Operational performance and rejection of trace organics, nutrients and total organic carbon was monitored during pilot-scale testing and compared to the performance of full-scale trains operating on the same feed water employing conventional RO membranes (i.e., TFC-HR, Koch Membrane Systems). During several occasions, challenge tests were conducted at pilot-scale to examine rejection of trace organics during periods of elevated feed concentrations. The best performing ULPRO membrane was also selected for testing over 5,000 hours in a 6.6 m^3/min (2.5-mgd) full-scale RO train to further verify findings from pilot-scale and to determine rejection behavior and operating costs under full-scale conditions. The study was assisted through state-of-the-art membrane characterization to describe the fouling behavior of NF and ULPRO membranes and its role on operation (e.g., flux decline) and rejection. Findings of this study were compiled into a model framework to describe and predict the rejection of organic micropollutants during NF or RO treatment, which is taking into account physicochemical properties of solutes and membranes as well as key operational conditions affecting solute rejection.

Membranes tested during this study using the membrane testing protocol displayed a wide range of operating characteristics including fouling tendency, permeate productivity, and feed pressure while displaying a wide range of rejection behavior in terms of TOC, ammonium, nitrate and select trace organic contaminant removal. Based upon the laboratory-scale testing, the candidate NF and ULPRO membranes can be divided into two categories: 1) Permeate quality similar to the benchmark RO membrane (TFC-HR) while operating at significantly lower pressure (ULPRO membranes: ESPA2, TFC-ULP, TMG10, XLE, and one NF membrane: NF-90); and 2) Lower pressure than the category 1 membranes, effectively rejecting TOC, selective for most trace organic contaminants, but were limited in their removal of monovalent ions, including nitrate and ammonium (NF membranes: NF-270, NF-200, and NF-4040). Three NF membranes (TFC-S, ESNA1-LF, and NE-90) exhibited a flux decline between 50 and 80 percent diminishing the benefits of operating at lower feed pressure. The selection of membranes for pilot-scale testing, therefore, became an exercise in balancing the benefits of lower pressure and energy requirements with constituent rejection and permeate quality. The TMG10 (Toray America) and NF-90 (Dow/Filmtec) membranes were selected as candidate membranes for pilot-scale testing based upon high specific fluxes after a period of fouling as compared to the 'benchmark' RO membrane (TFC-HR) and efficient rejection of TOC (<0.5 mg/L), nitrogen species (<10 mg-N/L), and organic trace contaminants. Although the category 2 membrane NF-4040 (Dow/Filmtec) did not meet the nitrogen and TOC water quality criteria, this membrane was also selected for pilot-scale testing due to moderate flux decline, efficient trace organic rejection (usually greater than 90 percent), and a significantly higher specific flux than the 'benchmark' TFC-HR membrane.

The TMG10 and NF-90 membranes tested on microfiltered non-nitrified wastewater (WBWRP) exhibited a relative constant temperature-corrected specific flux (TCSF) after initial membrane compaction and fouling of 0.03 L/m^2-hr-kPa (0.12 gfd/psi). Similar to observations made during membrane selection on laboratory-scale, the NF-90 exhibited a larger flux decline as compared to the TMG10 during pilot-scale testing. For comparison, the benchmark TFC-HR membrane employed at the WBWRP under similar conditions achieved a lower stabilized TCSF of approximately 0.022 L/m^2-hr-kPa (0.09 gfd/psi). During these test periods, the permeate TOC and ammonia concentrations of the TMG10 (NF-90) were below 0.2 mg-C/L (<0.3 mg-C/L) and below 2 mg-N/L (<3 mg-N/L), respectively. Comparing pilot-scale testing of the TMG10 to a period of full-scale operation indicated that the TMG10 performed similarly or better than the ESPA2 and TFC-HR in terms of TOC and ammonia permeate concentrations and rejection. Rejection of regulated organics, such as THMs, HAAs and NDMA, was also found to be similar to the 'benchmark' TFC-HR membrane with a poor to moderate rejection of chloroform and NDMA. Both, the TMG10 and NF-90 tested at pilot-scale and the TFC-HR and ESPA2 employed on full-scale were highly efficient in rejecting unregulated organics with non detectable concentrations or only traces of the target compounds in the permeates. In a side-by-side 3,000 hour test in a full-scale vessel at WBWRP, the TMG10 and the ESPA2 (serving as benchmark) membranes performed very similar regarding specific flux and product water quality (regarding TOC, nitrogen, and regulated and unregulated organics).

The NF-4040 membrane was tested at pilot-scale on microfiltered nitrified/denitrified wastewater (SWC) and exhibited a relative small initial flux decline followed by a very constant TCSF of 0.08 L/m^2-hr-Pa (0.33 gfd/psi) over 1,200 hours. As expected, during the 1,200 hours of testing the specific flux of the NF-4040 was more than four times higher as compared to the 'benchmark' RO membrane TFC-HR. At all times, TOC permeate concentrations of the NF-4040 were below 0.5 mg-C/L. Nitrate, however, was not rejected by the NF-4040 and negative rejections were observed in the NF-4040 membrane permeate. Of the unregulated trace organic compounds targeted, only traces of chlorinated flame retardants, carbamazepine and bisphenol-A were detected in the permeate of the NF-4040 membrane immediately after start-up. Rejection of trace organics, however, improved with membrane fouling and subsequent sampling campaigns revealed no detections in permeate samples. This was an interesting finding and especially important for NF membranes, because initial (organic) fouling will always occur where membranes are employed to treat surface water or secondary and tertiary treated effluents. Foulant precipitation and cake-layer formation result in a considerable change of membrane surface characteristics with respect to membrane hydrophobicity, surface charge and surface morphology, hence potentially affecting transport mechanisms of contaminants as compared to virgin membranes. The transport of hydrophilic ionic organic contaminants was especially hindered in NF membranes as a result of improved electrostatic exclusion, pore clogging and an increased adsorption capacity of fouled polyamide membranes.

Due to the relatively high fouling potential of secondary effluent feed water, initial high specific flux rates of the TMG10 and NF-90 as observed during both laboratory- and pilot-scale testing were reduced to values similar to the commonly employed ULPRO membrane, ESPA2, but exceeded the conventional RO membrane, TFC-HR. For a full-scale installation operating at a feed pressure of 896 kPa (130 psi) and 85 percent recovery, this relates to an electrical consumption of 0.37 kWh/m^3 (1.40 kWh/kgal) of permeate produced. Based on an example electrical cost of \$0.12/kWh, the operating cost of producing one m^3 of permeate utilizing the TMG10 or ESPA2 membrane would be approximately \$0.05. Pilot- and full-scale operational results appear to support the argument that the choice of membranes to achieve more significant cost savings associated with lower feed pressure is limited to low fouling *loose* NF membranes. During the testing period, the NF-4040 sustained pressures 3–4 times lower than the feed pressure of conventional RO membranes currently employed at full-scale facilities. These lower feed pressures would result in potential savings of \$0.03–0.08 per m^3 (\$0.1–0.3 per kgal) as compared to current installations using conventional RO membranes while treating reclaimed water with an IMS.

Findings of this research clearly demonstrated that a single model does not exist which is capable of reliably predicting the mass transport of organic micropollutants during high-pressure membrane treatment. The physicochemical properties of the solutes are the key factors determining rejection and if properly considered and put in context with relevant membrane properties, rejection can be qualitatively predicted. This qualitative model, a solute rejection diagram (Figure 7.1), highlights the complexity of membrane rejection performance while considering solutes possessing different physicochemical properties. Although operational conditions, such as flux or recovery, are not considered, the rejection diagram allows utilities to estimated rejection of individual compounds semi-quantitatively. One approach was developed to quantitatively describe and predict the rejection of non-ionic and ionic compounds of concern (Spiegler-Kedem model). With the exception of highly hydrophobic compounds, the Spiegler-Kedem model provided a high to moderate degree of accuracy for the rejection of non-ionic solutes by a conventional RO, an ULPRO and an NF membrane. For ionic solutes, the Spiegler-Kedem model underpredicted rejection as expected since electrostatic exclusion is not properly considered in the model.

RECOMMENDATIONS TO UTILITIES

This report provides guidance for utilities regarding the viability of low-pressure RO (ULPRO) and NF membranes to replace conventional RO membranes employed in integrated membrane systems (IMS) to treat source water of impaired quality. Findings of this study reveal that ULPRO membranes can consistently meet potable water quality requirements for treating source water of impaired quality with respect to TOC, total nitrogen, and both regulated and unregulated trace organic compounds. ULPRO membranes (such as ESPA2, TMG10, XLE, TFC-ULP) exhibited specific flux values that were slightly higher as compared to conventional RO membranes (TFC-HR). Because purchase cost and expected life of these membranes are similar, ULPRO membranes potentially offer lower operating expenses and therefore should be considered in membrane procurement or replacement programs of an IMS. Achieving significant cost savings associated with lower feed pressure, however, is limited to certain low fouling *loose* NF membranes (such as NF-200, NF-4040 or NF-270). While these membranes demonstrated efficient rejection of TOC and a high selectivity for a wide range of trace organics, they will result in lowered permeate water quality in terms of nitrogen species such as ammonia and nitrate. These membranes will likely not be capable of meeting stringent nitrogen standards and would require a tailored wastewater pretreatment to reduce ammonia and nitrate concentrations below the Federal Drinking Water Standard of 10 mg-N/L. Providing the nitrogen standard can be met prior to membrane treatment and given the selectivity and low operating cost of *loose* NF membranes, these membranes are attractive for the design of multiple barrier treatment trains, consisting of high-pressure membranes followed by an advanced oxidation process (such as UV/hydrogen peroxide or ozone) or groundwater recharge of impaired water followed by membrane treatment. Findings of this study underline the fact that proper pre-treatment of ULPRO and NF membranes is as critical as it is for conventional RO membranes. If not properly maintained, the benefit of operating NF and ULPRO membranes at lower feed pressure is quickly lost, where severe biofouling is not prevented and a constant permeate flux is maintained. The laboratory-scale testing protocol developed to select viable candidate ULPRO and NF membranes was a useful tool in separating the performance of different membranes. The candidate membranes selected using this protocol performed as expected during pilot- and full-scale testing. This testing protocol can, therefore, assist the water industry in better pre-selecting viable membranes for applications using a given impaired water source such as recycled water.

APPENDIX A
PHYSICAL-CHEMICAL PROPERTIES OF INDICATOR COMPOUNDS

Table A.1 Physical and Chemical Properties of Indicator Compounds

Compound	Tris(2-chloroethyl)phosphate (TCEP)	Tris(1-chloro-2-propyl)phosphate (TCPP)	Tris(1,3-dichloro-2-propyl) phosphate (TDCPP)
Formula	C6 H12 Cl3 O4 P	C9 H18 Cl3 O4 P	C9 H15 Cl6 O4 P
Chemical structure	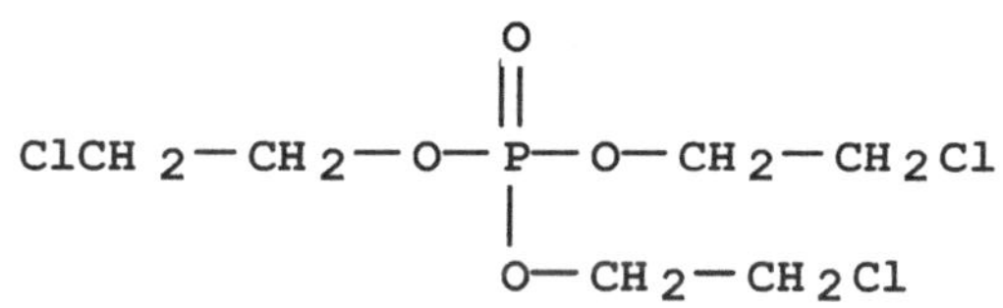	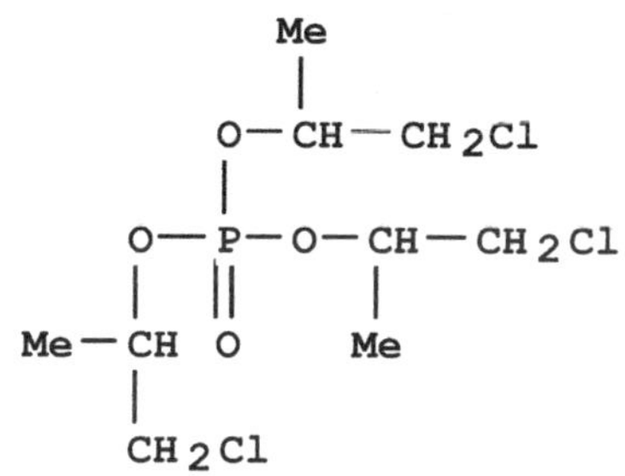	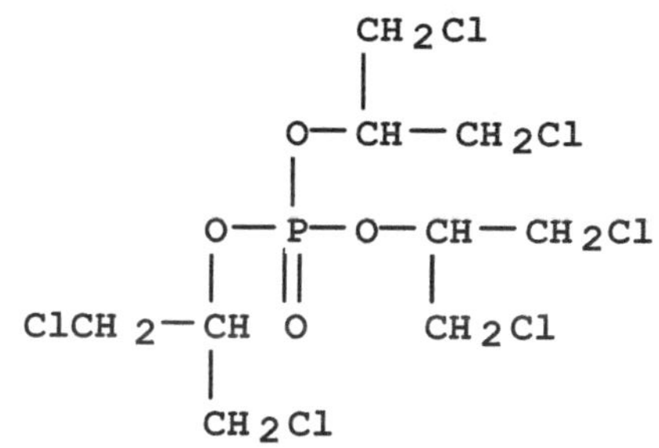
CAS	115-96-8	13674-84-5	13674-87-8
Molecular weight [g/mol]	285.49	327.57	430.9
Water solubility [mg/L]	7000	7	less than 1000
Charge (at neutral pH)	Neutral	Neutral	Neutral
Log KOW (ACD)	0.484±0.360	1.526±0.373	1.791±0.771
Log D pH 1	0.48	1.53	1.79
Log D pH 4	0.48	1.53	1.79
Log D pH 7	0.48	1.53	1.79
Log D pH 8	0.48	1.53	1.79
Log D pH 10	0.48	1.53	1.79
pK_a (ACD)	N/A	N/A	N/A

Note: ACD - Calculated using Advanced Chemistry Development (ACD) Software Solaris V4.67, source: SciFinder Scholar 2002.
References: Data from SRC PhysProp Database, source: http://www.Chemfinder.com

Table A.1 (continued)

Compound	Mecoprop	Dichlorprop	Diclofenac
Formula	C10 H11 Cl O3	C9 H8 Cl2 O3	C14 H11 Cl2 N O2
Chemical structure	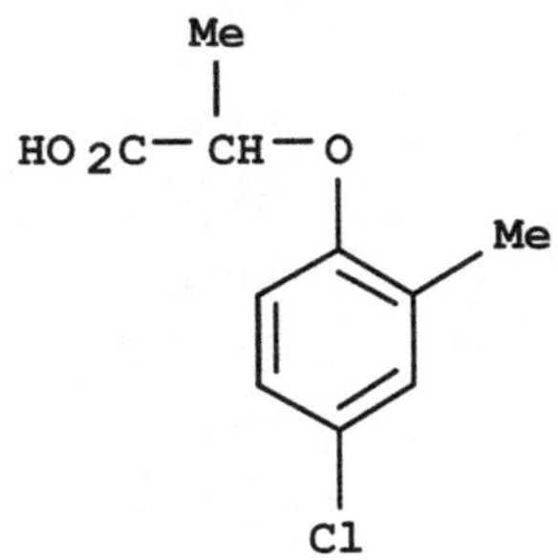	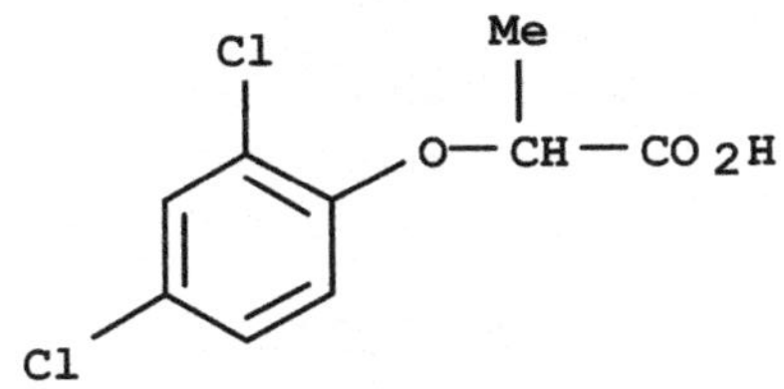	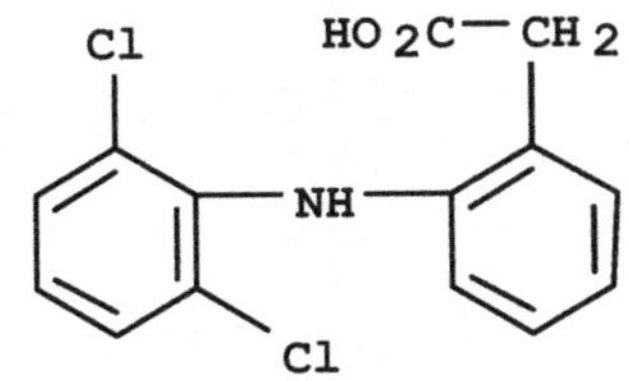
CAS	93-65-2	120-36-5	15307-86-5
Molecular weight [g/mol]	214.65	235.06	296.15
Water solubility [mg/L]	620	350	2.37
Charge (at neutral pH)	Negative	Negative	Negative
Log KOW (ACD)	2.835±0.268	2.945±0.279	3.284±0.361
Log D pH 1	2.83	2.94	3.28
Log D pH 4	1.96	1.93	3.06
Log D pH 7	-0.8	-0.78	0.48
Log D pH 8	-1.19	-1.1	-0.35
Log D pH 10	-1.26	-1.15	-0.81
pK_a (ACD)	3.18±0.20	3.03±0.20	4.18±0.20

Table A.1 (continued)

Compound	Propyphenazone	Ibuprofen	Ketoprofen
Formula	C14 H18 N2 O	C13 H18 O2	C16 H14 O3
Chemical structure	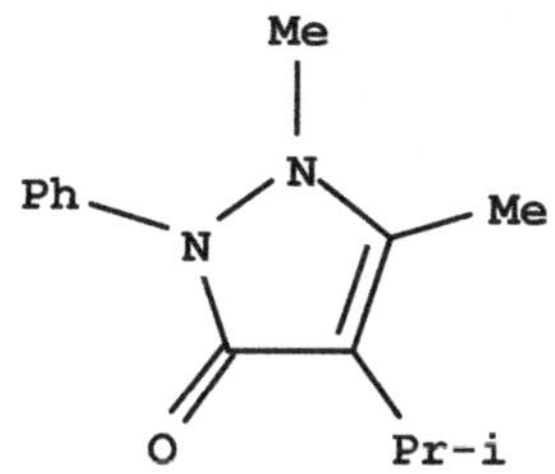	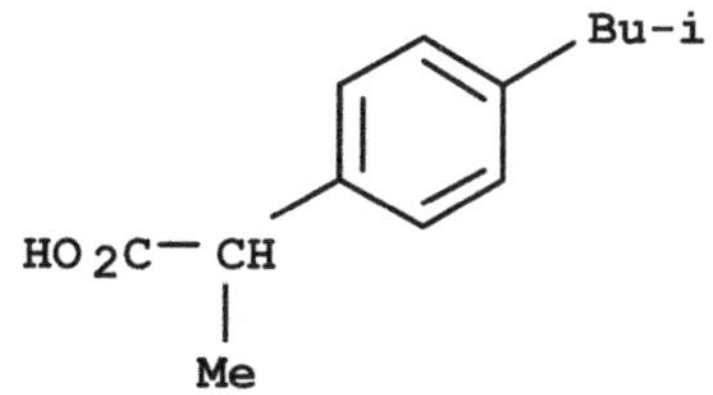	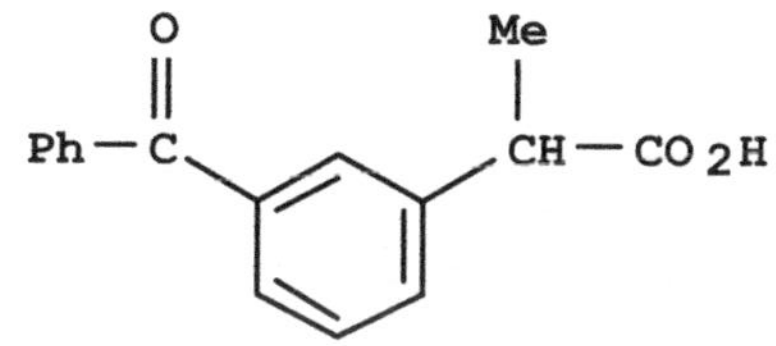
CAS	479-92-5	15687-27-1	22071-15-4
Molecular weight [g/mol]	230.31	206.28	254.28
Water solubility [mg/L]	hard to dissolve	21	51
Charge (at neutral pH)	Neutral	Negative	Negative
Log KOW (ACD)	1.737±0.335	3.722±0.227	2.814±0.326
Log D pH 1	0.36	3.72	2.81
Log D pH 4	1.73	3.58	2.61
Log D pH 7	1.74	1.15	0.07
Log D pH 8	1.74	0.25	-0.79
Log D pH 10	1.74	-0.36	-1.28
pK_a (ACD)	2.37±0.20	4.41±0.20	4.23±0.20

Table A.1 (continued)

Compound	Naproxen	Salicylic acid	Bisphenol A
Formula	C14 H14 O3	C7 H6 O3	C15 H16 O2
Chemical structure	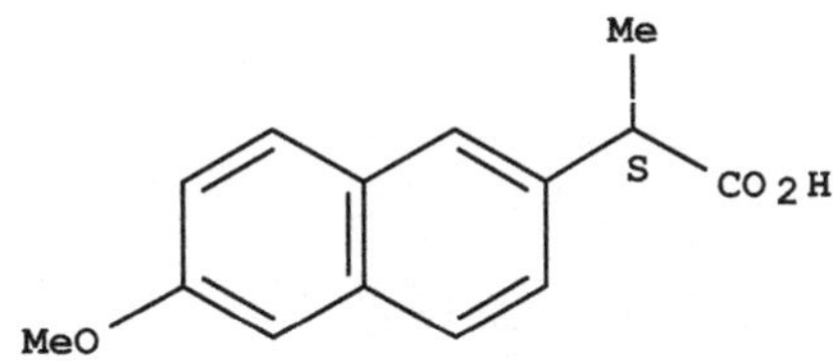	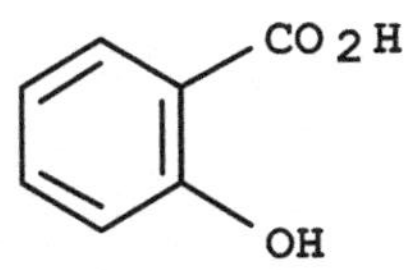	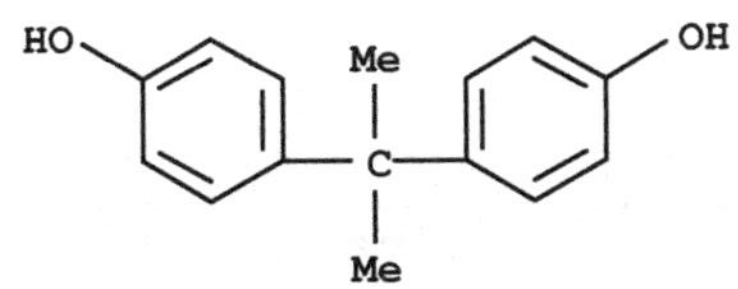
CAS	22204-53-1	69-72-7	80-05-7
Molecular weight [g/mol]	230.26	138.12	228.29
Water solubility [mg/L]	15.9	2240	120
Charge (at neutral pH)	Negative	Negative	Neutral
Log KOW (ACD)	2.998±0.239	2.061±0.247	3.431±0.232
Log D pH 1	3	2.06	3.43
Log D pH 4	2.85	1.03	3.43
Log D pH 7	0.41	-1.68	3.43
Log D pH 8	-0.48	-1.99	3.42
Log D pH 10	-1.09	-2.04	2.89
pK_a (ACD)	4.40±0.20	3.01±0.20	9.73±0.20

Table A.1 (continued)

Compound	Caffeine	Carbamazepine	Primidone
Formula	C8 H10 N4 O2	C15 H12 N2 O	C12 H14 N2 O2
Chemical structure	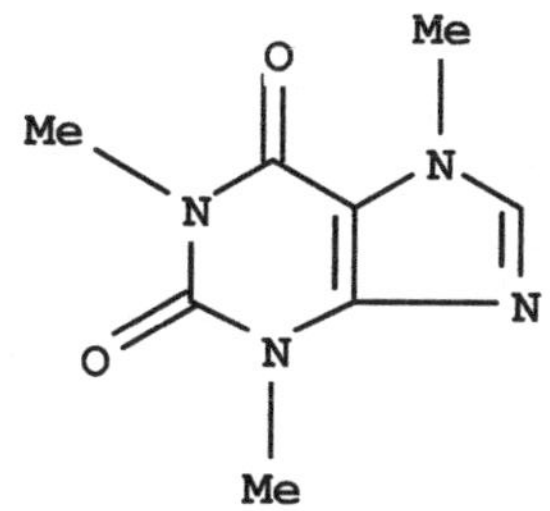	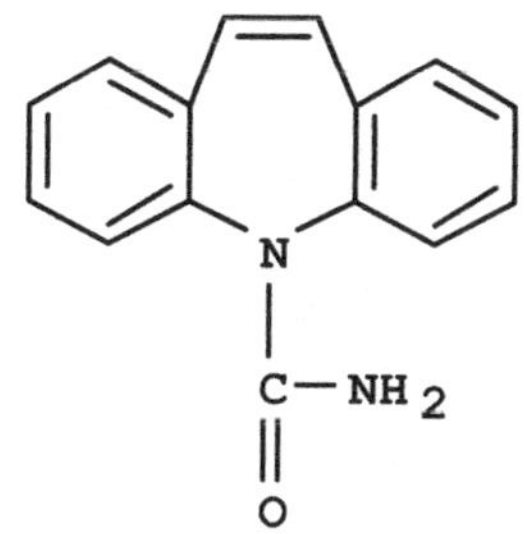	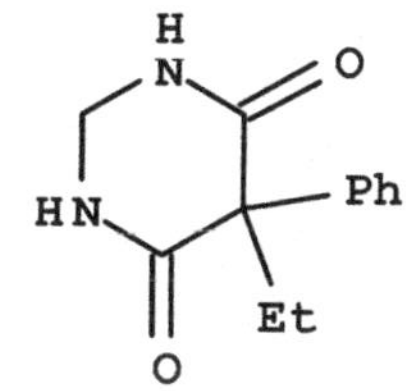
CAS	58-08-2	298-46-4	125-33-7
Molecular weight [g/mol]	194.19	236.27	218.25
Water solubility [mg/L]	2160	17.66	500
Charge (at neutral pH)	Neutral	Neutral	Neutral
Log KOW (ACD)	-0.081±0.351	2.673±0.376	-0.844±0.417
Log D pH 1	-0.62	2.66	-0.85
Log D pH 4	-0.08	2.67	-0.84
Log D pH 7	-0.08	2.67	-0.84
Log D pH 8	-0.08	2.67	-0.84
Log D pH 10	-0.08	2.67	-0.85
pKa (ACD)	1.39±0.70	13.94±0.20	12.26±0.40

Table A.1 (continued)

Compound	Fenofibrate	Gemfibrozil	Clofibric acid
Formula	C20 H21 Cl O4	C15 H22 O3	C10 H11 Cl O3
Chemical structure	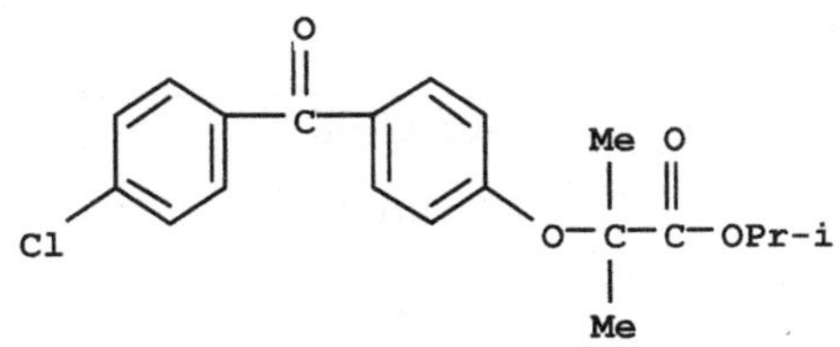	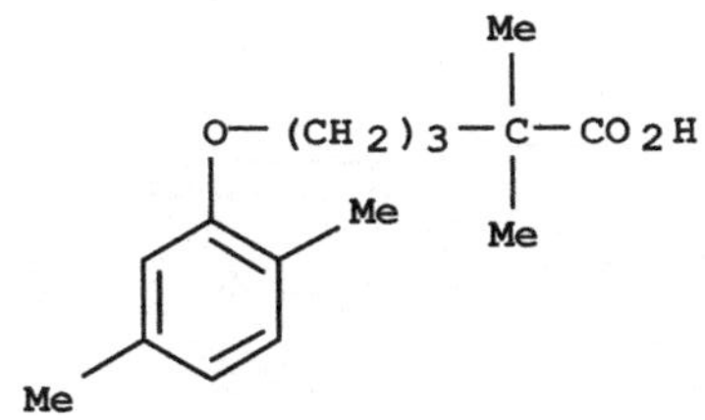	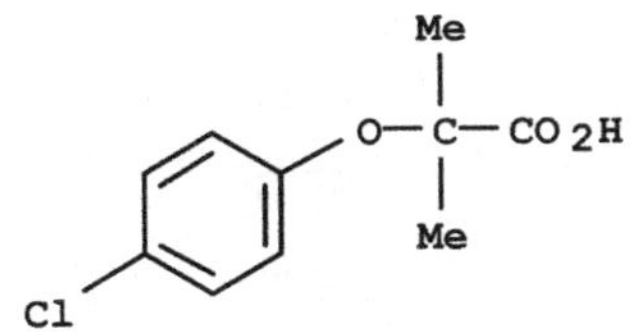
CAS	49562-28-9	25812-30-0	882-09-7
Molecular weight [g/mol]	360.83	250.33	214.65
Water solubility [mg/L]	less than 0.5	19	100 (in 1:6 EtOH:PBS)
Charge (at neutral pH)	Neutral	Negative	Negative
Log KOW (ACD)	4.804±0.388	4.387±0.487	2.724±0.272
Log D pH 1	4.8	4.39	2.72
Log D pH 4	4.8	4.31	1.85
Log D pH 7	4.8	2.14	-0.91
Log D pH 8	4.8	1.19	-1.3
Log D pH 10	4.8	0.32	-1.37
pKa (ACD)	N/A	4.75±0.20	3.18±0.20

Table A.1 (continued)

Compound	Phenacetine	Trischloroacetic acid	Dichloroacetic acid
Formula	C10 H13 N O2	C2 H Cl3 O2	C2 H2 Cl2 O2
Chemical structure	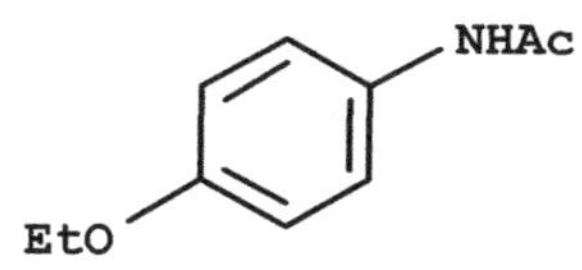	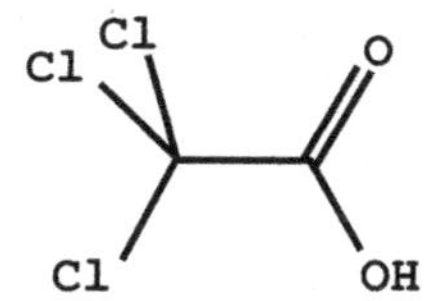	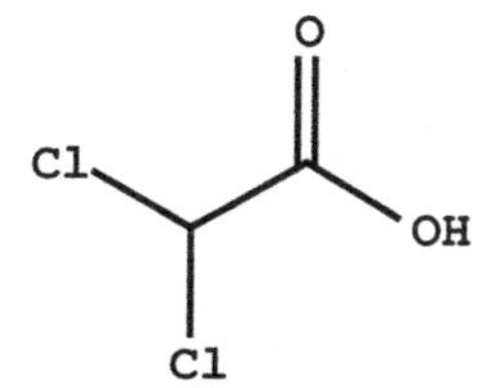
CAS	62-44-2	76-03-9	79-43-6
Molecular weight [g/mol]	179.22	163.39	128.94
Water solubility [mg/L]	766	44000	100 000
Charge (at neutral pH)	Neutral	Negative	Negative
Log KOW (ACD)	1.626±0.223	1.67±0.41	0.54±0.29
Log D pH 1	1.61	1.41	0.39
Log D pH 4	1.63	-1.2	-2.07
Log D pH 7	1.63	-2.42	-3.54
Log D pH 8	1.63	-2.43	-3.56
Log D pH 10	1.63	-2.43	-3.56
pKa (ACD)	N/A	1.10±0.20	1.37±0.20

Table A.1 (continued)

Compound	Bromoform	Chloroform	Methylamine
Formula	C H Br3	C H Cl3	C H5 N
Chemical structure	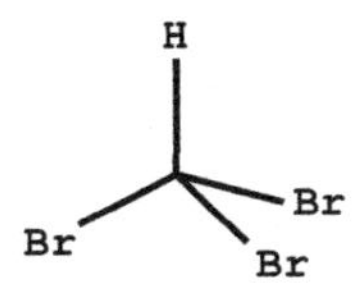	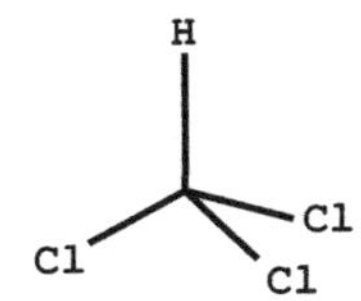	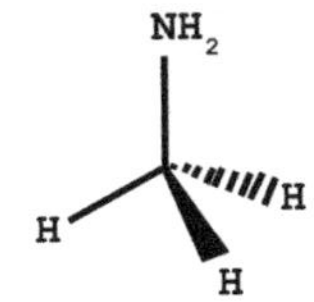
CAS	75-25-2	67-66-3	74-89-5
Molecular weight [g/mol]	252.73	119.3779	31
Water solubility [mg/L]	3100	7950	1.08E+08
Charge (at neutral pH)	Neutral	Neutral	Positive
Log KOW (ACD)	2.42±0.33	1.97	-0.66±0.20
Log D pH 1	2.42	1.97	-3.76
Log D pH 4	2.42	1.97	-3.76
Log D pH 7	2.42	1.97	-3.66
Log D pH 8	2.42	1.97	-3.19
Log D pH 10	2.42	1.97	-1.41
pKa (ACD)	N/A	N/A	10.66±0.20

Table A.1 (continued)

Compound	2-naphthol	Estriol	17b-estradiol
Formula	C15 H25 NO3	C18 H24 O3	C18 H24 O2
Chemical structure	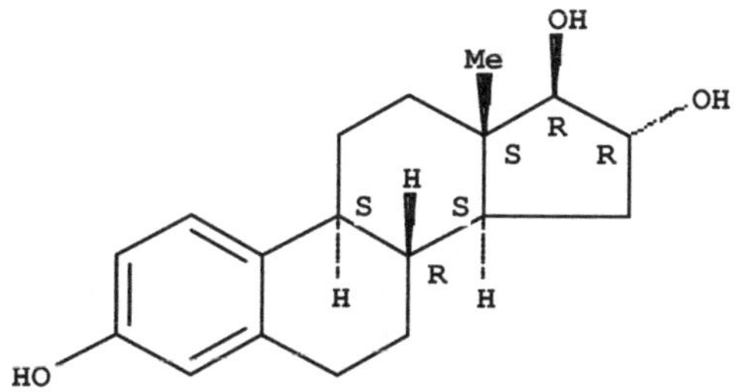	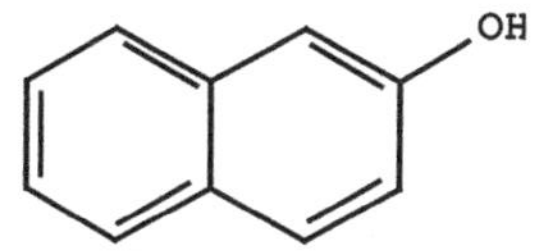	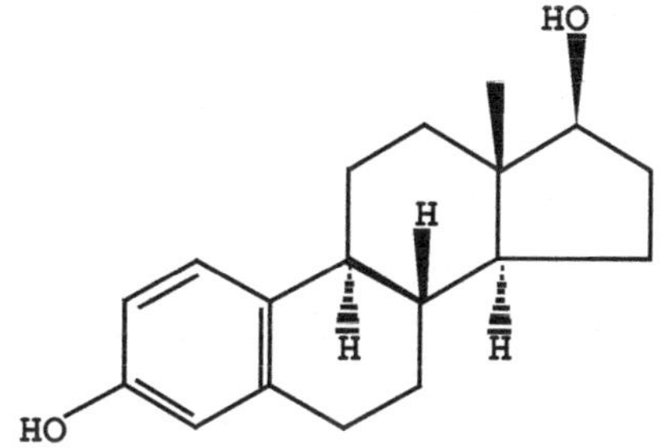
CAS	37350-58-6	50-27-1	50-28-2
Molecular weight [g/mol]	144.17	288.38	272.39
Water solubility [mg/L]	755	441	3.6
Charge (at neutral pH)	Neutral	Neutral	Neutral
Log KOW (ACD)	2.71	2.944±0.281	4.13±0.25
Log D pH 1	2.71	2.94	4.13
Log D pH 4	2.71	2.94	4.13
Log D pH 7	2.71	2.94	4.13
Log D pH 8	2.7	2.94	4.13
Log D pH 10	2.15	2.78	3.98
pKa (ACD)	9.57±0.20	10.35±0.70	10.37±0.20

Table A.1 (continued)

Compound	Testosterone	N-Nitrosodimethylamine (NDMA)	Naphthalene
Formula	C19 H28 O2	C2 H8 N2O	C10H8
Chemical structure	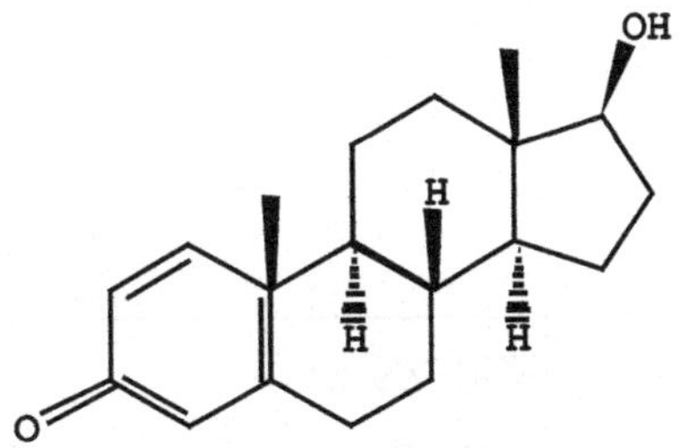	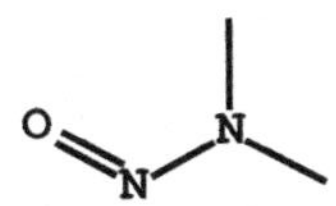	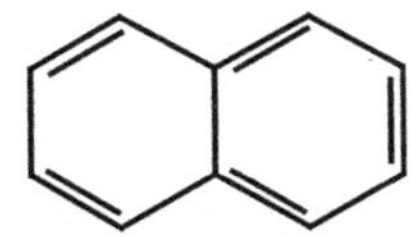
CAS	58-22-0	62-75-9	91-20-3
Molecular weight [g/mol]	288.42	74.08	128
Water solubility [mg/L]	23.4	1.00E+06	31
Charge (at neutral pH)	Neutral	Neutral	Neutral
Log KOW (ACD)	3.48±0.28		3.3
Log D pH 1	3.47		3.3
Log D pH 4	3.47		3.3
Log D pH 7	3.47		3.3
Log D pH 8	3.47		3.3
Log D pH 10	3.47		3.3
pKa (ACD)	N/A	3.56	N/A

Table A.1 (continued)

Compound	Methanol	Ethylene Glycol	1,3-Propanediol	1,4-Butanediol	Diethylene Glycol	Triethylene Glycol	Diethylene Glycol Monobutyl Ether	Tetra ethylene Glycol	Tetra ethylene Glycol Dimethyl Ether
Formula	CH_4O	$C_2H_6O_2$	$C_3H_80_2$	$C_4H_{10}O_2$	$C_4H_{10}O_3$	$C_6H_{14}O_4$	$C_8H_{18}O_3$	$C_8H_{18}O_5$	$C_{10}H_{22}O_5$
Chemical Structure	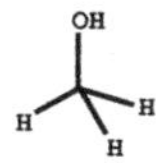				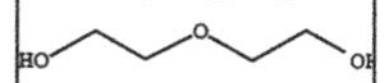	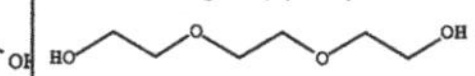	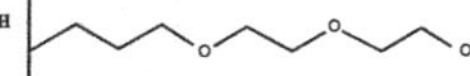	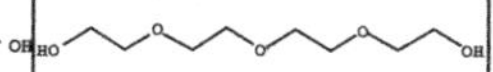	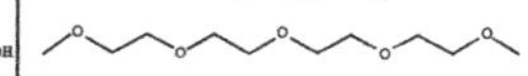
CAS	67-56-1	107-21-1	504-63-2	110-63-4	111-46-6	112-27-6	112-34-5	112-60-7	143-24-8
Molecular Weight [(g/mol)	32	62	76	90	106	150	162	194	222
Water solubility, [mg/L]	1.00E+06	1.00E+06	1.00E+06	1.00E+06	1.00E+06	1.00E+6	1.00E+06	Not available	1.00E+06
Log Kow	-0.77	-1.36	-1.04	-0.83	-1.47	Not available	0.56	Not available	-0.77
pKa	15.3	15.1	n/a	14.5	n/a	Not available	n/a	Not available	15.3

Table A.1 (continued)

Characteristics	Isopropanol	1,2-Propanediol	1,3-Butanediol	Glycerin	Neopentyl Glycol
Formula	C_3H_8O	$C_3H_8O_2$	$C_4H_{10}O_2$	C3H8O3	$C_5H_{12}O_2$
Chemical Structure	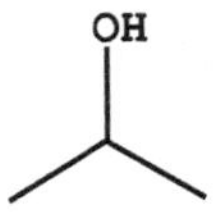	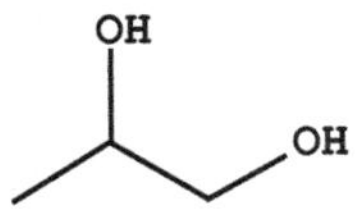	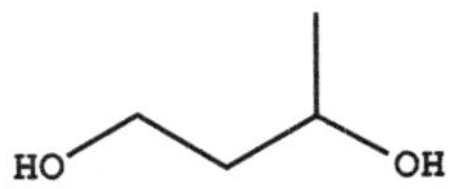	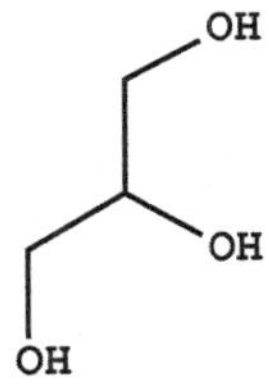	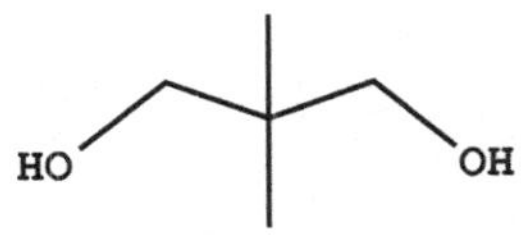
CAS	67-63-0	57-55-6	107-88-0	56-81-5	126-30-7
Molecular Weight, [g/mol]	60	76	90	92	104
Water solubility, [mg/L]	1.0E+06	1.0E+06	Not available	1.0E+06	Not available
Log Kow	0.05	-0.92	Not available	-1.76	Not available
pKa	17.1	14.9	Not Available	14.4	Not available

Table A.1 (continued)

Characteristics	Phenol	Benzyl Alcohol	Dextrose
Formula	C_6H_6O	C_7H_8O	$C_6H_{12}O_6$
Chemical Structure	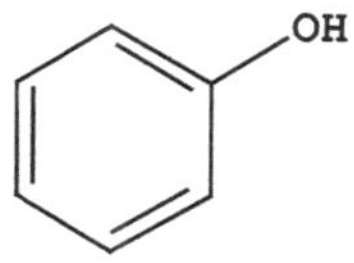	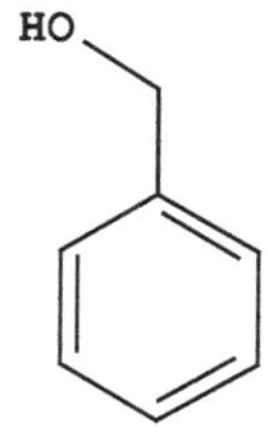	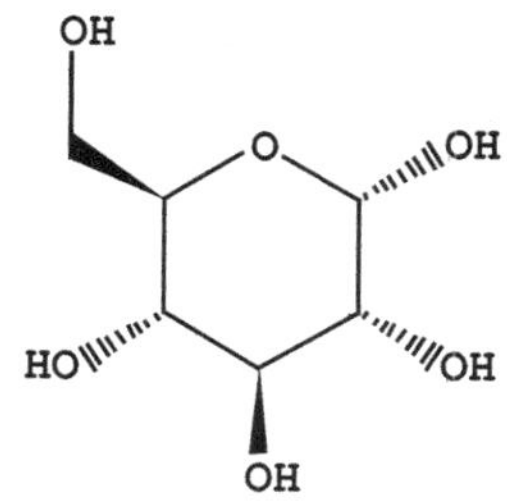
CAS	108-95-2	100-51-6	50-99-7
Molecular Weight [g/mol]	94	108	180
Water solubility [mg/L]	8.28E+04	4.29E+04	1.20E+06
Log Kow (EXP)	1.46	1.1	-3.24
pKa	9.99	15.4	12.9

APPENDIX B
LABORATORY-SCALE TESTING

Pressure and Specific Flux

ULPRO Membranes

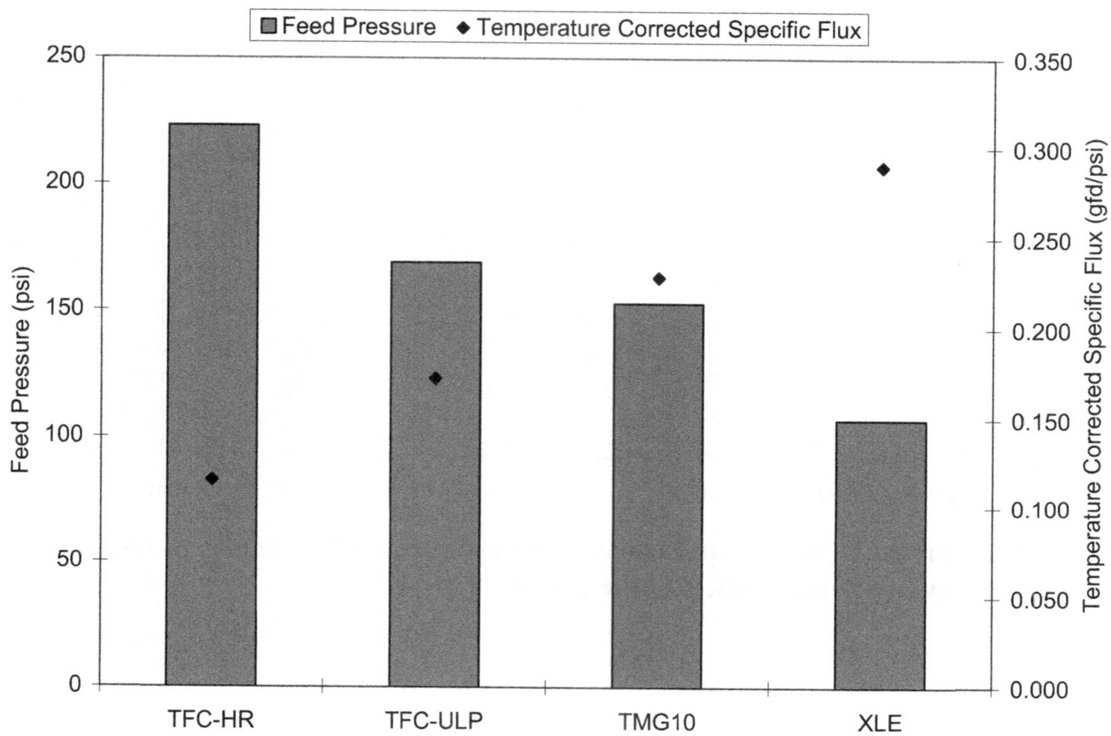

Figure B.1 Feed pressure and temperature corrected specific flux values of tested ULPRO membranes versus the 'benchmark'RO membrane TFC-HR

NF Membranes

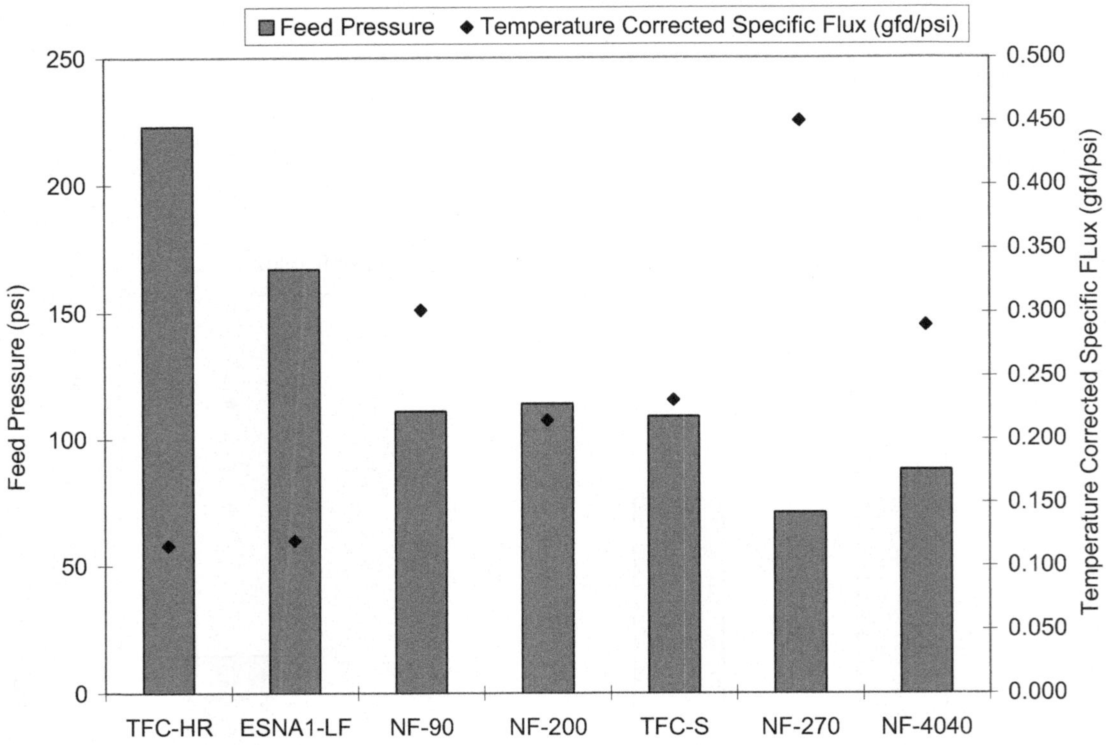

Figure B.2 Feed pressure and temperature corrected specific flux values of tested NF membranes versus the'benchmark'RO membrane TFC-HR

TOC and Nitrogen Rejection

ULPRO Membranes

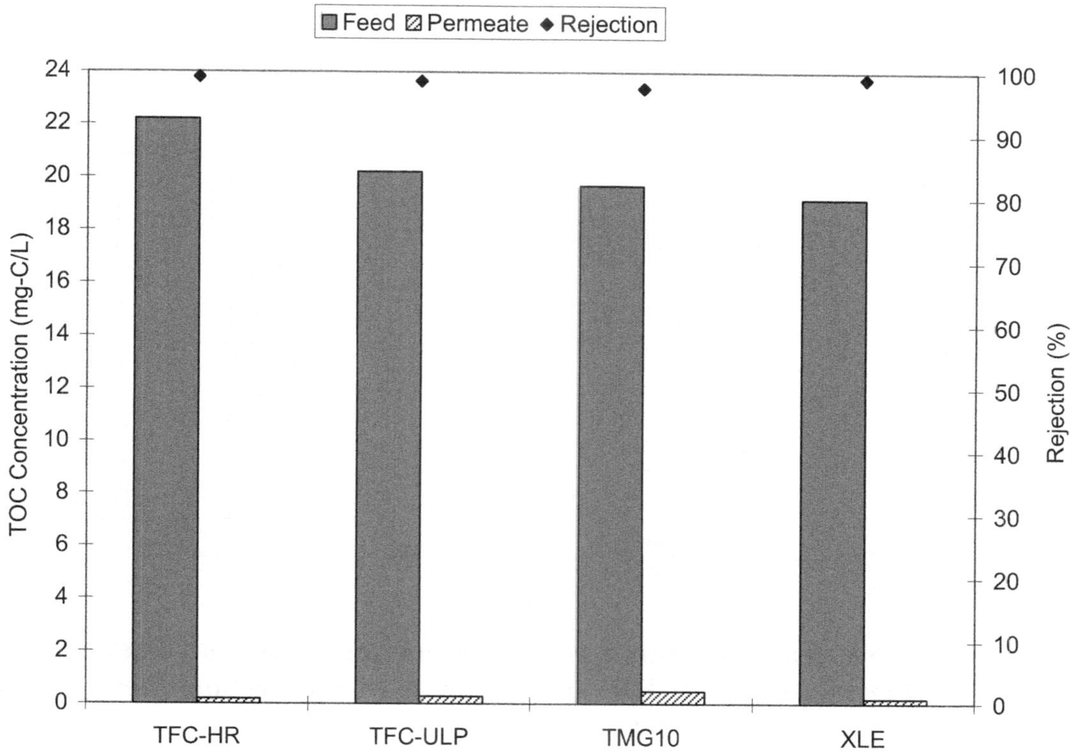

Figure B.3 Feed and permeate TOC concentrations and TOC rejection values for 'benchmark'RO membrane (TFC-HR) and candidate ULPRO membranes during 80 percent recovery experiments

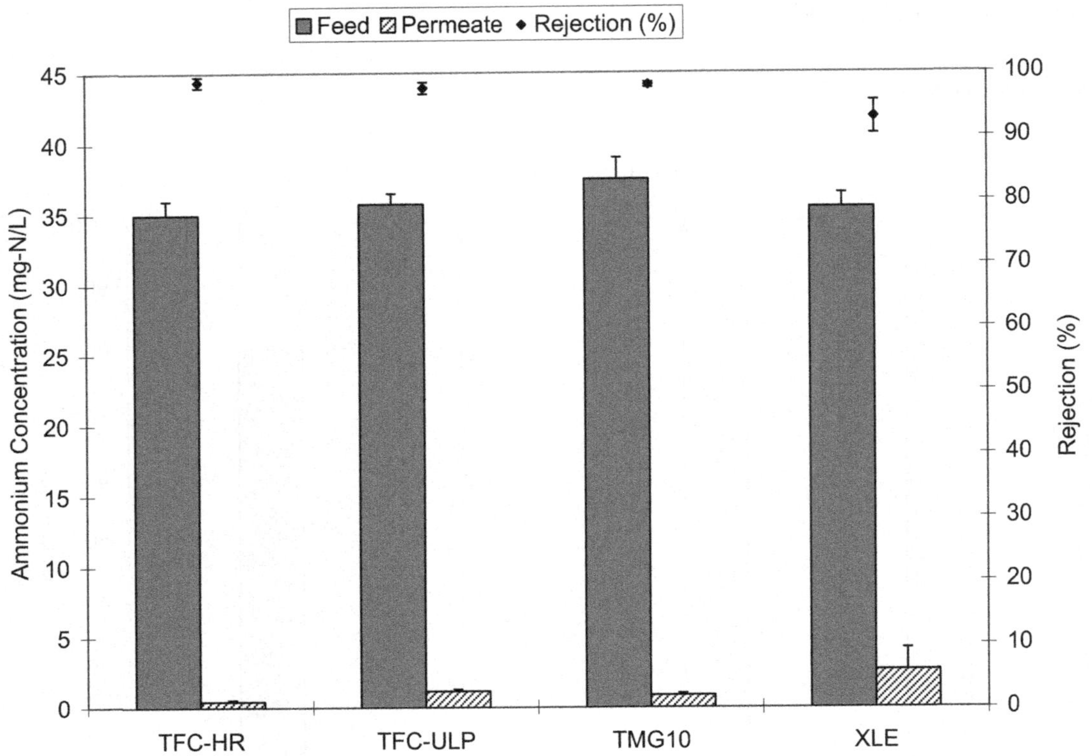

Figure B.4 Feed and permeate ammonium concentrations and rejection values for'benchmark'RO membrane (TFC-HR) and candidate ULPRO membranes during 30 percent recovery experiments (error bars represent the average deviation (n=2))

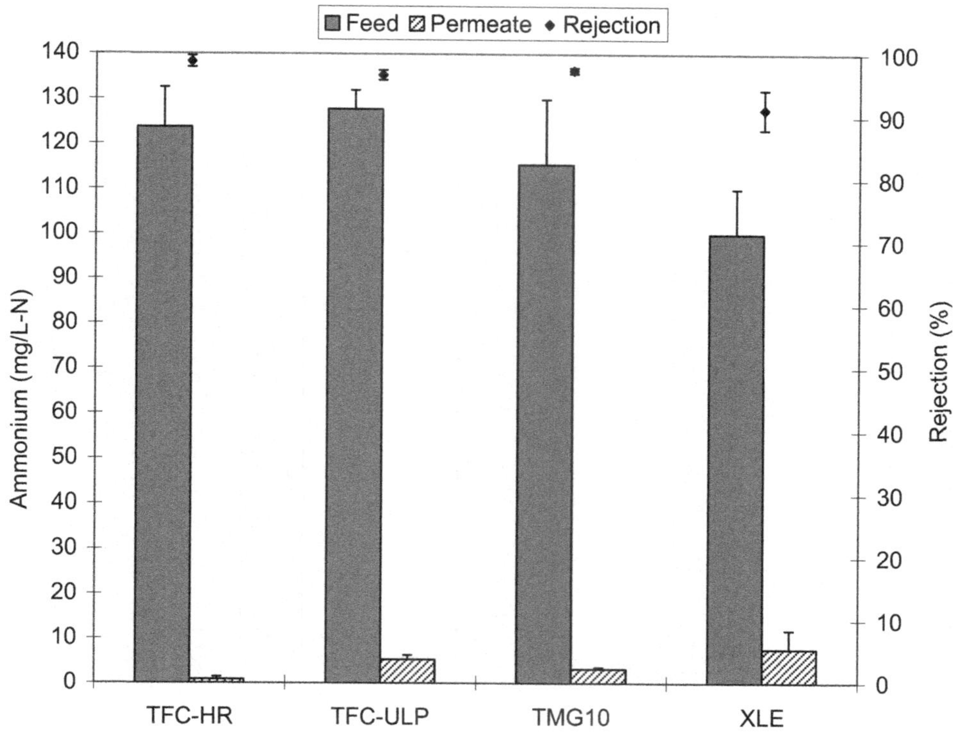

Figure B.5 Feed and permeate ammonium concentrations and rejection values for'benchmark'RO membrane and candidate ULPRO membranes during 80 percent recovery experiments (error bars represent the average deviation (n=2))

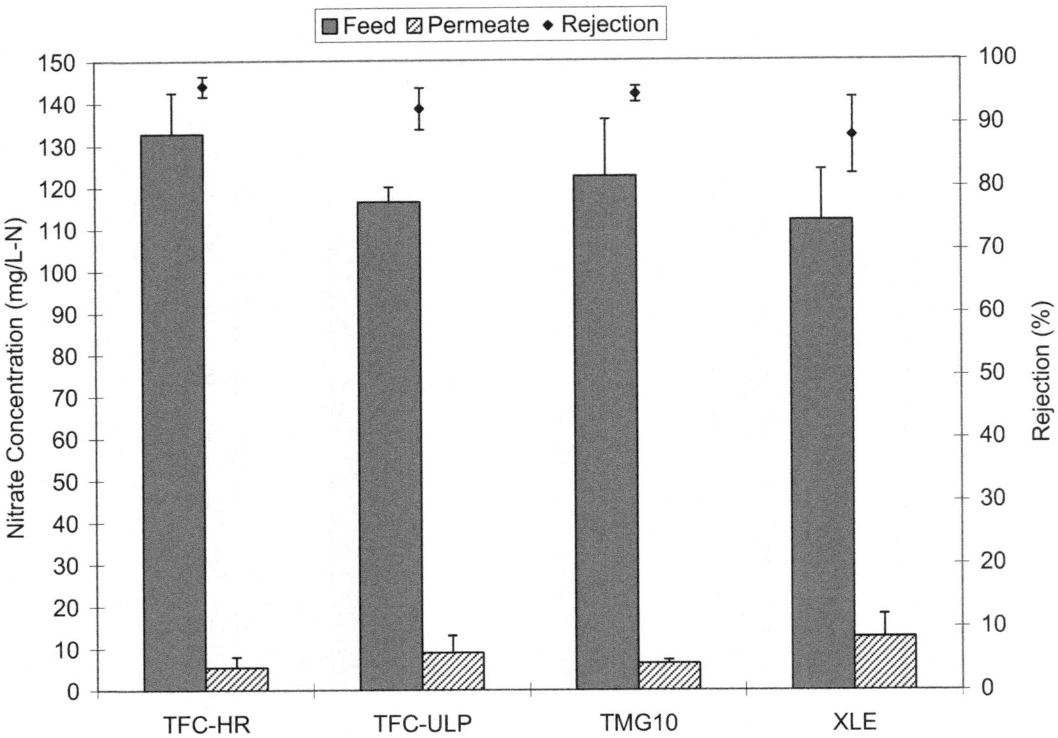

Figure B.6 Feed and permeate nitrate concentrations and rejection values for 'benchmark'RO membrane and candidate ULPRO membranes during 80 percent recovery experiments (error bars represent the average deviation (n=2))

NF Membranes

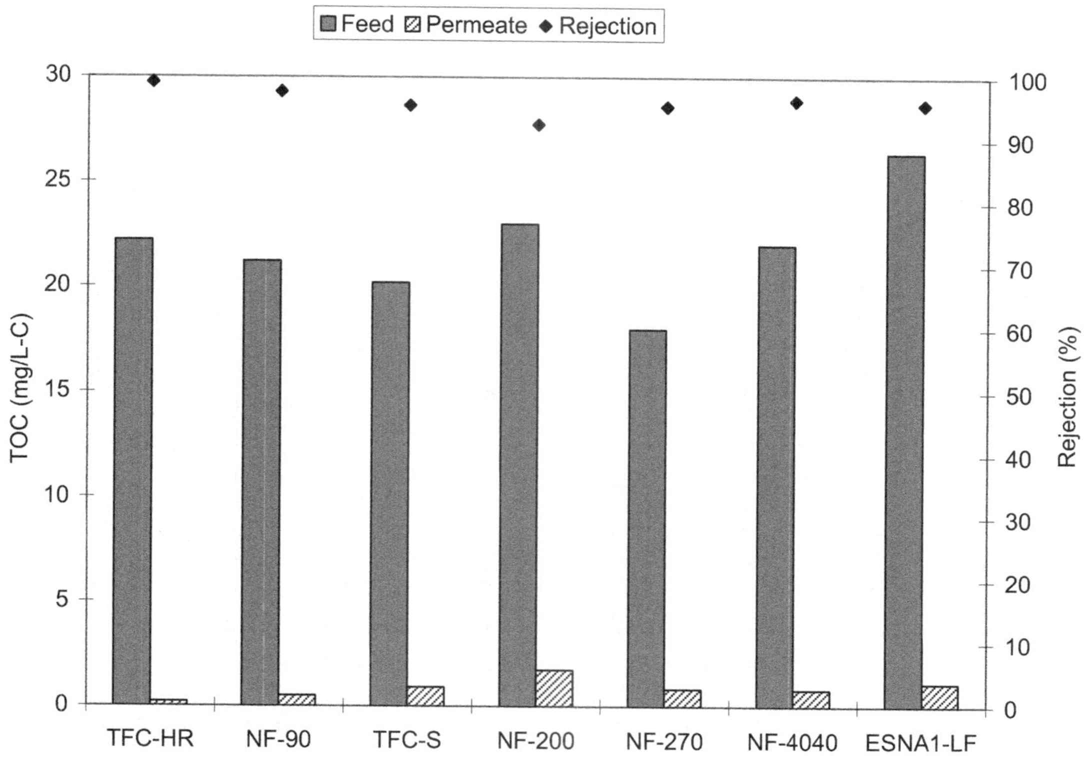

Figure B.7 Feed and permeate TOC concentrations and rejection values for 'benchmark'RO membrane and candidate NF membranes during 80 percent recovery experiments

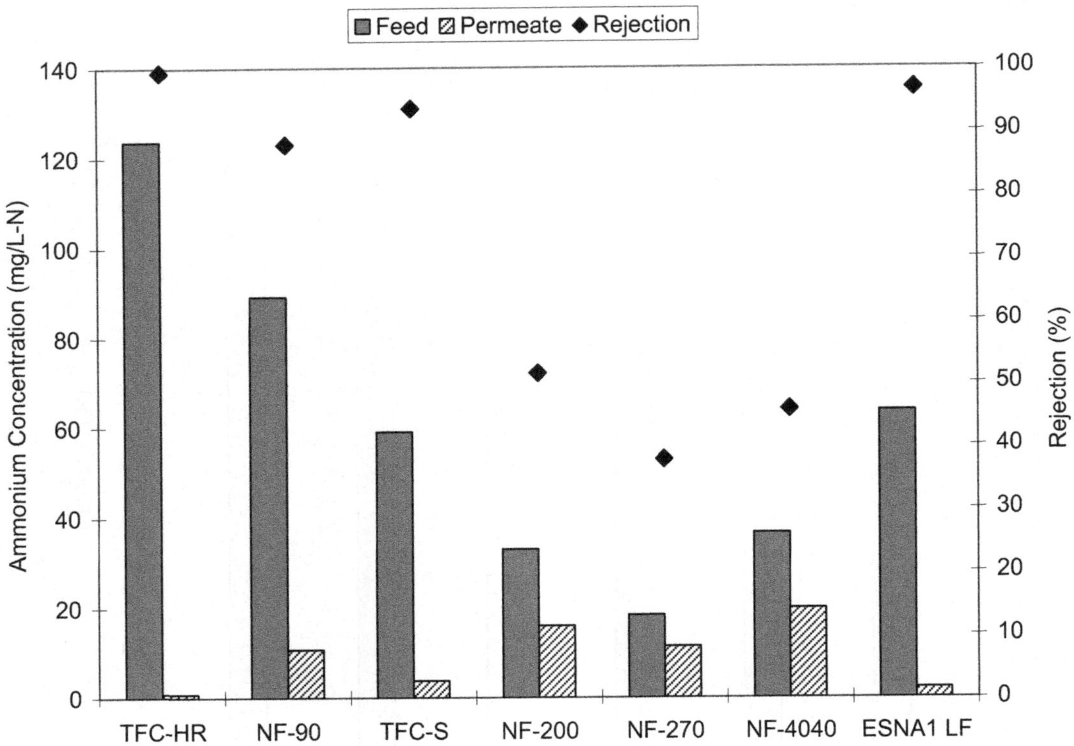

Figure B.8 Feed and permeate ammonium concentrations and rejection values for'benchmark'RO membrane and candidate NF membranes during 80 percent recovery experiments

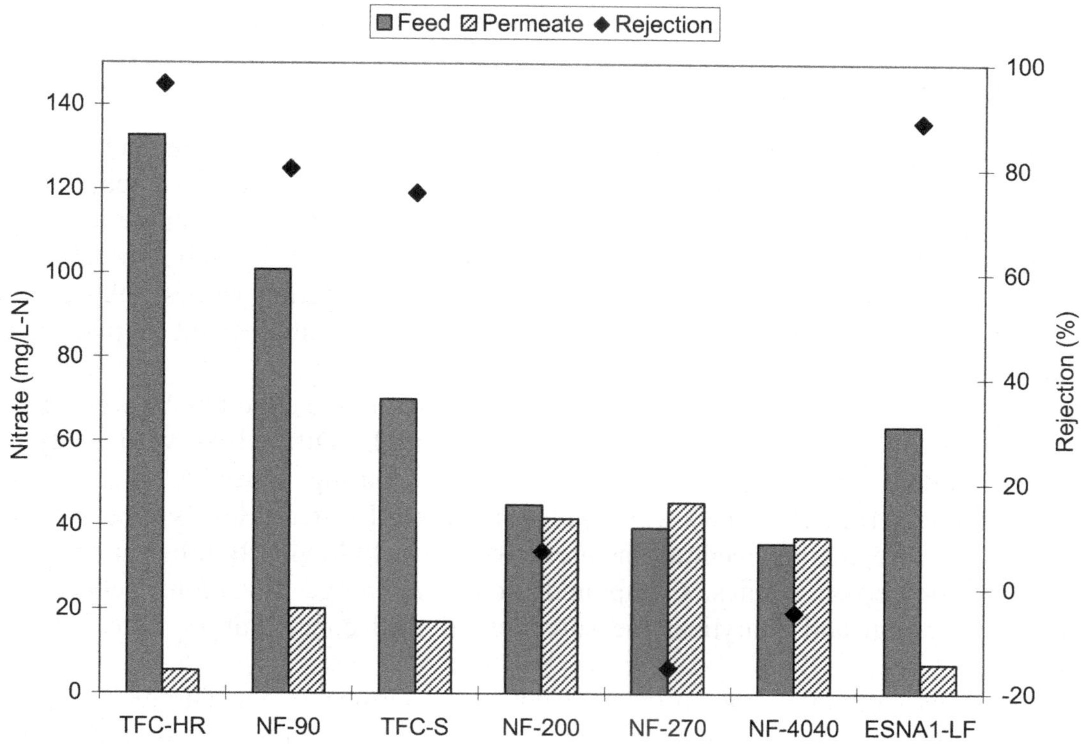

Figure B.9 Feed and permeate nitrate concentrations and rejection values for 'benchmark'RO membrane and candidate NF membranes during 80 percent recovery experiments

Unregulated Organic Rejection

ULPRO Membranes

The following figures present trace organic contaminant results for 80 percent recovery experiments. When tabulating permeate concentrations for the figures, it was necessary to distinguish between quantifiable concentrations of a solute (presented as a numeric value), detectable but not quantifiable concentrations of a solute (presented as n.q.), and non-detectable concentrations of a solute (presented as n.d.). When computing rejection values, 99.9 percent was given arbitrarily for solutes that were not detected in permeate samples and 99 percent for solutes that were detected but not quantified.

Trace organic feed concentrations for initial ULPRO membrane experiments were spiked into type II water at levels of approximately 200 to 300 ng/L. Difficulties with analytical methods required certain experiments to be repeated and during repeated trace organic contaminant rejection experiments solutes were spiked into type II water at levels approximately 500 to 600 ng/L. Therefore, feed concentrations presented in the following figures vary between 200 and 700 ng/L for certain solutes. Compounds for which large deviations between the nominal feed concentration and analytical measurement of feed concentration were observed were not tabulated.

Non-ionic trace organic contaminant feed and permeate concentrations during 'benchmark' TFC-HR membrane experiments and subsequent rejection are summarized in Figure B.10. Two solutes were quantified in permeate samples taken during 80 percent recovery experiments including the small molecular weight pharmaceutical residue phenacetine and the hydrophobic plasticizer bisphenol-A. Bisphenol-A concentrations in feed water was measured at significantly higher concentrations than were expected during trace organic contaminant rejection experiments with certain membranes. It is theorized that bisphenol-A is used in membrane manufacturing and can leach into concentrate or permeate streams, which were recycled back to the feed container during these experiments (Schäfer et al. 2005). The rejection of all non-ionic solutes, however, during 80 percent recycle experiments was 90 percent or greater indicating that traditional RO membranes are efficient in removing/reducing concentrations of nonionic organic solutes (in the size range investigated) present in wastewater effluents.

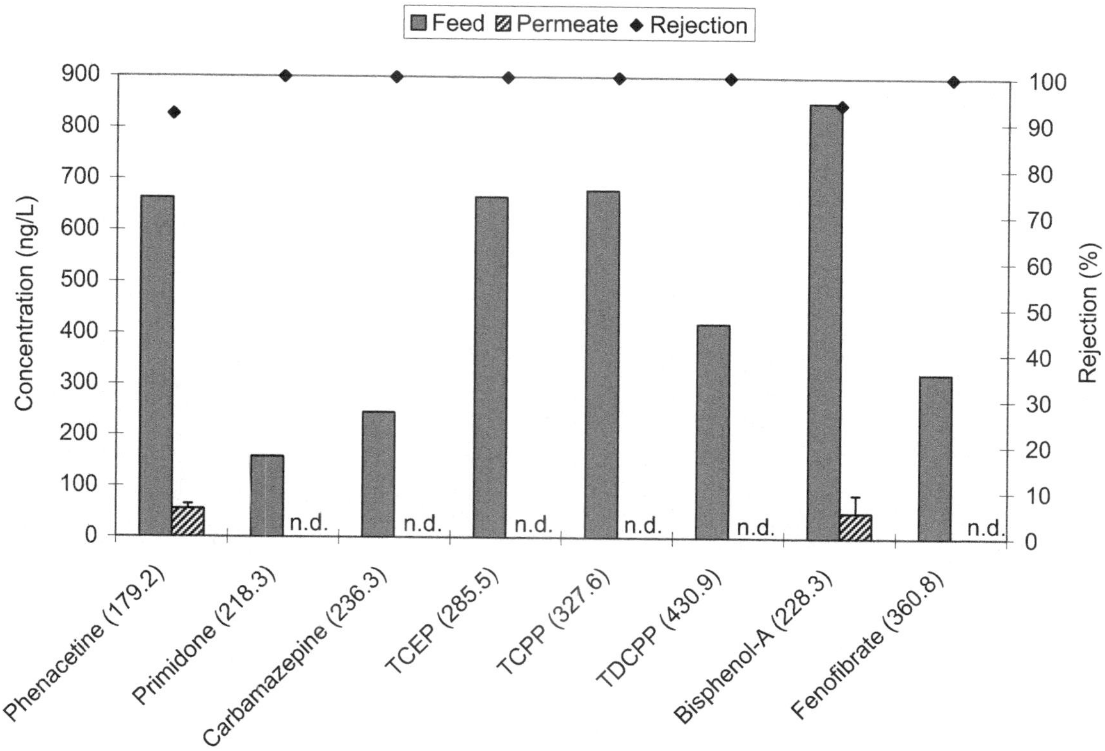

Figure B.10 Non-ionic target trace organic contaminant feed and permeate concentrations and rejection values for the'benchmark'RO TFC-HR membrane during 80 percent recovery experiments
(value in parenthesis is molecular weight of individual compound; permeate concentration error bars represent the average deviation (n=4) , n.d. and n.q. denote not detected and non-quantifiable, respectively)

Ionic trace organic contaminant feed and permeate concentrations during 'benchmark' TFC-HR membrane experiments and subsequent rejection are summarized in Figure B.11. Of the ionic solutes employed in rejection experiments, only clofibric acid and dichloroprop were detected in any permeate samples. Although it is generally understood that ionic solutes are usually well removed by charged RO and NF membranes due to electrostatic and size exclusion (Bellona et al. 2004), the low detection levels for ionic organic solutes using the Tier 2 analyte GC/MS method made these solutes easier to detect in matrix free permeate samples than non-ionic solutes. Therefore, permeate concentrations of ionic trace organic contaminants during RO experiments were less than 5 ng/L resulting in rejection values greater than 98 percent based on feed concentrations of 300 ng/L.

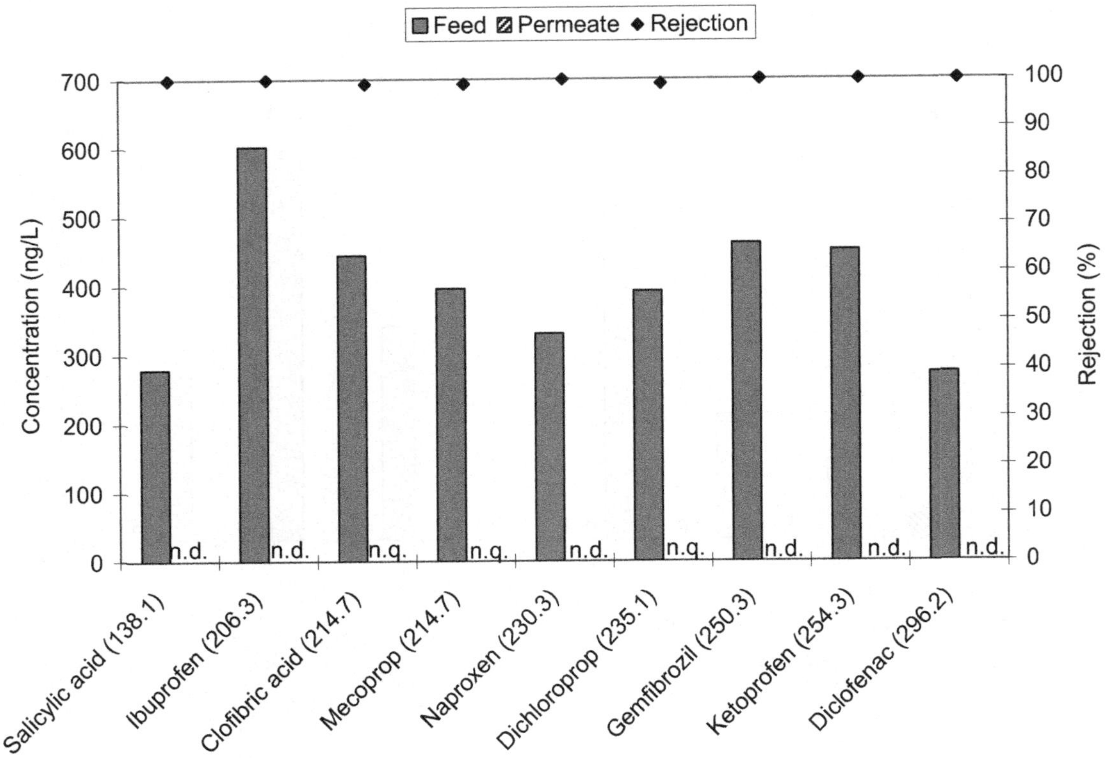

Figure B.11 Ionic target trace organic contaminant feed and permeate concentrations and rejection values for the'benchmark'RO TFC-HR membrane during 80 percent recovery experiments (value in parenthesis is molecular weight of individual compound, n.d. and n.q. denote not detected and non-quantifiable, respectively)

Non-ionic trace organic contaminant rejection by the TFC-ULP membrane is presented in Figure B.12. Non-ionic solute rejection was observed to be similarly efficient as the TFC-HR membrane although the TFC-ULP membrane operated at lower feed pressure and higher specific flux. It was observed, however, that phenacetine rejection was lower than the 'benchmark' RO membrane (TFC-HR) and traces of three additional solutes were detected in permeate samples including TCEP, carbamazepine and propyphenazone. Ionic trace organic rejection experimental results (Figure B.13) further demonstrated the high removal efficiencies of RO membranes for these solutes since only trace concentrations of naproxen were detected in any permeate samples analyzed.

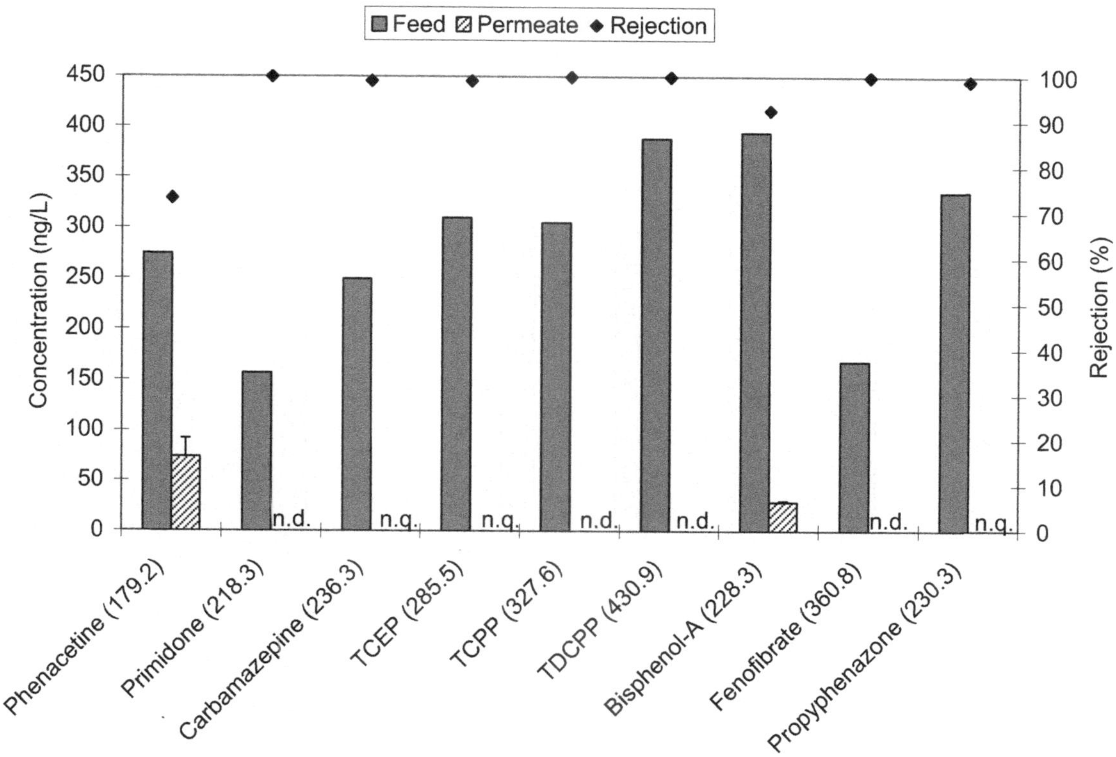

Figure B.12 Non-ionic target trace organic contaminant feed and permeate concentrations and rejection values for the TFC-ULP membrane during 80 percent recovery experiments (value in parenthesis is molecular weight of individual compound; permeate concentration error bars represent the average deviation (n=4) , n.d. and n.q. denote not detected and non-quantifiable, respectively)

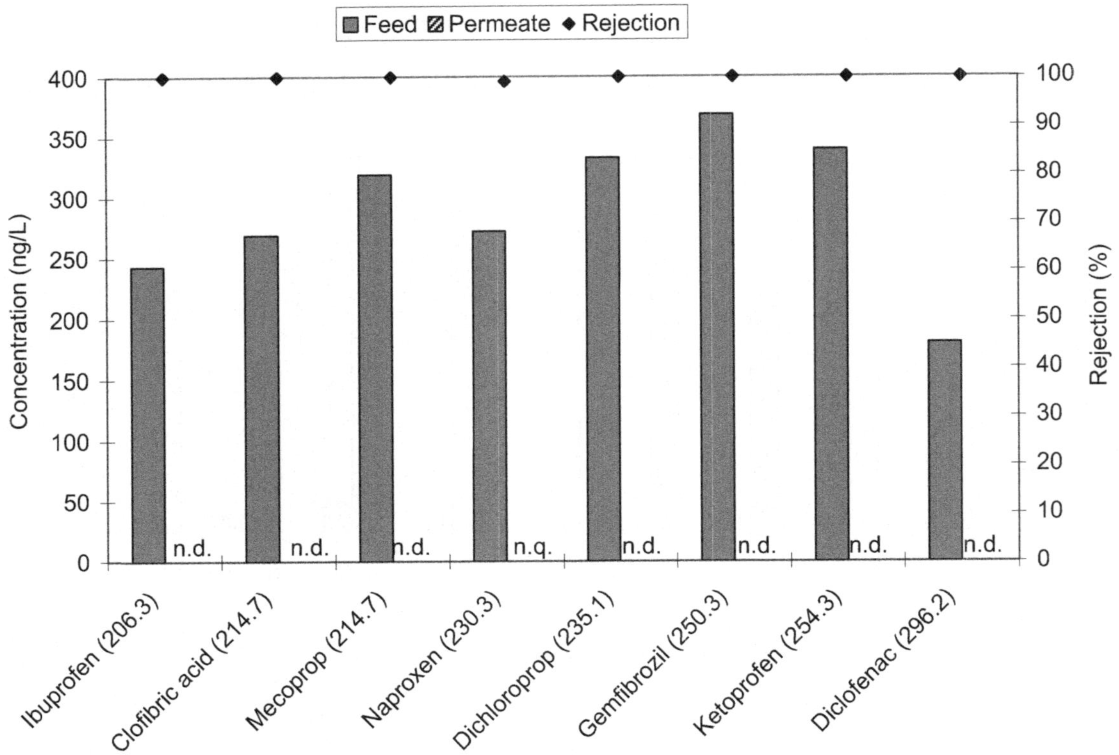

Figure B.13 Ionic target trace organic contaminant feed and permeate concentrations and rejection values for the TFC-ULP membrane during 80 percent recovery experiments (value in parenthesis is molecular weight of individual compound; permeate concentration error bars represent the average deviation (n=4), n.d. and n.q. denote not detected and non-quantifiable, respectively)

The rejection of non-ionic trace organic solutes by the TMG10 membrane is presented in Figure B.14. Bisphenol-A concentrations were found to be at concentrations an order of magnitude higher than what was spiked into the feed water, and therefore are not presented. This was likely due to bisphenol-A leaching from the virgin membrane material as described above. In addition to the quantification of phenacetine, trace concentrations of the chlorinated flame retardant TCEP were detected in permeate samples taken during 80 percent recycle TMG10 membrane experiments. Additionally, caffeine was quantified in feed water samples, which, was not the case for the TFC-HR and TFC-ULP membranes. Caffeine is a difficult solute to quantify with the GC/MS method employed and quantification depends highly upon the age of the capillary column. Therefore, the quantification of caffeine did not occur with all membranes tested.

Ionic organic contaminants were also quantified at low concentrations in permeate samples taken during 80 percent recycle experiments including clofibric acid, naproxen, dichloroprop and diclofenac (Figure B.15). Unlike the non-ionic trace organic contaminants, where it appears that molecular size plays a role in permeation (i.e., phenacetine), there appears to be no correlation between molecular weight and the rejection of ionic trace organic contaminants. Ionic trace organic contaminant rejection by the TMG10 membrane was nevertheless considered efficient, since it achieved rejections higher than 90 percent for all of the contaminants employed in the study.

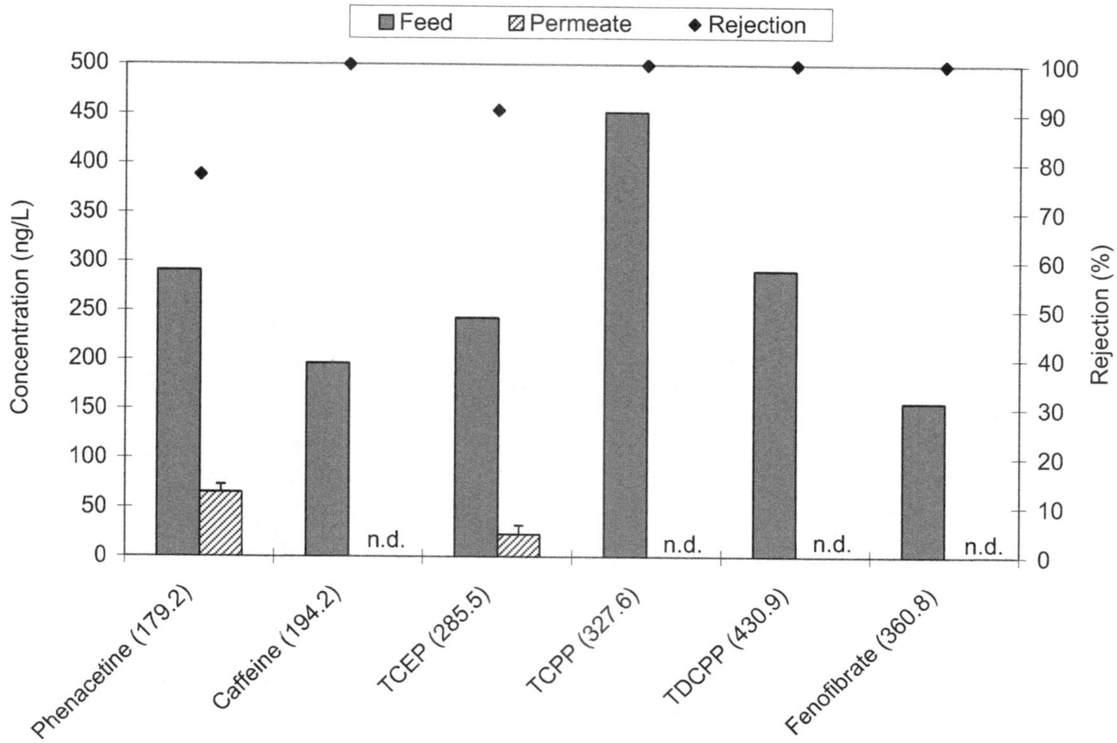

Figure B.14 Non-ionic target trace organic contaminant feed and permeate concentrations and rejection values for the TMG10 membrane
during 80 percent recovery experiments
(value in parenthesis is molecular weight of individual compound; permeate concentration error bars represent the average deviation (n=4) , n.d. and n.q. denote not detected and non-quantifiable, respectively)

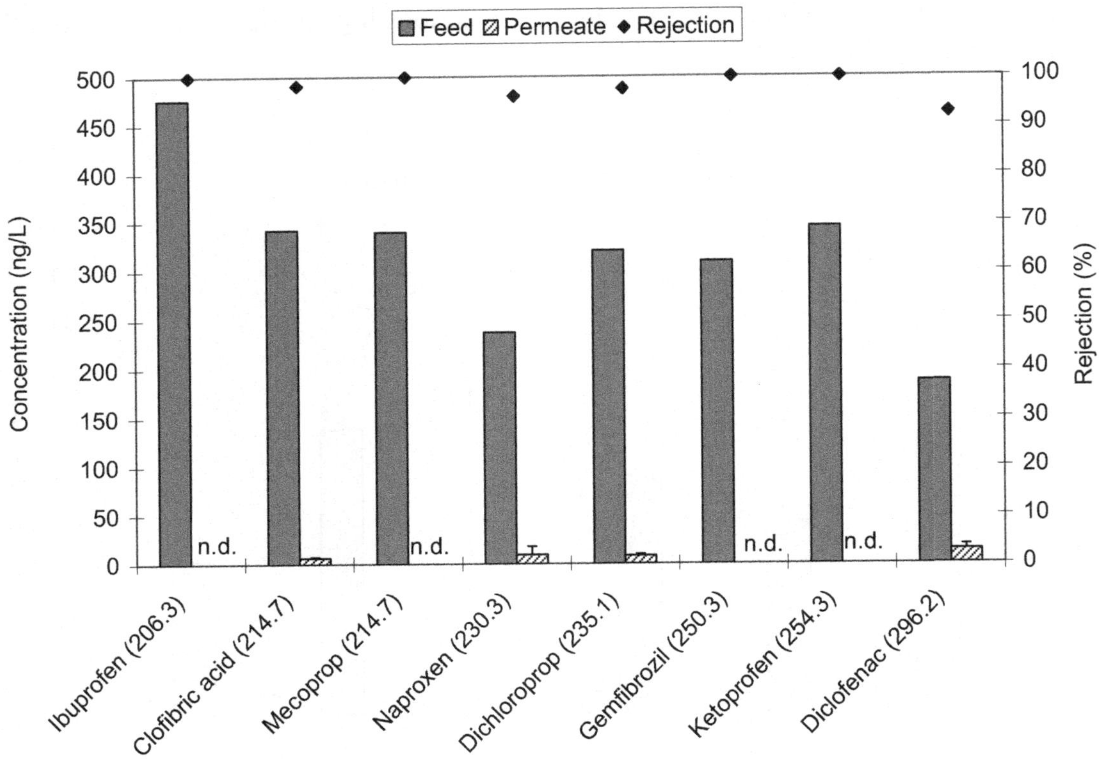

Figure B.15 Ionic target trace organic contaminant feed and permeate concentrations and rejection values for the TMG10 membrane during 80 percent recovery experiments (value in parenthesis is molecular weight of individual compound; permeate concentration error bars represent the average deviation (n=4) , n.d. and n.q. denote not detected and non-quantifiable, respectively)

The most permeable ULPRO membrane tested during this study, the XLE, also displayed the most permeate detections of trace organic contaminants. Non-ionic organic contaminants quantified in permeate samples included phenacetine, TCEP and TCPP (Figure B.16). Similarly to TMG10 membrane results, bisphenol-A was quantified at concentrations approximately 10 times higher than were spiked and are not presented in Figure B.16.

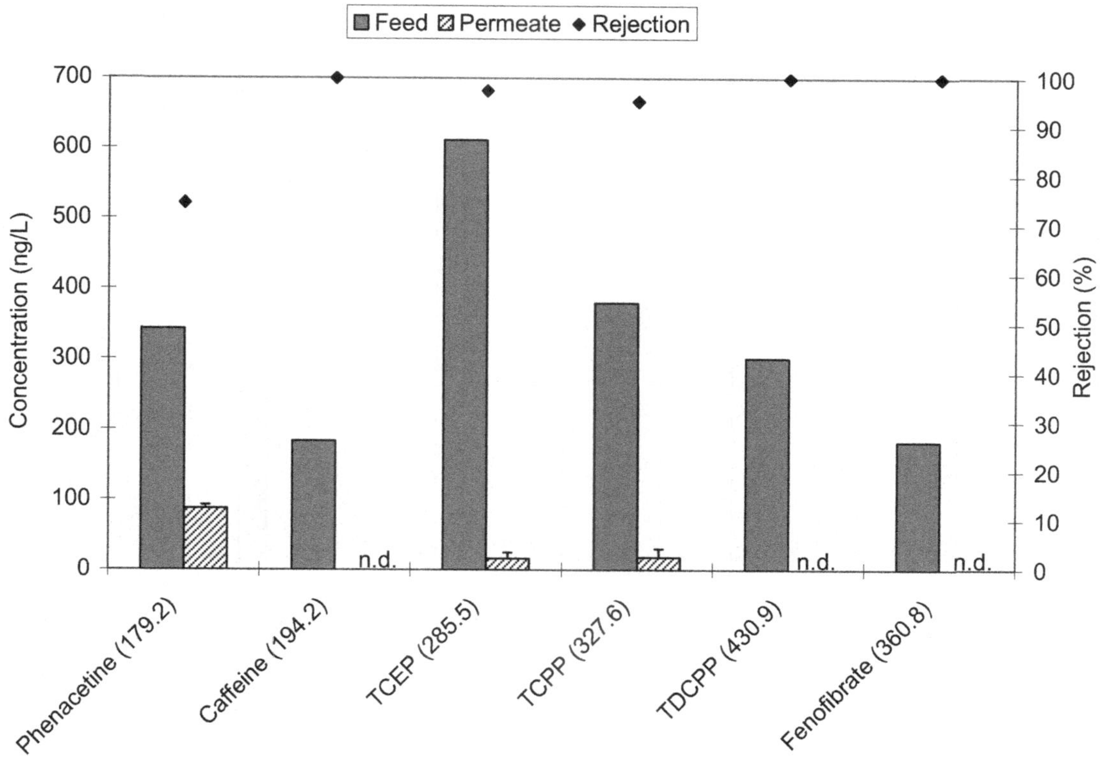

Figure B.16 Non-ionic target trace organic contaminant feed and permeate concentrations and rejection values for the XLE membrane during 80 percent recovery experiments (value in parenthesis is molecular weight of individual compound; permeate concentration error bars represent the average deviation (n=4) , n.d. and n.q. denote not detected and non-quantifiable, respectively)

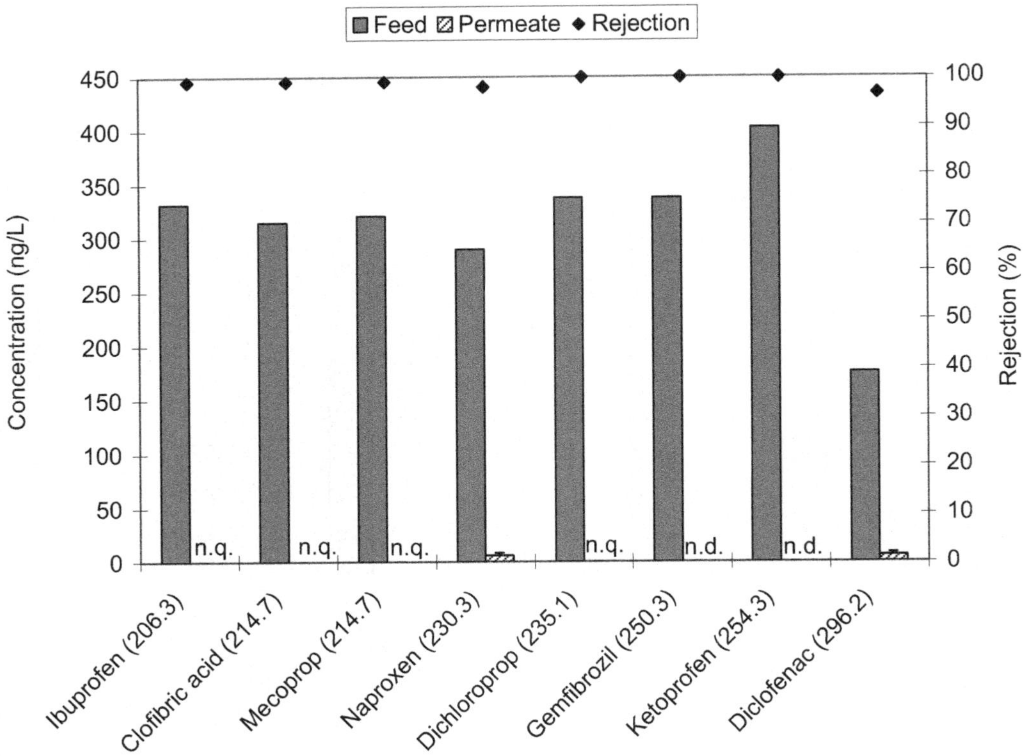

Figure B.17 Ionic target trace organic contaminant feed and permeate concentrations and rejection values for the XLE membrane during 80 percent recovery experiments (value in parenthesis is molecular weight of individual compound; permeate concentration error bars represent the average deviation (n=4) , n.d. and n.q. denote not detected and non-quantifiable, respectively)

The RO and ULPRO membranes TFC-HR, TFC-ULP, TMG10, and XLE were found to efficiently reject trace organics employed in this study and permeate concentrations were generally below 10 ng/L or non-quantifiable with exception of the low molecular weight pharmaceutical residue phenacetine. Permeate phenacetine concentrations were generally 80 percent less than feed concentrations, but indicate that small uncharged molecular weight compounds are less efficiently removed by ULPRO and to a lesser extent by RO membranes. Ionic trace organic contaminants, however, were found to be well removed by the RO and ULPRO membranes, which is theorized to be due to contributions of steric hindrance and electrostatic exclusion to the rejection of these solutes.

Table B.1 summarizes the candidate ULPRO membrane trace organic contaminant rejection results during both 30 percent and 80 percent recovery experiments by summarizing the number of feed and permeate detections and quantifications and the percentage of the compounds detected in feed water samples that were also detected in permeate samples. Of the ULPRO membranes tested, the TFC-ULP membrane experiments were conducted with the most number of organic solutes due to the final establishment of the full analytical method during testing of this membrane. The TFC-ULP and TMG10 membranes were found to have similar number of permeate detections and percentage of solutes in the feed water that were detected in permeate samples although the TFC-ULP membrane had the fewest number of permeate quantifications. The XLE membrane exhibited the most number and highest percentage of permeate organic solute detections, which is assumed to be a reflection of the membranes high productivity and solvent permeability.

Table B.1
Summary of candidate ULPRO trace organic contaminant rejection experiments

Membrane	TFC-ULP		TMG10		XLE	
Feed water recovery during experiment (percent)	30	80	30	80	30	80
Feed quantifications (#)	17	17	14	14	14	14
Permeate detections and quantifications (#)	4	6	4	6	9	9
Permeate quantifications (#)	2	2	3	6	5	5
Compounds detected perm/feed (percent)	24	35	21	43	64	64

NF Membranes

The rejection of the candidate NF membranes was in general similar to the ULPRO membranes tested, where larger molecular weight and charged solutes were better rejected than small molecular weight uncharged organic solutes. In general, however, the number of permeate quantifications of trace organic solutes increased with increasing permeability and permeate productivity.

Feed and permeate concentrations of non-ionic trace organic contaminants and subsequent rejection values for the NF-90 membrane during 80 percent recovery experiments are presented in Figure B.18. The NF-90 membrane was found to perform the best in terms of the lowest number of trace organic permeate quantifications out of the NF membranes considered for the study. In general, the NF-90 membrane behaved similarly to the XLE membrane and had quantifiable concentrations of phenacetine, TCEP and TCPP in permeate samples.

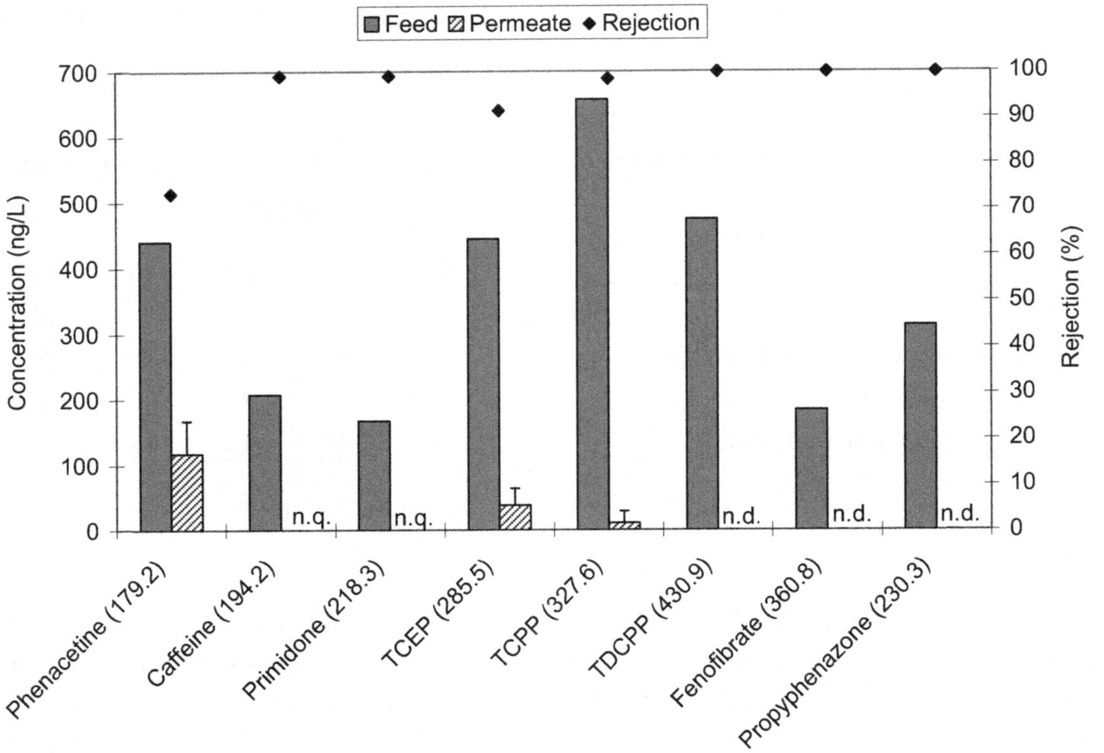

Figure B.18 Non-ionic target trace organic contaminant feed and permeate concentrations and rejection values for the NF-90 membrane during 80 percent recovery experiments (value in parenthesis is molecular weight of individual compound; permeate concentration error bars represent the average deviation (n=4) , n.d. and n.q. denote not detected and non-quantifiable, respectively)

Ionic trace organic solute rejection results for the NF-90 membrane during 80 percent recovery experiments are presented in Figure B.19. Although all of the solutes besides ketoprofen were detected in NF-90 membrane permeate samples, the rejection of these solutes was greater than 90 percent resulting in permeate concentrations of less than 20 ng/L. Trace organic rejection results indicate that the NF-90 membrane can efficiently reject (> 90 percent) most trace organic contaminants employed during this study, but that small molecular weight non-ionic solutes present in wastewater effluents are difficult to remove.

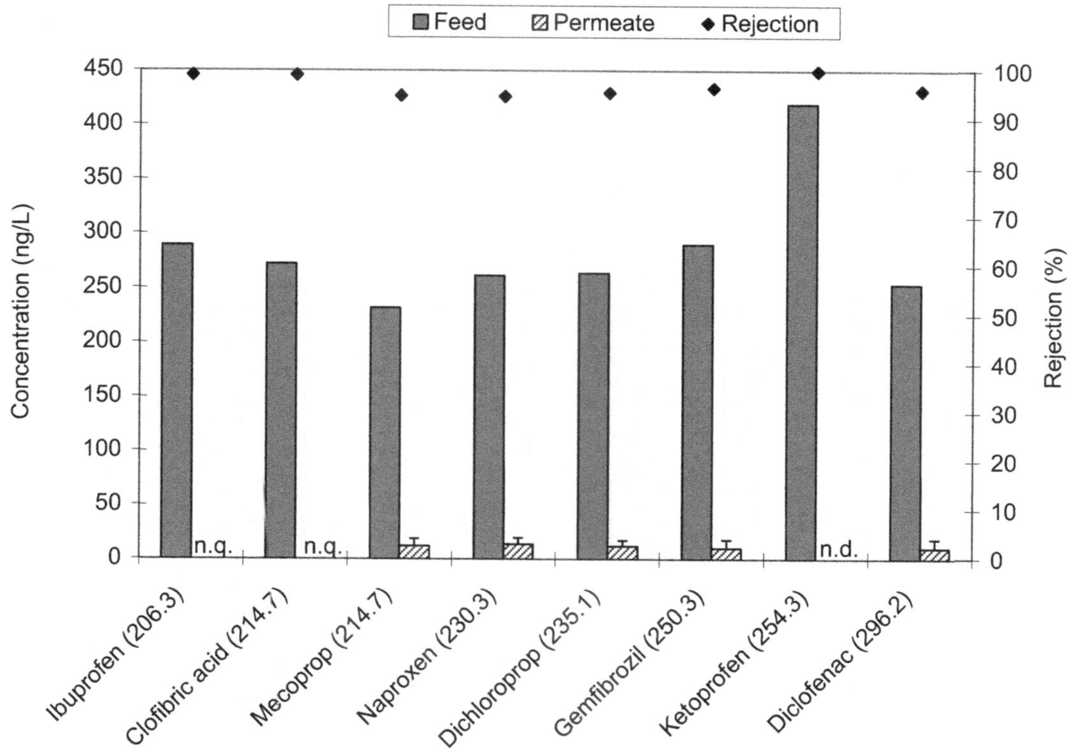

Figure B.19 Ionic target trace organic contaminant feed and permeate concentrations and rejection values for the NF-90 membrane during 80 percent recovery experiments (value in parenthesis is molecular weight of individual compound; permeate concentration error bars represent the average deviation (n=4) , n.d. and n.q. denote not detected and non-quantifiable, respectively)

Similar to NF-90 membrane results, the TFC-S membrane was found to only moderately reject the smallest non-ionic trace organic solute phenacetine, while efficiently removing ionic trace organic solutes (Figures B.20 and B.21). Increased concentrations of TCEP and TCPP as well as the quantification of carbamazepine in permeate samples; however, indicate that the TFC-S membrane is not as efficient in removing non-ionic trace organic solutes as the NF-90 membrane. Ionic trace organic solutes quantified in TFC-S membrane permeates were at low concentrations, resulting in rejection values for these solutes greater than 90 percent for all of the compounds considered (Figure B.21).

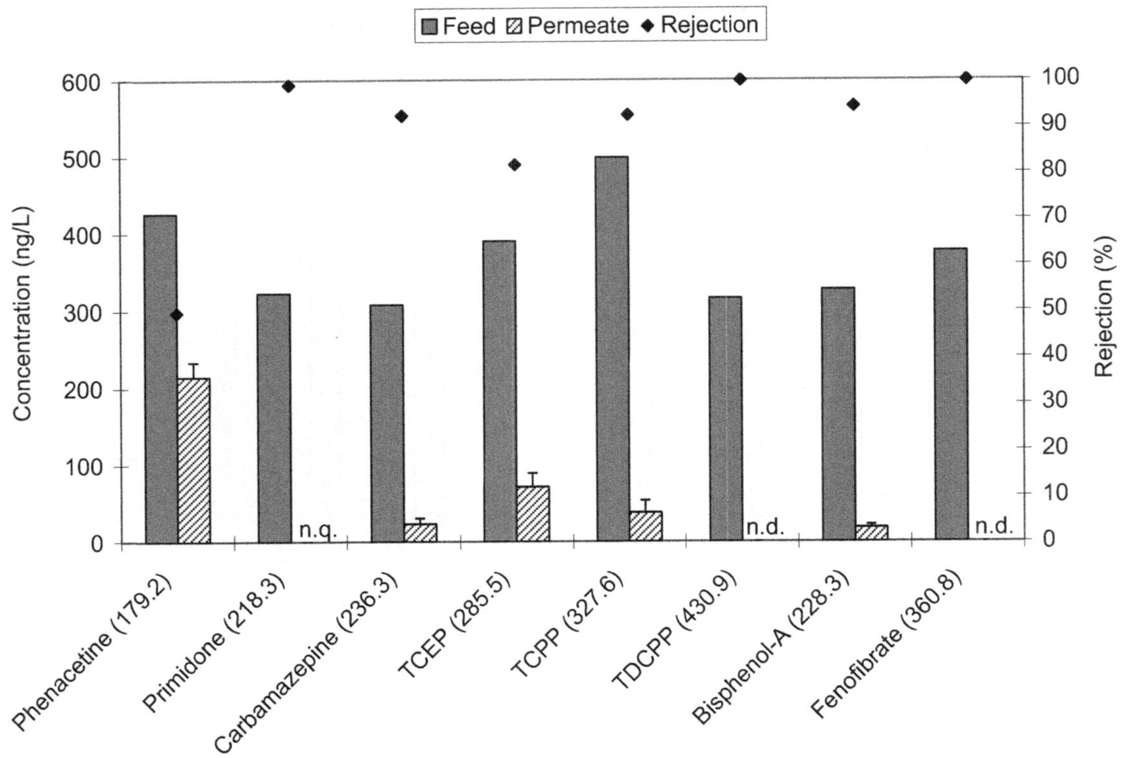

Figure B.20 Non-ionic target trace organic contaminant feed and permeate concentrations and rejection values for the TFC-S membrane during 80 percent recovery experiments (value in parenthesis is molecular weight of individual compound; permeate concentration error bars represent the average deviation (n=4) , n.d. and n.q. denote not detected and non-quantifiable, respectively)

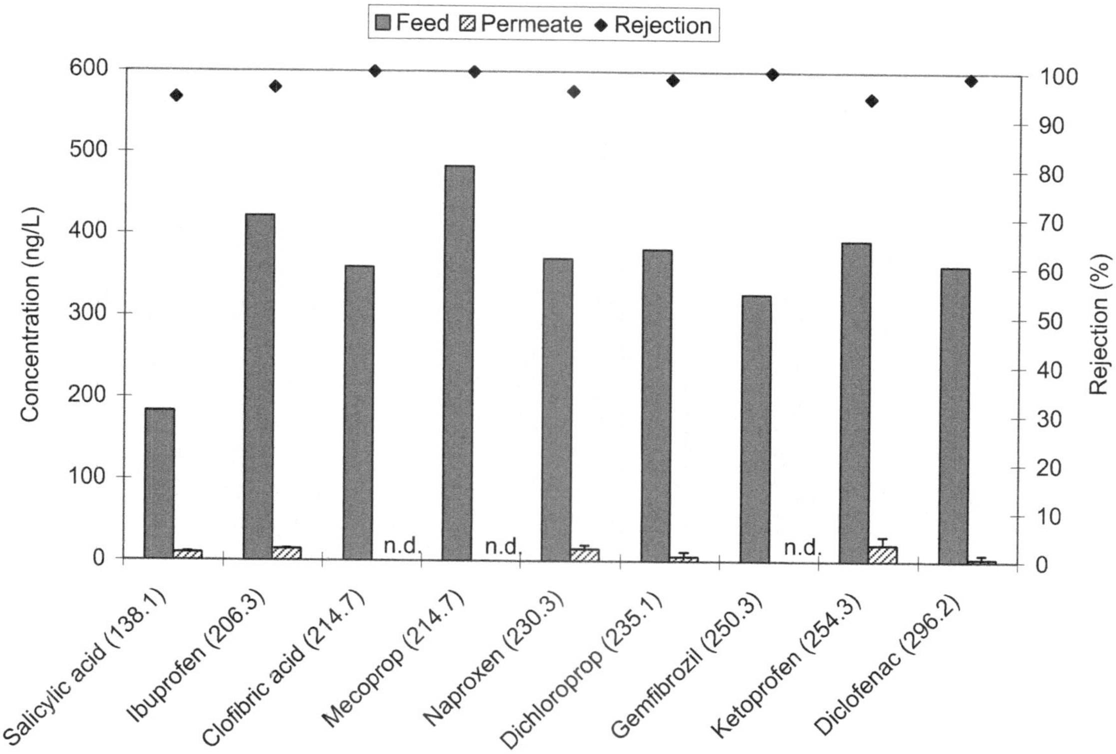

Figure B.21 Ionic target trace organic contaminant feed and permeate concentrations and rejection values for the TFC-S membrane during 80 percent recovery experiments (value in parenthesis is molecular weight of individual compound; permeate concentration error bars represent the average deviation (n=4) , n.d. and n.q. denote not detected and non-quantifiable, respectively)

The rejection of non-ionic and ionic trace organic solutes by the NF-200 membrane is presented in Figure B.22 and B.23, respectively. Similar to the other membranes tested, phenacetine and bisphenol-A were detected at the highest concentrations in permeate samples taken during 80 percent recovery experiments. In addition to the solutes detected in TFC-S membrane permeate samples, the hydrophilic solute caffeine was also detected in permeate samples taken during rejection experiments. Besides the solutes bisphenol-A and phenacetine, however, the rejection of quantifiable non-ionic trace organic contaminants by the NF-200 membrane was greater than 90 percent. Ionic trace organic rejection was observed to be greater than 97 percent for all of the solutes employed in the study, indicating that the NF-200 membrane can reject ionic trace organic solutes as efficiently as the ULPRO membranes employed. Bellona and Drewes (2005) determined that the NF-200 membrane could efficiently reject (>90 percent) small molecular weight ionic organics such as acetic acid (molecular weight of 60 Daltons) due to electrostatic exclusion as a result of a relatively high membrane surface charge. Other studies have determined that the rejection of charged solutes is a function of membrane permeability (often given as a pore size) and membrane surface charge and that certain NF membranes are effective in removing charged solutes (Bowen and Mohammed 1998).

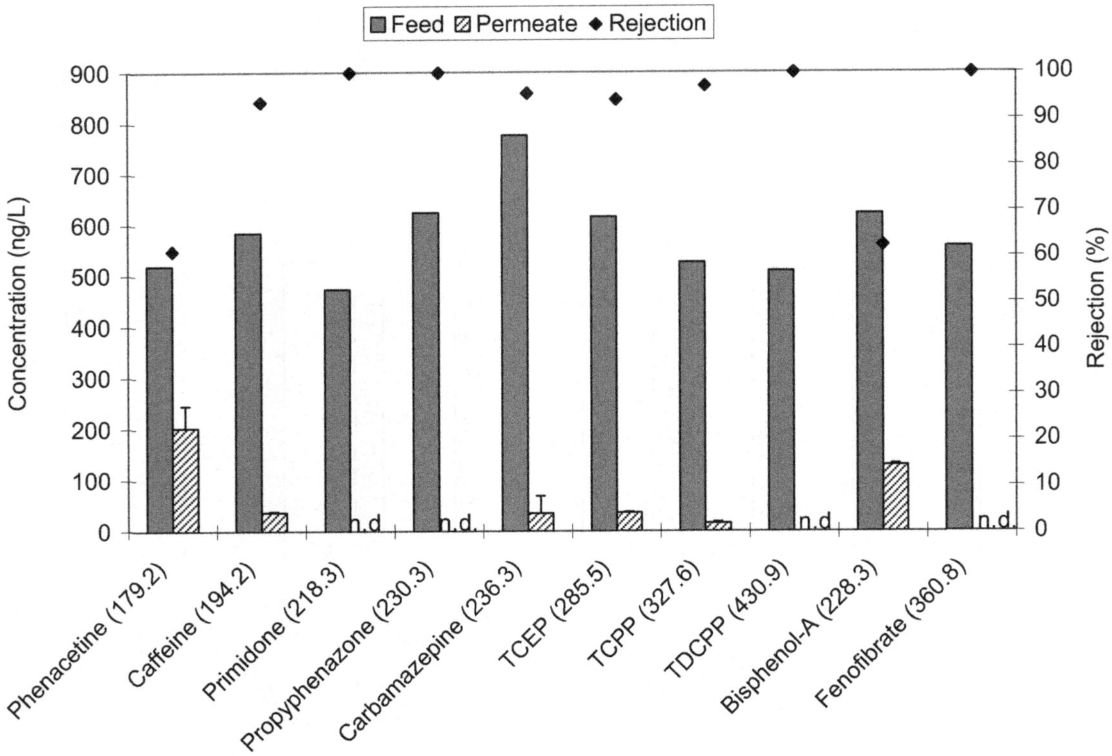

Figure B.22 Non-ionic target trace organic contaminant feed and permeate concentrations and rejection values for the NF-200 membrane during 80 percent recovery experiments (value in parenthesis is molecular weight of individual compound; permeate concentration error bars represent the average deviation (n=4) , n.d. and n.q. denote not detected and non-quantifiable, respectively)

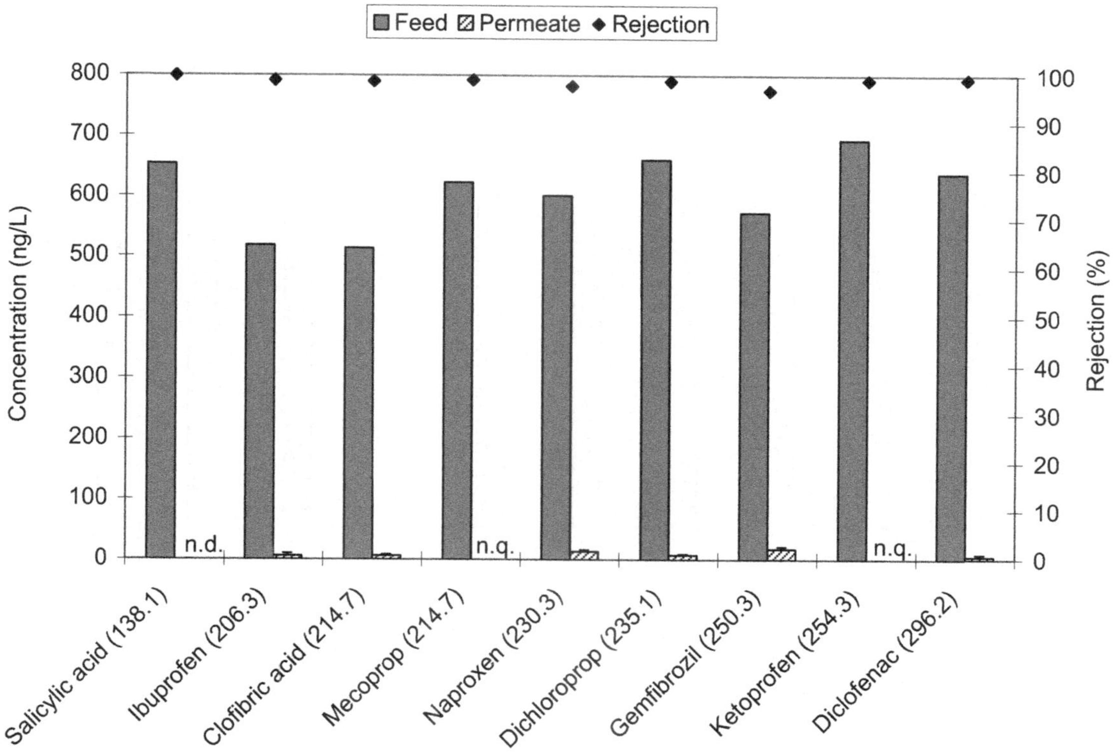

Figure B.23 Ionic target trace organic contaminant feed and permeate concentrations and rejection values for the NF-200 membrane during 80 percent recovery experiments (value in parenthesis is molecular weight of individual compound; permeate concentration error bars represent the average deviation (n=4) , n.d. and n.q. denote not detected and non-quantifiable, respectively)

The rejection of trace organic contaminants by the most permeable and productive NF membrane, the NF-270, is presented in Figures B.24 and B.25. Although the greatest percentage of solutes that were detected in feed water were detected in permeate samples taken during 80 percent recovery rejection experiments, permeate concentrations of non-ionic and ionic trace organic contaminants were similar to other NF membranes tested, with the exception of carbamazepine, which was detected at concentrations of approximately 100 ng/L. Ionic trace organic rejection experiments displayed rejection efficiencies similar to other membranes tested, where all of the solutes employed in the study were rejected greater than 90 percent during 80 percent recovery experiments.

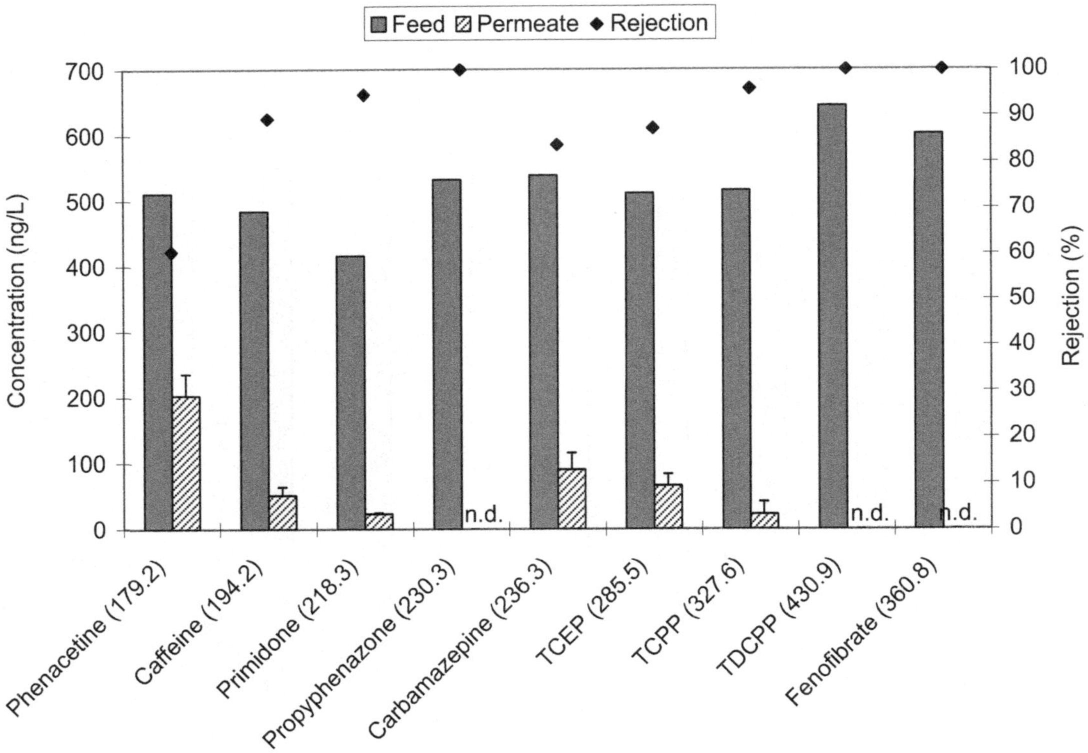

Figure B.24 Non-ionic target trace organic contaminant feed and permeate concentrations and rejection values for the NF-270 membrane during 80 percent recovery experiments (value in parenthesis is molecular weight of individual compound; permeate concentration error bars represent the average deviation (n=4) , n.d. and n.q. denote not detected and non-quantifiable, respectively)

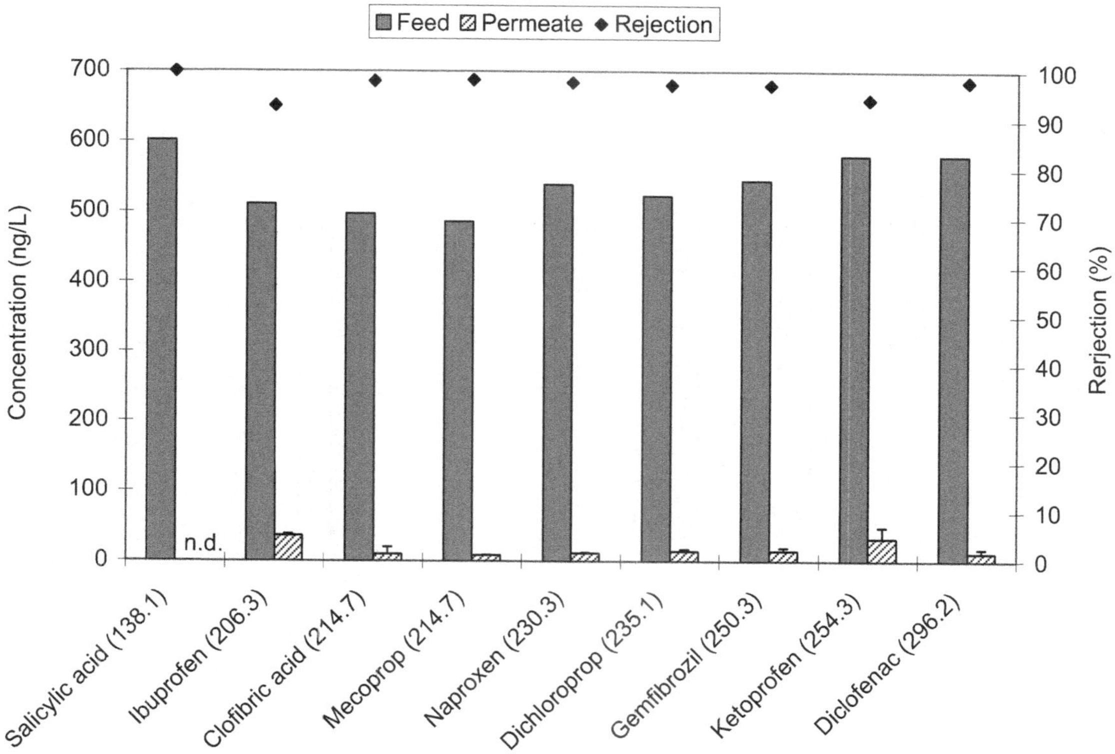

Figure B.25 Ionic target trace organic contaminant feed and permeate concentrations and rejection values for the NF-270 membrane during 80 percent recovery experiments (value in parenthesis is molecular weight of individual compound; permeate concentration error bars represent the average deviation (n=4) , n.d. and n.q. denote not detected and non-quantifiable, respectively)

The ESNA1-LF membrane trace organic contaminant results exhibited a similar rejection efficiency as compared to the NF-270 membrane, although the ESNA1-LF was not nearly as permeable and required significantly higher feed pressures than the other NF membranes (Figures B.26 and B.27).

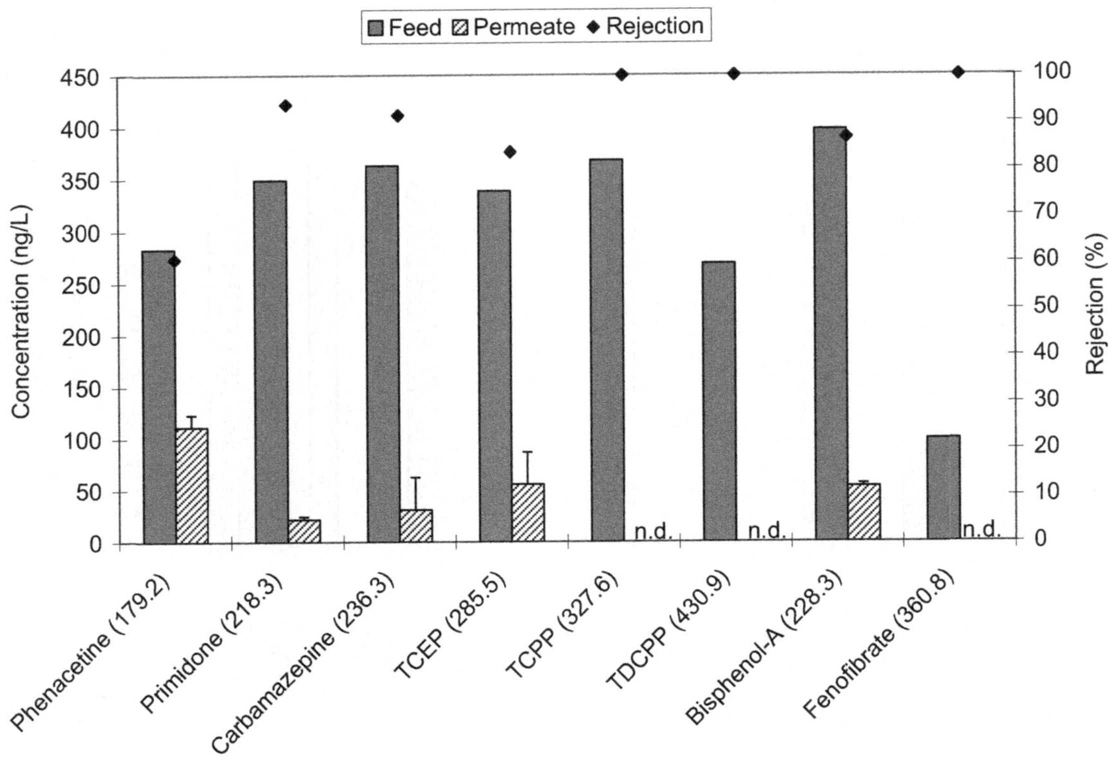

Figure B.26 Non-ionic target trace organic contaminant feed and permeate concentrations and rejection values for the ESNA1-LF membrane during 80 percent recovery experiments (value in parenthesis is molecular weight of individual compound; permeate concentration error bars represent the average deviation (n=4) , n.d. and n.q. denote not detected and non-quantifiable, respectively)

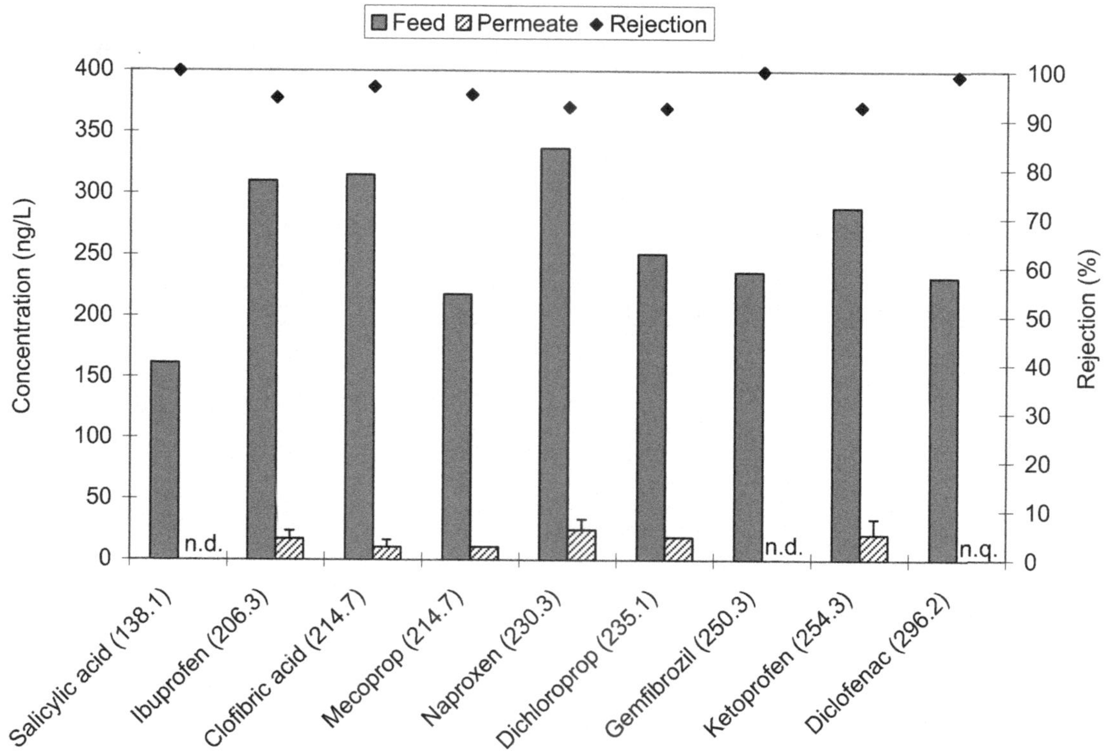

Figure B.27 Ionic target trace organic contaminant feed and permeate concentrations and rejection values for the ESNA1-LF membrane during 80 percent recovery experiments (value in parenthesis is molecular weight of individual compound; permeate concentration error bars represent the average deviation (n=4) , n.d. and n.q. denote not detected and non-quantifiable, respectively)

A summary of candidate NF membrane trace organic contaminant rejection experiments is presented in Table B.2. In general, the NF membranes employed during this study were observed to have a higher percentage of the solutes quantified in feed water in permeate samples than the RO and ULPRO membranes. With the exception of phenacetine, however, permeate concentrations were low resulting in rejection values greater than 80 percent. Although ionic trace organic solutes were more often detected in NF permeate samples, concentrations were below 40 ng/L, resulting in rejection values greater than 90 percent. As expected, the limitations of NF membranes were observed to be the removal of small molecular weight non-ionic organic solutes such as phenacetine, caffeine, and TCEP.

Table B.2
Summary of candidate NF trace organic contaminant experimental results

Membrane	NF-90		TFC-S		NF-200		NF-270		ESNA1-LF	
Feed water recovery during experiment (percent)	30	80	30	80	30	80	30	80	30	80
Feed detections (#)	16	16	17	17	18	18	18	18	17	17
Permeate detections and quantifcations (#)	9	12	10	13	11	12	15	14	9	12
Permeate Quantifications (#)	3	8	10	12	11	12	15	14	6	11
Compounds detected perm/feed (percent)	56	75	59	76	61	67	83	78	53	71

APPENDIX C
FULL-SCALE RO MEMBRANE PERFORMANCE

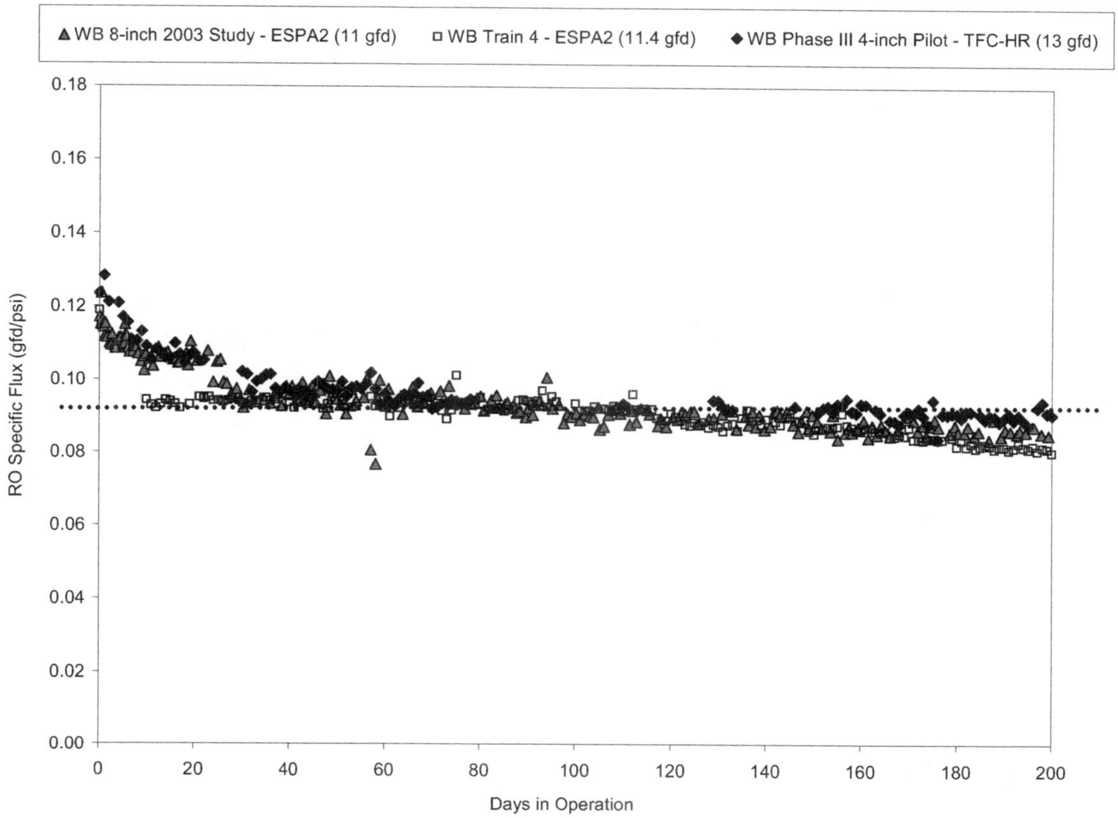

Figure C.1 Temperature corrected specific flux of the ESPA2 and TFC-HR membranes on pilot- and full-scale at the WBWRP
(dashed line shows stabilized TCSF of 0.09 gfd/psi)

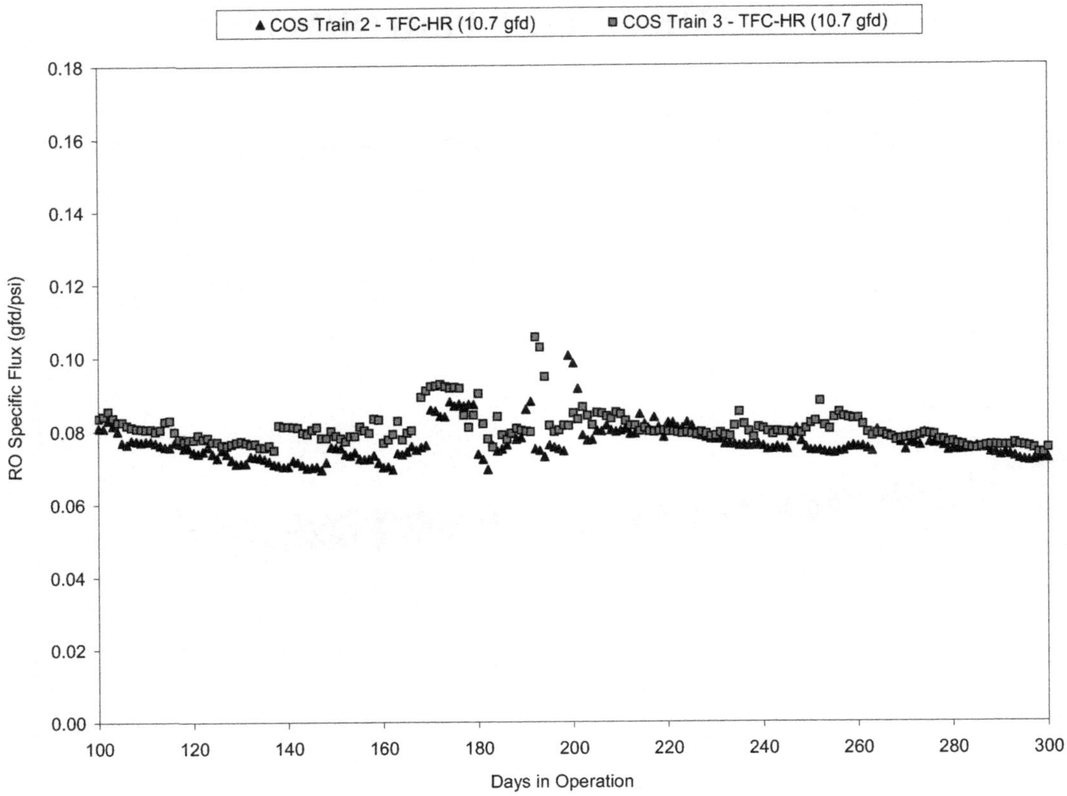

Figure C.2 Temperature corrected specific flux of the TFC-HR membrane on full-scale at the SWC

APPENDIX D
PILOT-SCALE TESTING

TOC and Nitrogen Removal by the TMG10 and NF-90 Membrane Treating Non-nitrified Feed Water

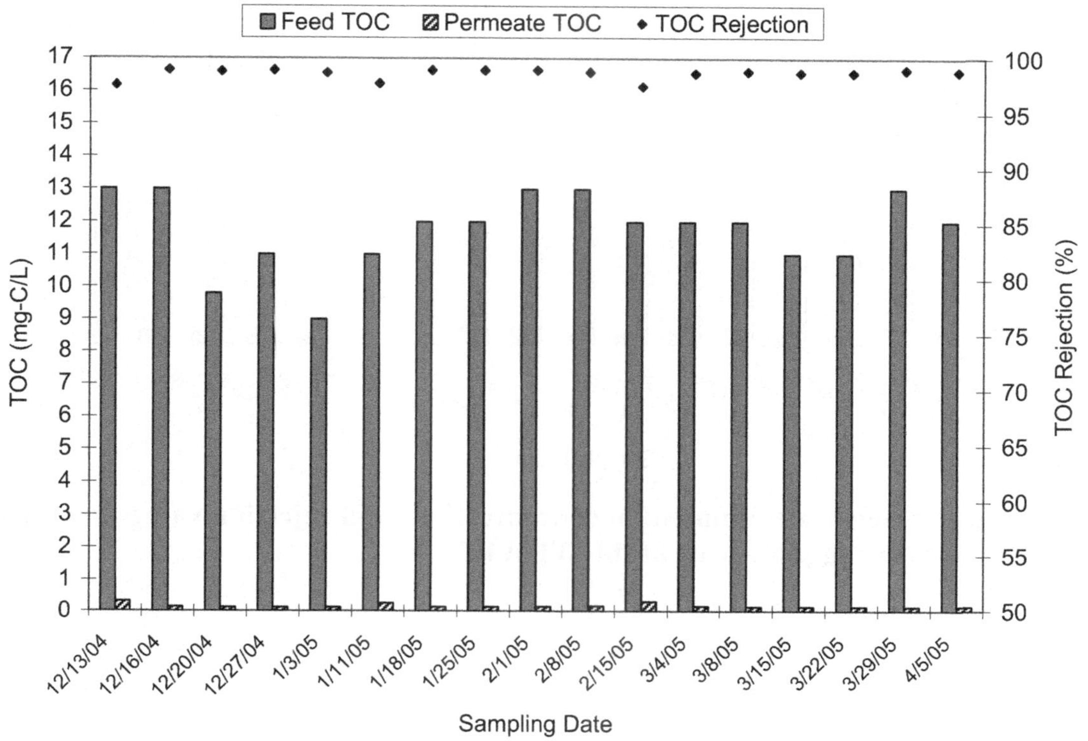

Figure D.1 Feed and permeate TOC concentrations and rejection values for the TMG10 membrane during pilot-scale testing at the WBWRP

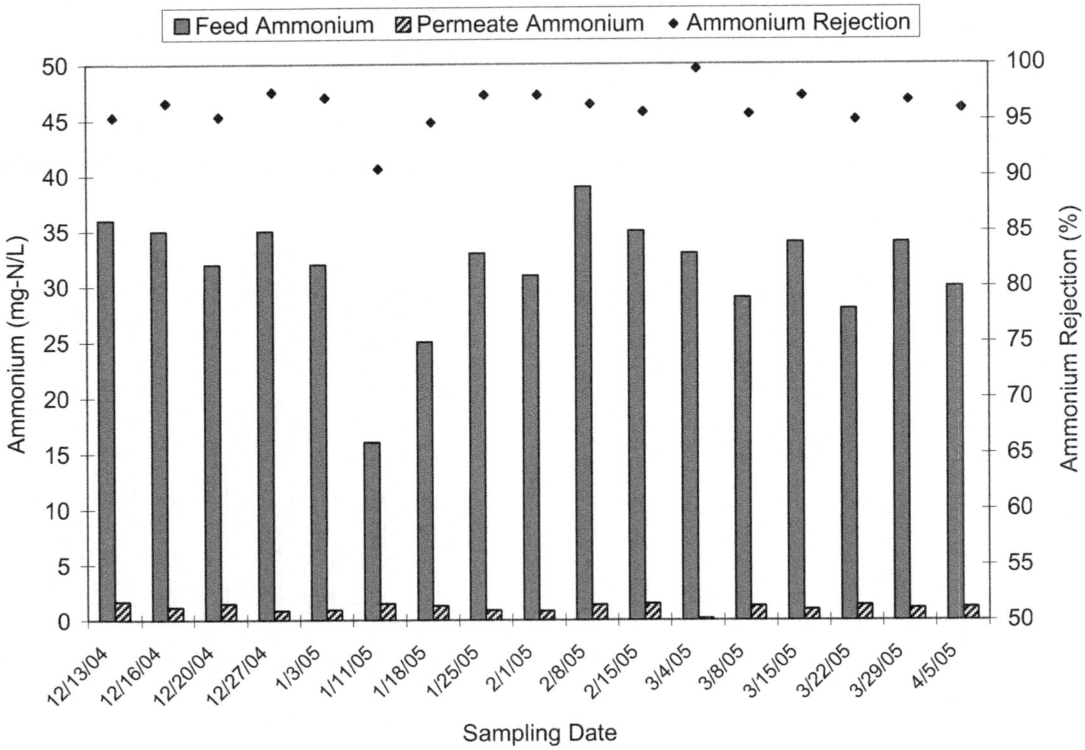

Figure D.2 Feed and permeate ammonium concentrations and rejection values for the TMG10 membrane during pilot-scale at the WBWRP

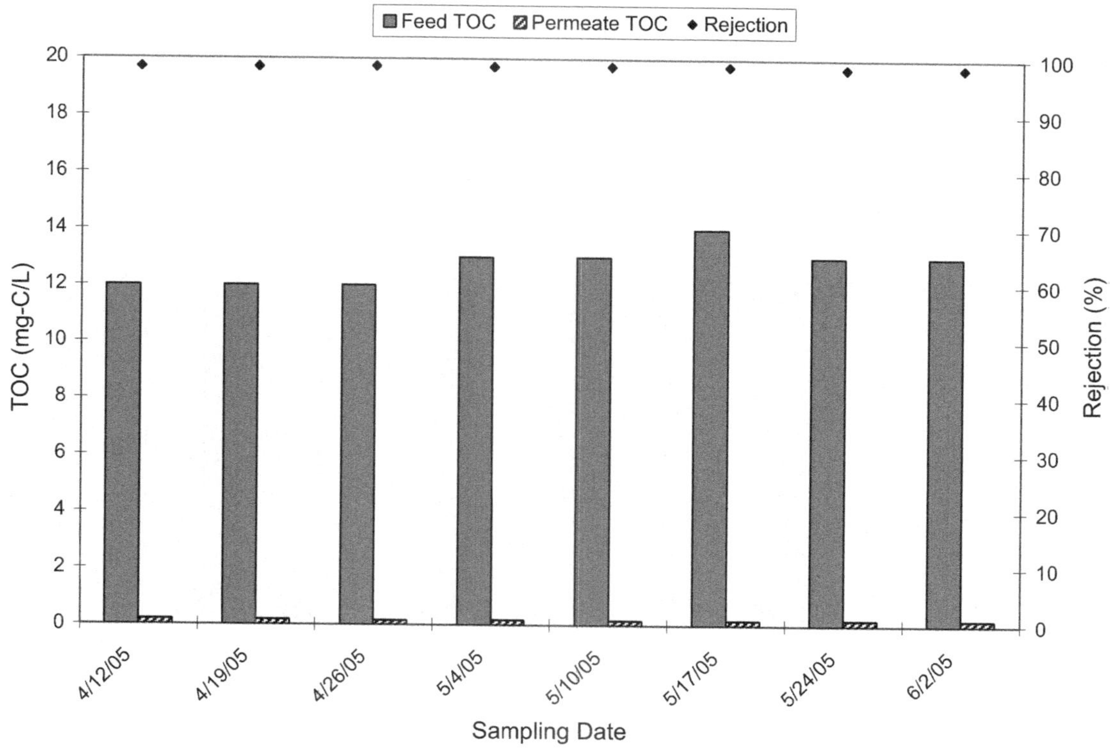

Figure D.3 Feed and permeate TOC concentrations and rejection values for the NF-90 membrane during pilot-scale testing at the WBWRP

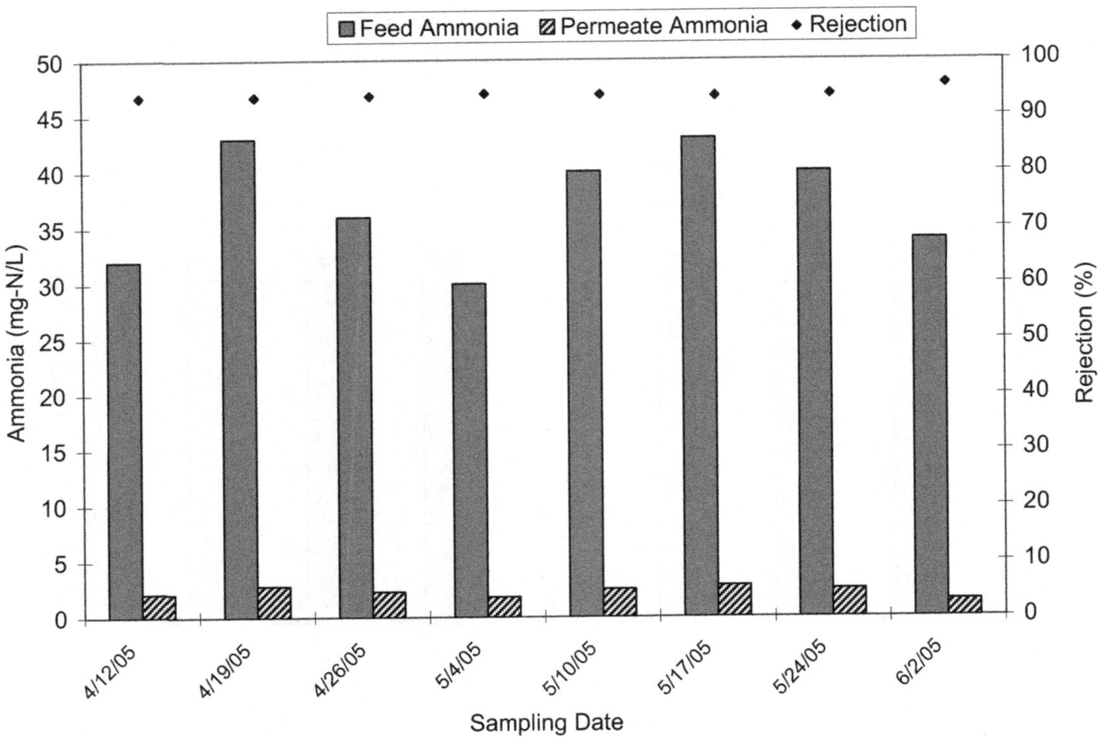

Figure D.4 Feed and permeate ammonium concentrations and rejection values for the NF-90 membrane during pilot-scale testing at the WBWRP

TOC Removal by the NF-4040 Membrane Treating Nitrified Feed Water

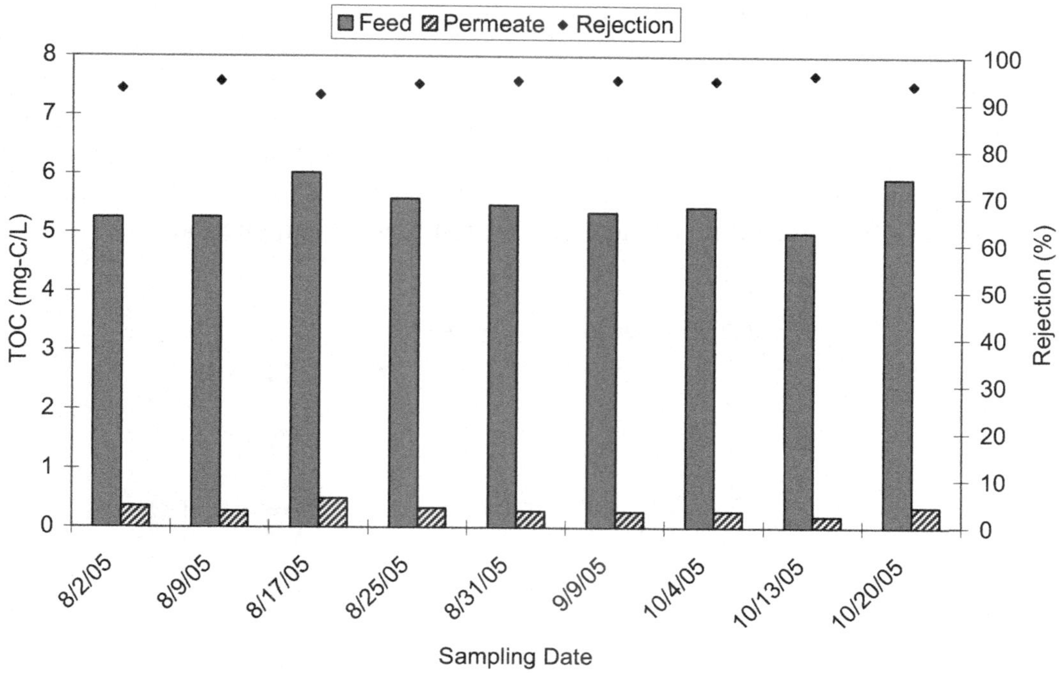

Figure D.5 Feed and permeate TOC concentrations and rejection values during NF-4040 membrane pilot-scale testing at the SWC

Water Quality Performance of the TMG10 Membrane Treating Nitrified/Denitrified Wastewater

Rejection of TOC and Nitrogen

During TMG10 membrane testing at pilot-scale, samples for total organic carbon (TOC) and nitrate were taken biweekly and analyzed by City of Scottsdale staff at the SWC. Due to pilot-skid downtime as a result of mechanical problems, samples were collected less frequently than originally planned. Summarized results for feed and combined permeate TOC samples taken weekly over the test period and corresponding TOC rejection values are presented in Figure D.6. Combined permeate TOC concentrations were below 0.3 mg-C/L resulting in TOC rejection values above 97 percent during the TMG10 membrane testing period.

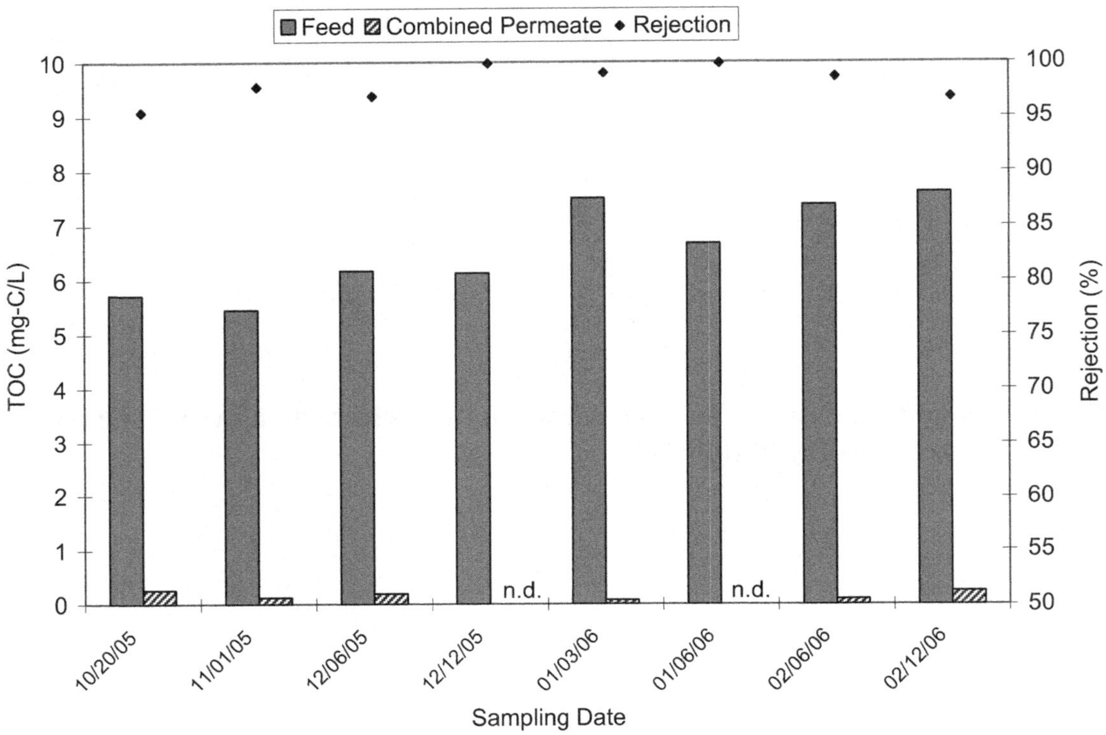

Figure D.6 Feed and permeate TOC concentrations and rejection values for the TMG10 membrane during pilot-scale testing at the SWC

Feed and permeate nitrate concentrations and subsequent rejection values for the TMG10 membrane testing period at the SWC are presented in Figure B.7. Permeate nitrate concentrations were below 1 mg-N/L and rejection was greater than 90 percent during all sampling campaigns. Feed water nitrate concentrations were variable throughout testing and ranged between 3 and 11 mg-N/L indicating that either the denitrifying efficiency of the wastewater process at the SWC varied during the testing period or raw wastewater quality changed over the testing period.

Table D.1 presents TMG10 membrane pilot-scale TOC and nitrate feed and permeate concentrations as average values during testing and average rejection values versus SWC full-scale TFC-HR TOC and nitrate data. During testing at the SWC, the TMG10 membrane exhibited slightly higher rejection for TOC and nitrate as compared to the TFC-HR membrane. Nitrate and TOC permeate concentrations were below 0.6 mg-N/L and 0.3 mg-C/L, respectively for both membranes during testing.

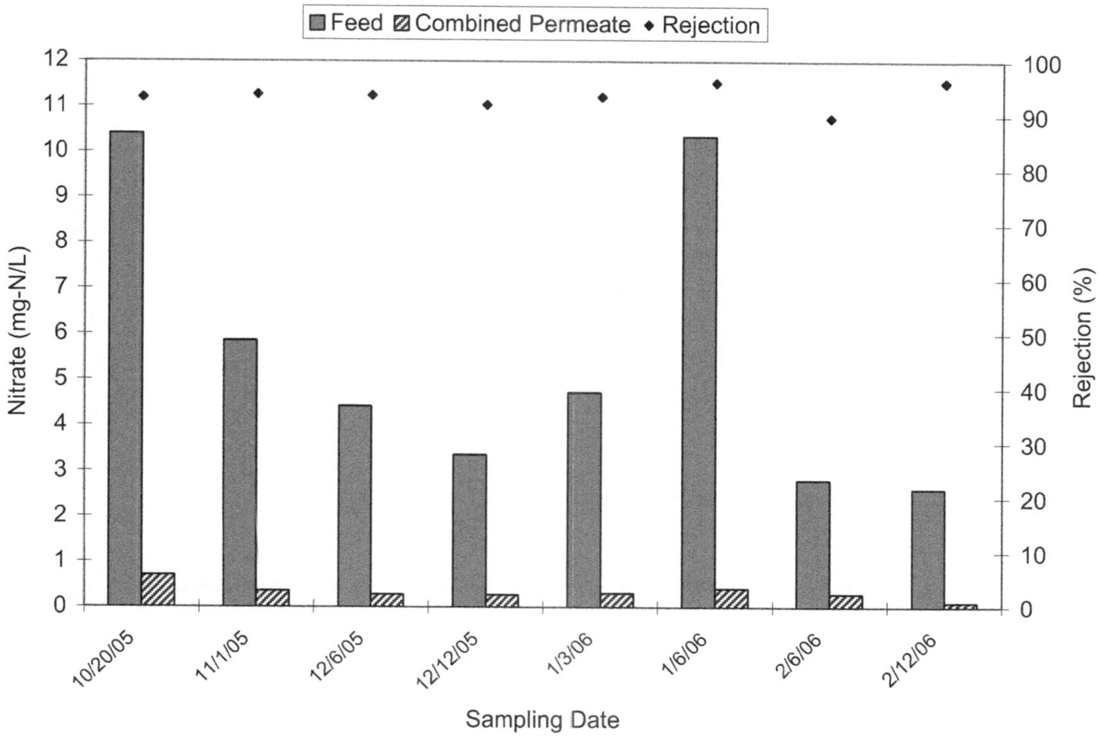

Figure B.7 Feed and permeate nitrate concentrations and rejection values for the TMG10 membrane during pilot-scale testing at the SWC

Table D.1
TOC, and nitrate concentrations and rejection values for TMG10 membrane on pilot-scale and the TFC-HR membrane on full-scale at the SWC

	Unit	Pilot-scale TMG10 Average 10/20/05-2/12/06	Full-scale TFC-HR Average 9/04 - 2/05
Recovery	Percent	82	85
TOC			
Feed TOC	mg-C/L	6.6	5.74
Permeate TOC	mg-C/L	0.12	0.23
TOC Rejection	percent	98.2	96.0
Nitrate			
Feed Nitrate	mg-N/L	5.6	7.13
Permeate Nitrate	mg-N/L	0.3	0.54
Nitrate Rejection	percent	94.6	92.4
Nitrate Rejection – F/B Average	percent	97.4	N/A

Rejection of Regulated Organics

Disinfection By-products. Summarized chloroform results from two sampling campaigns performed during pilot-scale TMG10 membrane testing and one sampling campaign performed on a full-scale treatment train employing the TFC-HR membrane are presented in Figure D.8. SWC feed water chloroform concentrations during TMG10 membrane testing were observed to be significantly lower than WBWRP feed water concentrations (Figure 4.4). Lower concentrations of chloroform in the SWC feed water as opposed to WBWRP feed water is theorized to be a result of lower chlorine dosing during the disinfection process, which may also explain low monochloramine concentrations as well. Since free ammonium was quantified in multiple feed water samples (data not shown) it appears that free chlorine is in shortage as compared to ammonium resulting in inadequate chloramine concentrations for the effective control of biofoulant formation, but may result in reduced DBP formation. The rejection of chloroform during the two sampling campaigns was between 60 and 80 percent for the TMG10 membrane and 60 percent for the TFC-HR membrane. The increased rejection of chloroform by the TMG10 at the SWC versus the WBWRP is hypothesized to be a result of membrane fouling creating an additional barrier that hinders partitioning of chloroform (Xu et al. 2006a).

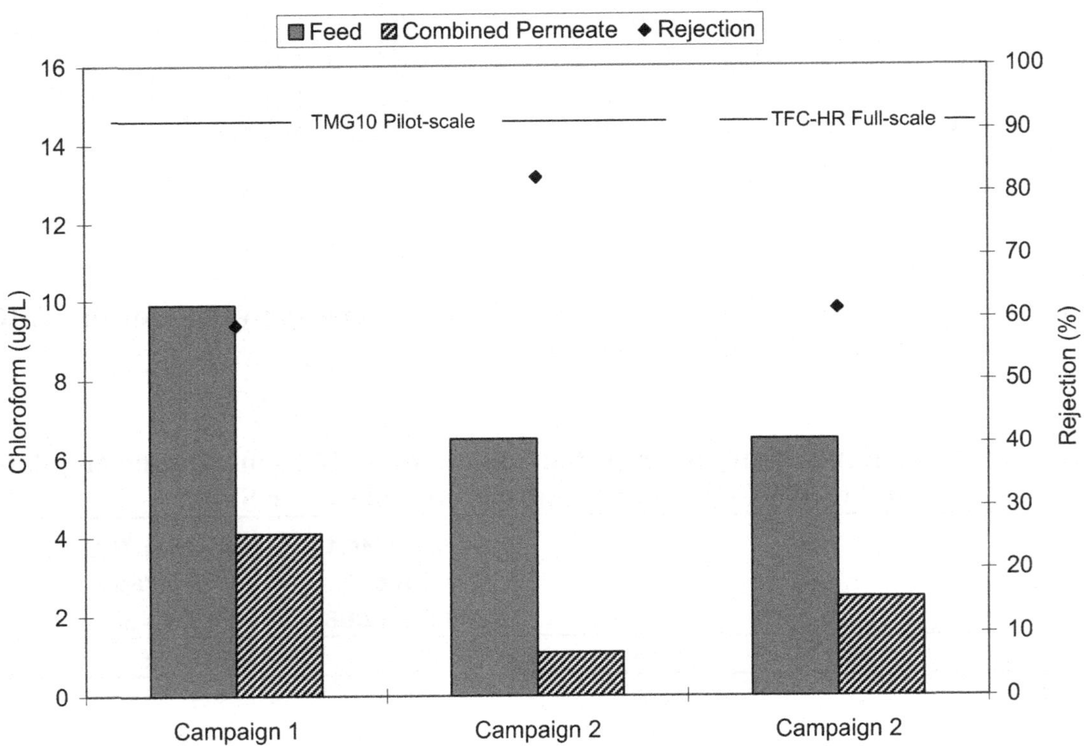

Figure D.8 Feed and permeate chloroform concentrations and rejection values for the TMG10 (pilot-scale) and TFC-HR (full-scale) membranes at the SWC

Summarized bromoform and TCAA feed and permeate concentrations for the TMG10 (pilot-scale) and TFC-HR (full-scale) membranes are presented in Table D.2 (Note: DCAA analysis failed for these samples). Bromoform was detected in one TMG10 feed sample and was not detected in permeate samples taken for the TMG10 and TFC-HR membranes. Similarly to chloroform results, TCAA concentrations were observed to be significantly lower in SWC feed water than WBWRP feed water and as a consequence resulted in no permeate detections for both the TMG10 and TFC-HR membranes.

Table D.2
Feed and permeate bromoform and trichloroacetic acid concentrations for the TMG10 (pilot-scale) and TFC-HR (full-scale) membranes at the SWC

	Bromoform		Dichloroacetic acid		Trichloroacetic acid	
	Feed µg/L	Permeate µg/L	Feed µg/L	Permeate µg/L	Feed µg/L	Permeate µg/L
TMG10 Pilot-scale						
Campaign 1	2.6	n.d.	N/A	N/A	6.4	n.d.
Campaign 2	n.d.	n.d.	N/A	N/A	11.9	n.d.
TFC-HR Full-scale						
Campaign 2	n.d.	n.d.	N/A	N/A	11.9	n.d.

N/A – not available
n.d. – not detected

Rejection of Unregulated Organics

During the period of pilot-scale testing at the SWC, which includes the period of August 2, 2005 until February 12, 2006, samples were collected for the quantification of unregulated trace organic contaminants listed in Table 2.1. SWC nitrified/denitrified feed water was analyzed monthly by the GC/MS method described in Chapter 2 (Tier 2 target compounds). Sampling campaigns were conducted for the analysis of the trace organic contaminants in SWC feed water as well as to determine the efficiency of the membranes employed at the SWC (at pilot- and full-scale) in removing organic contaminants. Approximately 30 samples were collected during SWC testing and the results were used to construct the box and whisker plot presented in Figure D.9.

Compared to WBWRP non-nitrified feed water (Figure 4.6) the range of the quantified pharmaceutically active compounds was observed to be more narrow in the SWC nitrified/denitrified feed water (Figure D.9). The median concentration of quantified pharmaceutically active compounds was below 500 ng/L with salicylic acid, gemfibrozil, naproxen, and carbamazepine most frequently detected in SWC feed water samples. Ibuprofen, ketoprofen, and primidone were quantified less frequently and concentrations were closer to the method limit of quantification. The chlorinated flame retardants TCEP, TCPP and TDCPP were quantified at similar concentrations to those observed in WBWRP feed water samples.

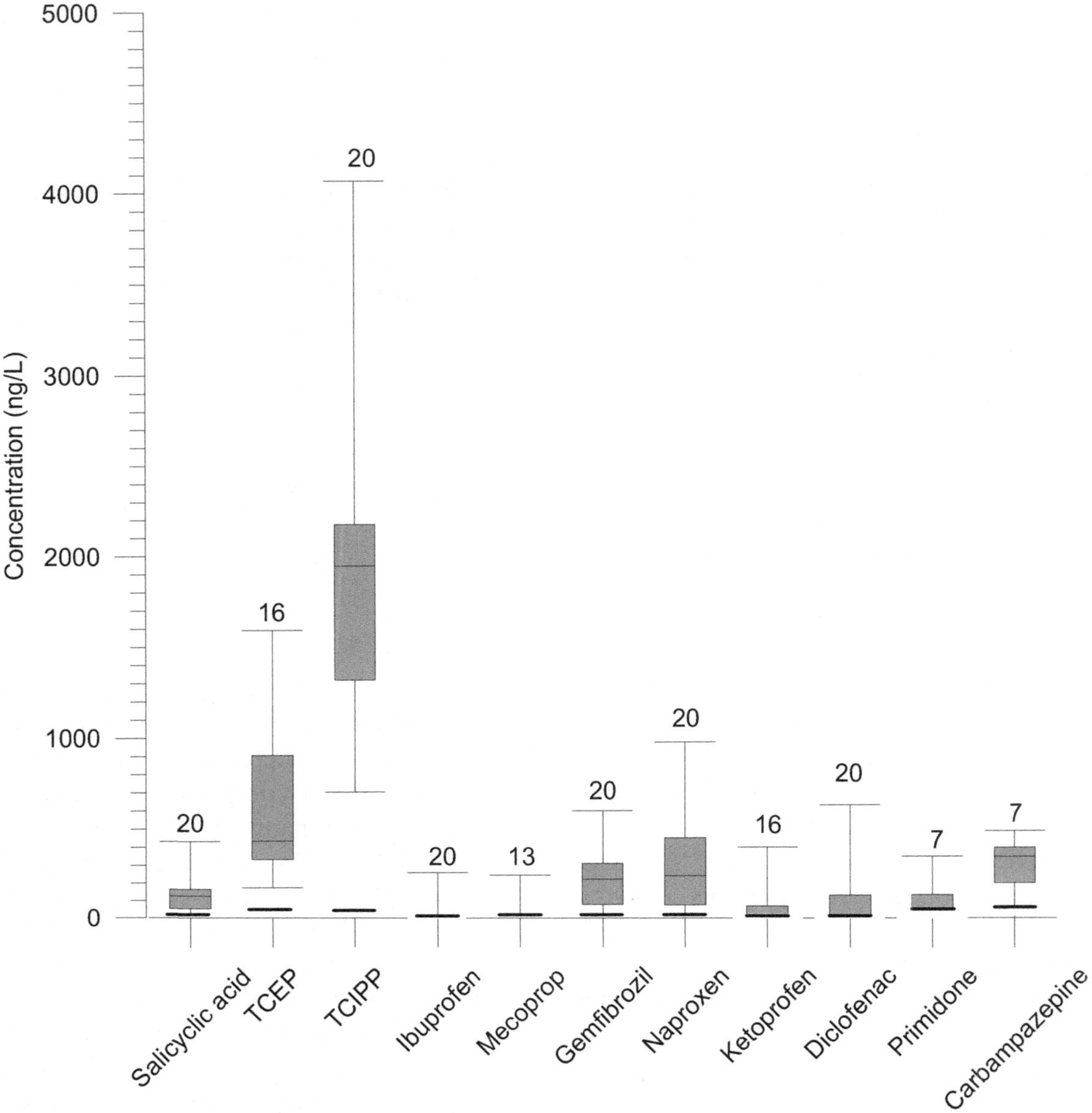

Figure D.9 Box and whisker plot of quantified trace organic contaminants present in SWC feed water between 8/2/2005 and 2/12/2006

Summarized results of trace organic contaminant analysis of pilot-scale TMG10 and full-scale TFC-HR membrane permeate samples are presented in Table D.3. No trace organic contaminants were quantified in any of the permeate samples collected from a full-scale TFC-HR membrane treatment train. Only naproxen was detected in one permeate sample from the TMG10 membrane, while all other solutes identified in SWC feed water were not detected in any permeate samples taken during TMG10 membrane testing.

Table D.3
Summarized TMG10 (pilot-scale) and TFC-HR (full-scale) membrane trace organic contaminant results during testing at SWC

	TMG10 Pilot-Scale	TFC-HR Full-scale
Samples (n)	8	3
Compound	ng/L	ng/L
Salicyclic acid	n.d.	n.d.
TCEP	n.d.	n.d.
TCIPP	n.d.	n.d.
Ibuprofen	n.d.	n.d.
Mecoprop	n.d.	n.d.
Gemfibrozil	n.d.	n.d.
Naproxen	n.d. (7) - Detect (1)	n.d.
Ketoprofen	n.d.	n.d.
Diclofenac	n.d.	n.d.
Primidone	n.d.	n.d.
Carbamazepine	n.d.	n.d.

n.d. – not detected

APPENDIX E
PROCESS AND INSTRUMENTATION DIAGRAM OF FULL-SCALE TEST VESSEL

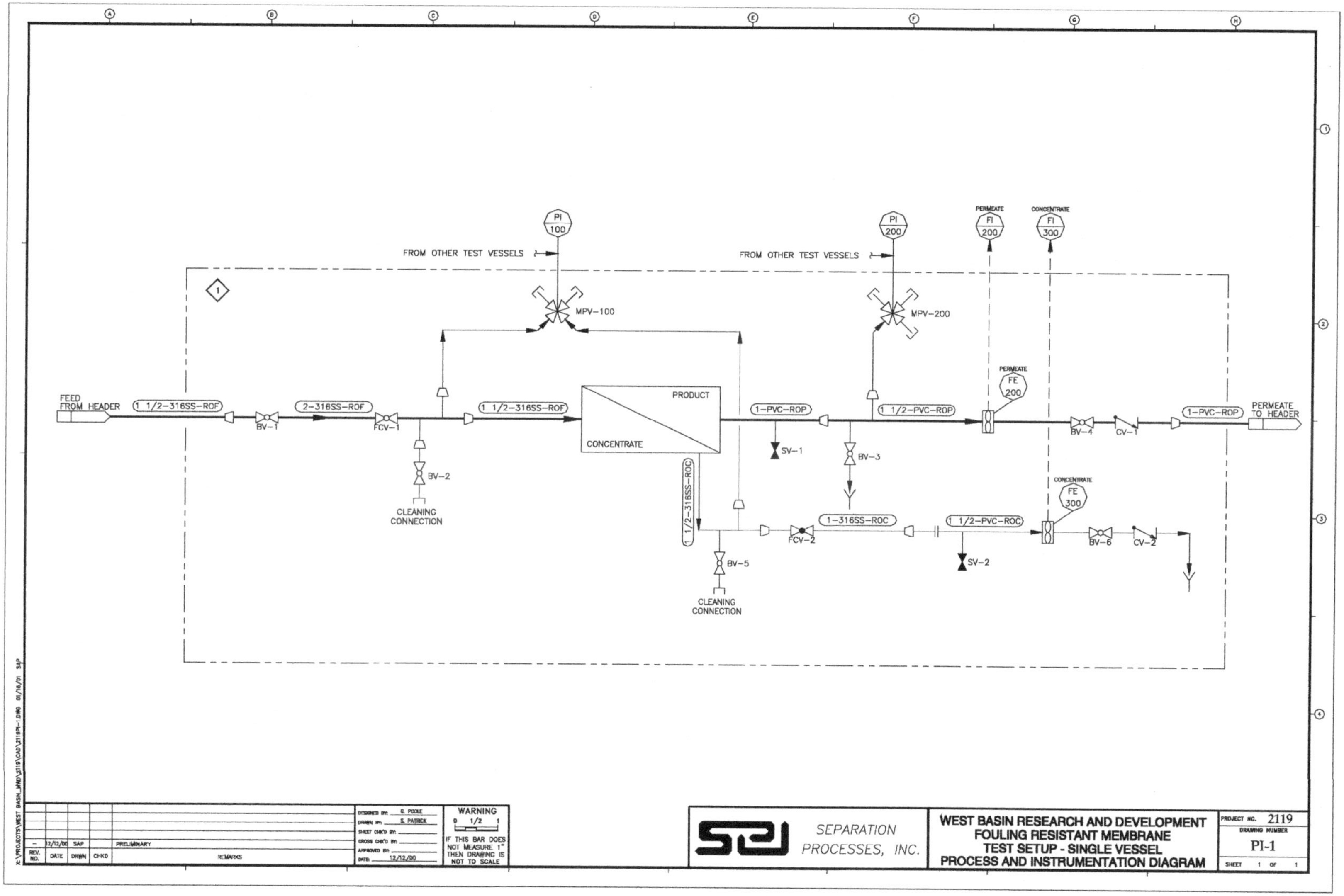

Figure E.1 Process and Instrumentation Diagram for a representative test vessel

APPENDIX F
MEMBRANE AUTOPSY RESULTS

NF-90 Membrane Autopsy

The NF-90 membrane was tested in the pilot-scale membrane skid over a course of 1,300 hours from April 7, 2005 to June 3, 2005, at the WBWRP, for treatment of microfiltered non-nitrified secondary effluent. Upon completion of the pilot-scale investigation, the lead element of the first stage and the tail-end element of the last stage were removed from the skid and soaked in 1 percent sodium metabisulfite solution for preservation. After shipping to the CSM laboratory, the elements were prepared for a membrane autopsy.

Membrane Surface Morphology and Elemental Composition

The ESEM micrographs of NF-90 virgin membrane specimen are shown in Figure F.1. The NF-90 virgin membrane contained a network of valleys and ridges, which was more apparent at magnification of x20,000 (Figure F.1(b)). Similar to the TMG10 virgin membrane, the EDS scan revealed the oxygen, carbon, and sulfur, which are representing the elemental composition in the virgin NF-90 membrane (Figure F.2).

After 1,300 hours of operation, no obvious foulants were observed by ESEM analysis for both lead and tail-end NF-90 elements (Figure F.3). Similar to the TMG10 membrane operated at WBWRP, the EDS analysis exhibited very small amounts of sodium and chlorine present in the fouled membrane (Figure F.4). Sodium was likely introduced by the preservation solution sodium metabisulfite, which was not completely removed by DI water rinsing prior to the characterization. Chlorine was attributed to the addition of chloramines in the feed water to control biogrowth. No other additional elements were detected by the EDS analysis indicating that the microfiltered secondary effluent feeding the NF-90 was not supersaturated in any constituents.

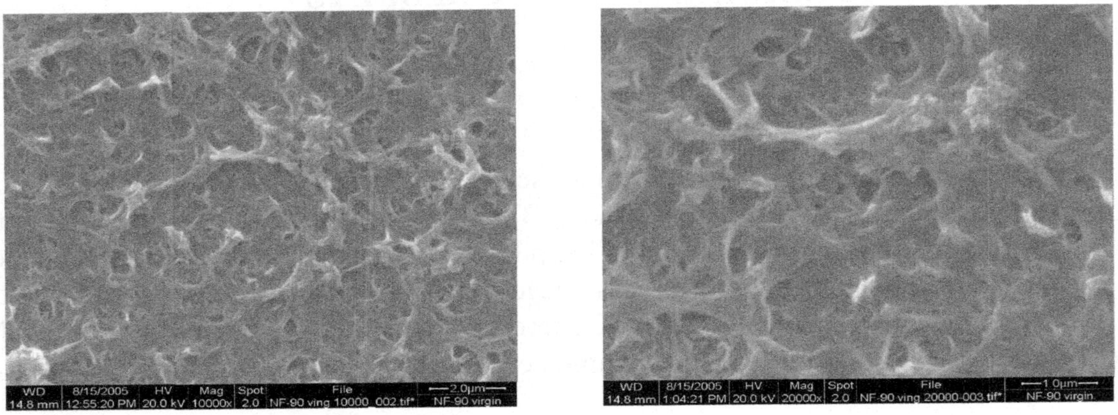

(a) (b)

Figure F.1 ESEM micrographs of NF-90 virgin membrane (a) x10,000 image; (b) x20,000 image

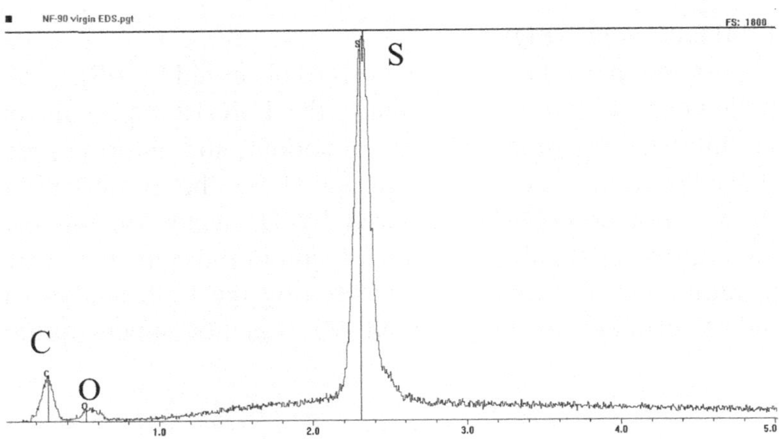

Figure F.2 EDS spectrum of NF-90 virgin membrane

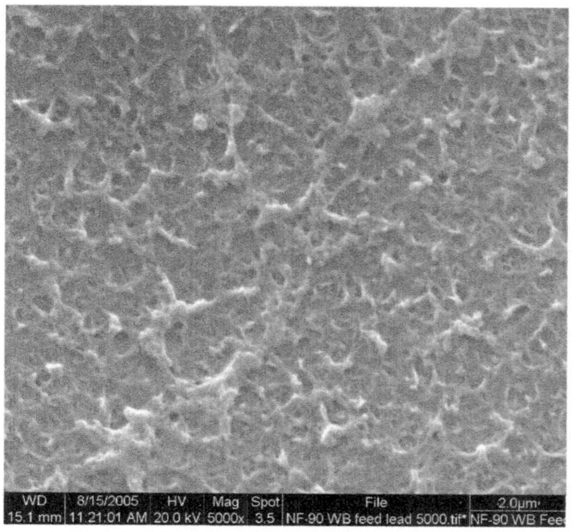

(a)

(b)

(c)

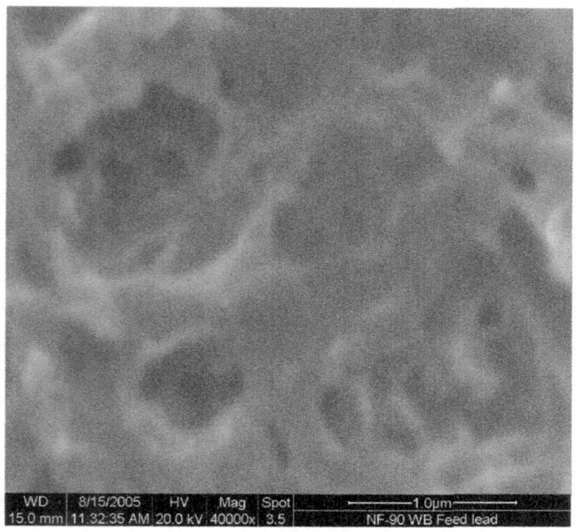

(d)

Figure F.3 ESEM micrographs of the feed-end lead-element NF-90 membrane specimen at different magnifications (a) x5,000; (b) x10,000; (c) x20,000; (d) x40,000

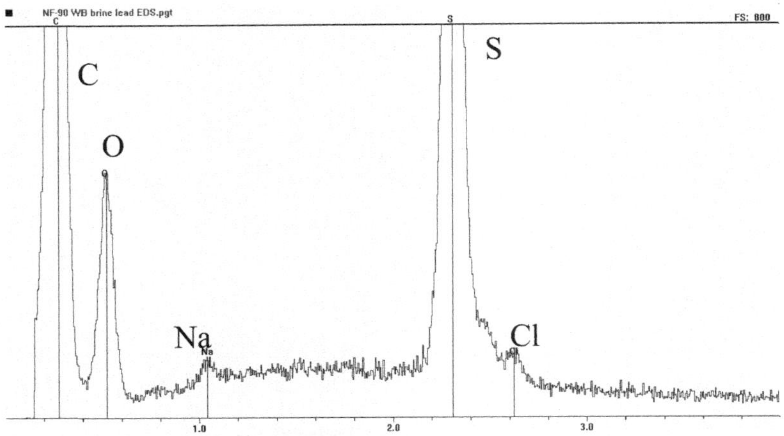

Figure F.4 EDS spectrum of the brine-end of lead-element NF-90 membrane specimen

ATR-FTIR Measurement

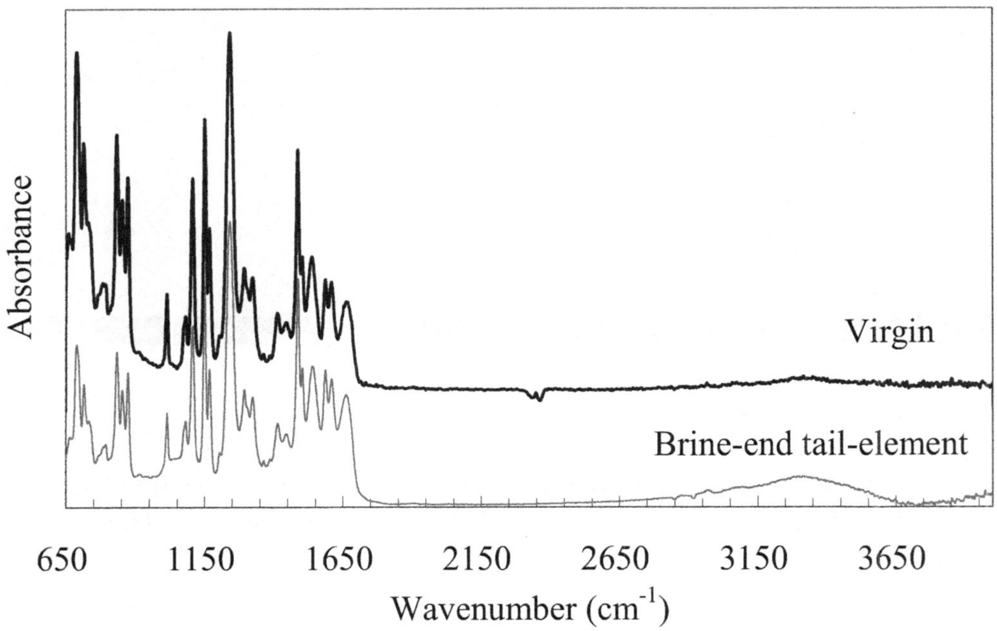

Figure F.5 ATR-FTIR spectra of virgin and fouled NF-90 membrane specimens
(spectra are shifted for better visualization)

NF-4040 Membrane Autopsy

The NF4040 membrane was tested in the pilot-scale membrane skid over a course of 1,600 hours from Aug 2, 2005 to Oct. 21, 2005, at the SWC. Upon completion of the pilot-scale investigation, a lead element and a tail-end element were removed from the skid and soaked in 1 percent sodium metabisulfite solution for preservation. After shipping to the CSM laboratory, the membrane elements were unrolled, cut into small pieces, and rinsed carefully with DI water (type II) prior to autopsy analysis.

Membrane Surface Morphology and Elemental Composition

The ESEM micrographs of NF-4040 virgin membrane specimen exhibited a very smooth active layer (Figure F.6). Similar to the TMG10 and NF-90 virgin membranes, the EDS scan detected oxygen, carbon, and sulfur atoms are representing the elemental composition in the virgin NF-4040 membrane (see spectra as shown in Figure F.2).

After 1,600 hours of operation, the ESEM micrographs of the lead NF-4040 element showed microbial colonies developed on the feed-end membrane surface (Figure F.7). Some white precipitate grains varying in size between 0.1-10 μm were observed on the membrane surface. By performing EDS spot scanning on the grains, the major constituents of the grains were found as barium (Figure F.8(a)). The EDS screen scanning did not detect other scalants in the fouling layer (Figure F.8(b)). The brine-end of the lead element displayed similar fouling layer as the feed-end with combination of microbial colonies and precipitate grains (Figure F.9). The EDS spot scanning showed high amount of aluminum and calcium in the inorganic precipitates with trace amount of barium (Figure F.10).

The feed-end and brine-end membrane specimens of the tail-end NF-4040 element exhibited a similar thick fouling layer consisting of a combination of a biomass layer and inorganic scalants (Figures F.11 and F.12). The EDS scanning detected large amount of calcium, aluminum, and phosphorus elements as inorganic scalants (Figures F.13 and F.14).

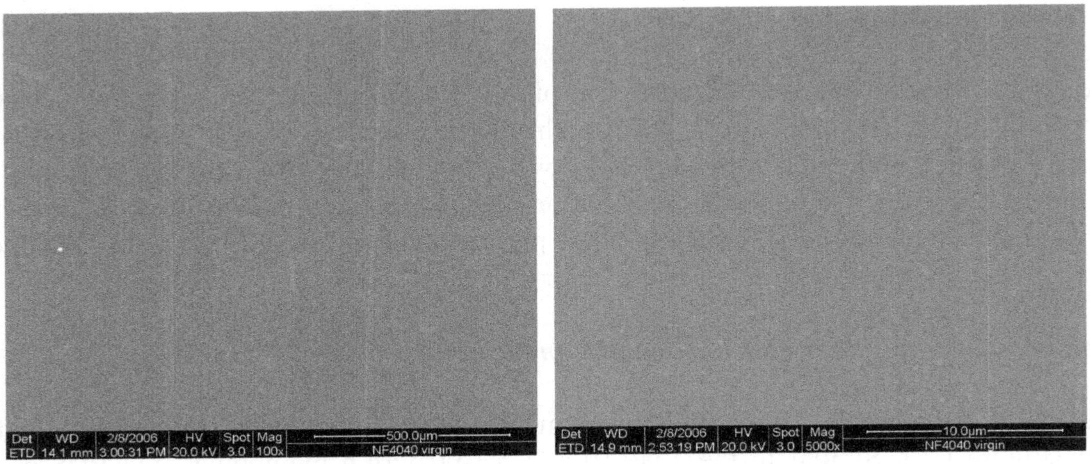

(a) (b)

Figure F.6 ESEM micrographs of virgin NF-4040 membrane specimen at different magnifications (a) x100; (b) x5,000

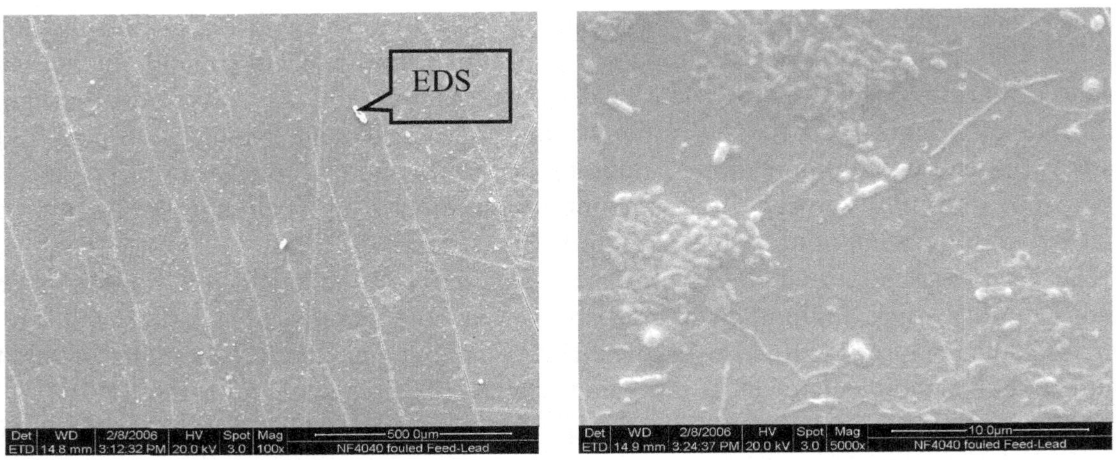

(a) (b)

Figure F.7 ESEM micrographs of the feed-end lead-element NF-4040 membrane specimen at different magnifications (a) x100 inorganic scaling formed with biofouling; (b) x5,000 microbial colonies

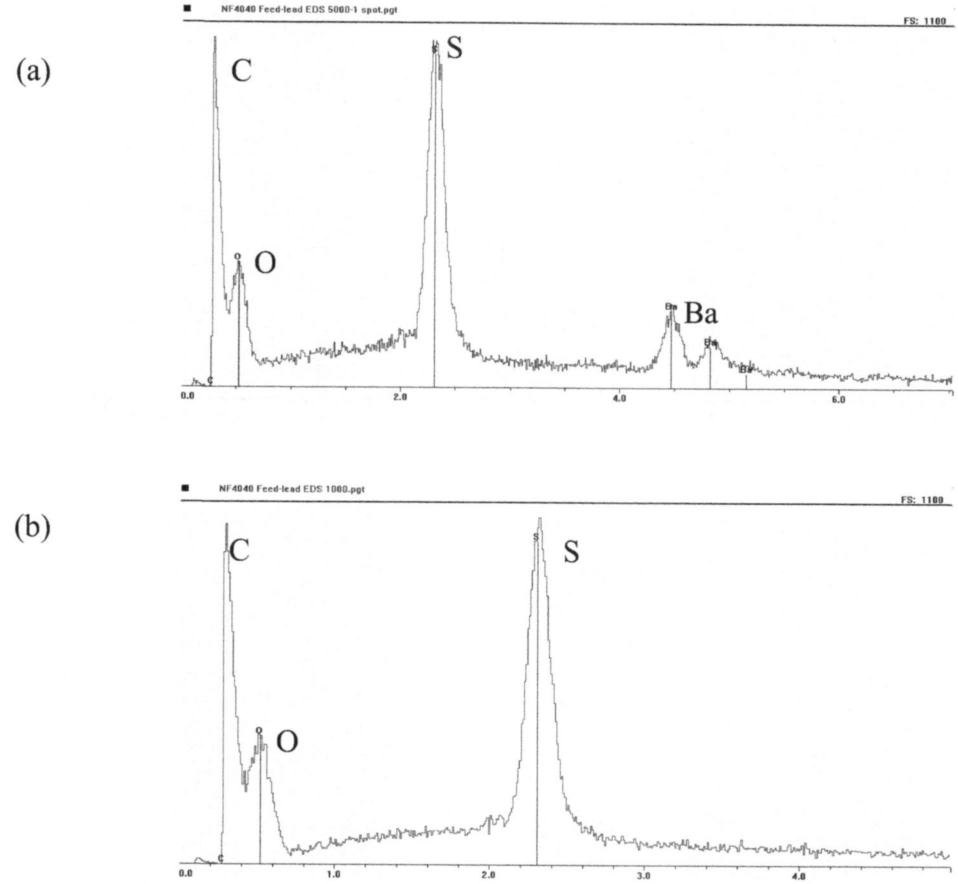

Figure F.8 EDS spectra of the feed-end lead-element NF-4040 membrane specimen (a) spot scanning on white grains; (b) screen scanning

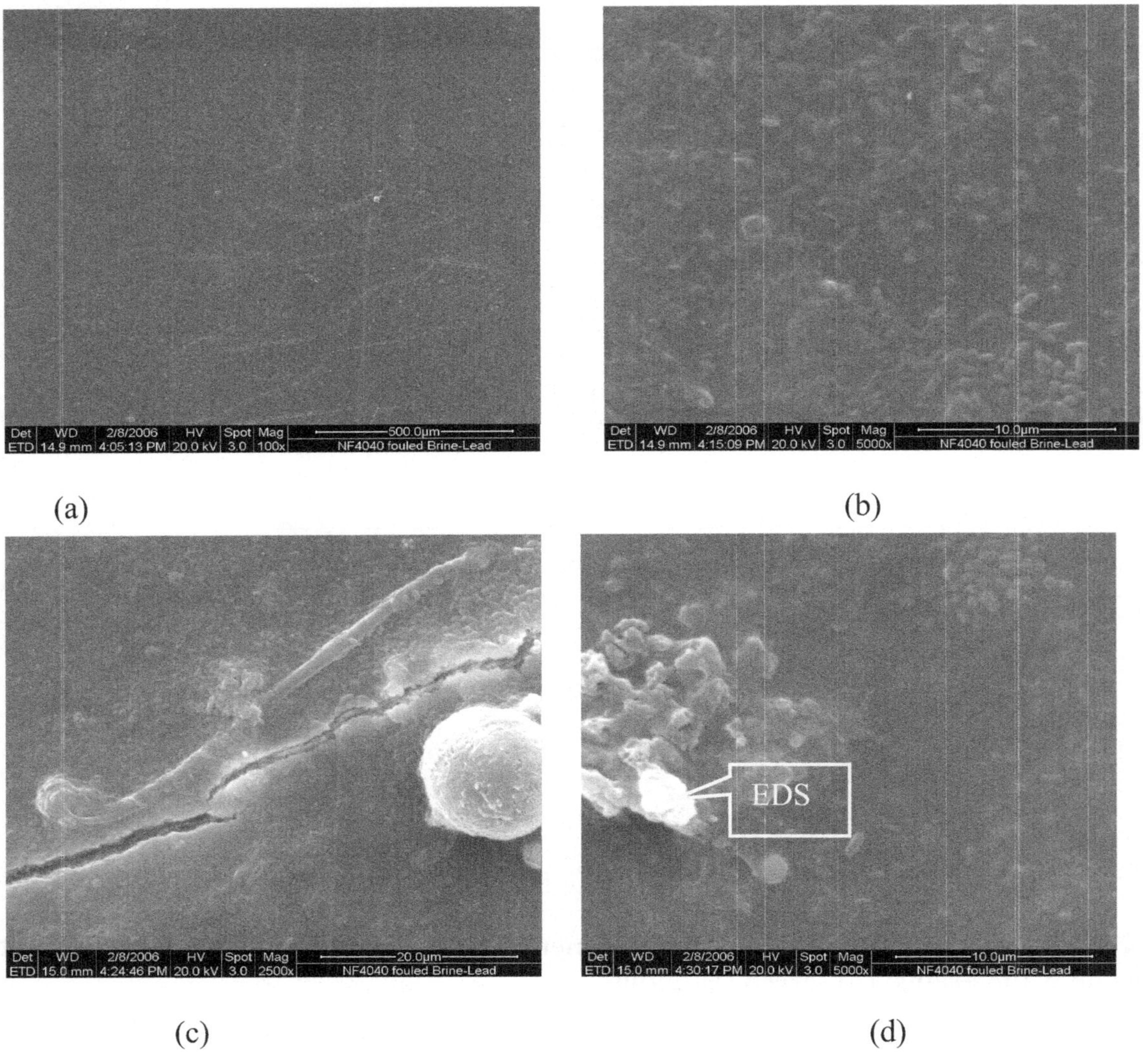

Figure F.9 ESEM micrographs of the brine-end lead-element NF-4040 membrane specimen at different magnifications (a) x100 inorganic scaling formed with biofouling; (b) x5,000 microbial colonies; (c) x5,000 white crystal of inorganic fouling; (d) x5,000 inorganic scaling and microbial colonies

(a)

(b)

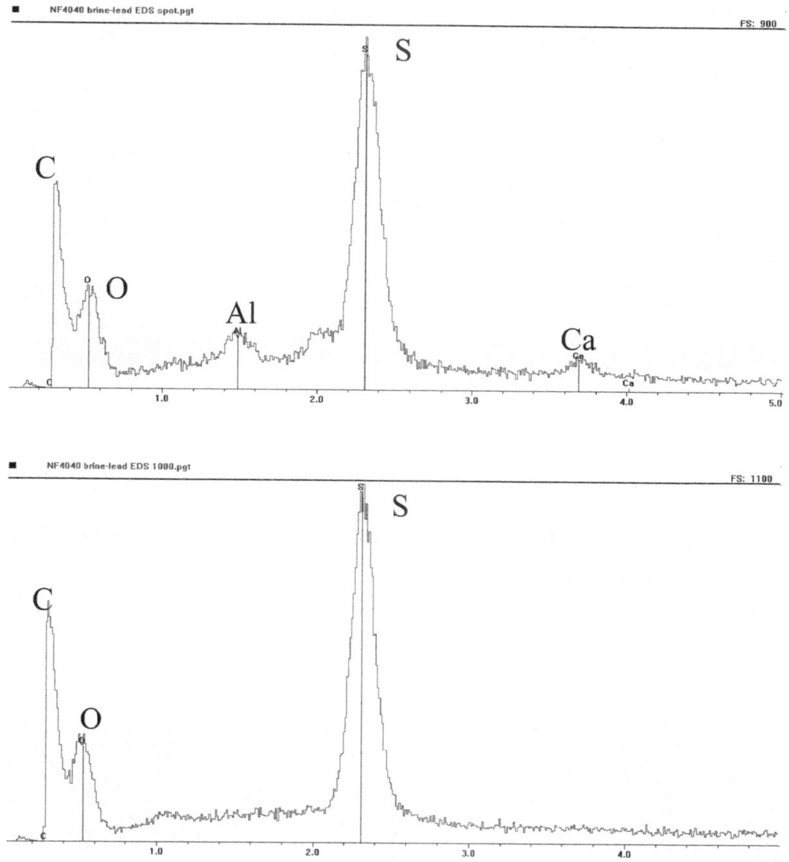

Figure F.10 EDS spectra of the brine-end lead-element NF-4040 membrane specimen (a) spot scanning on white grains; (b) screen scanning

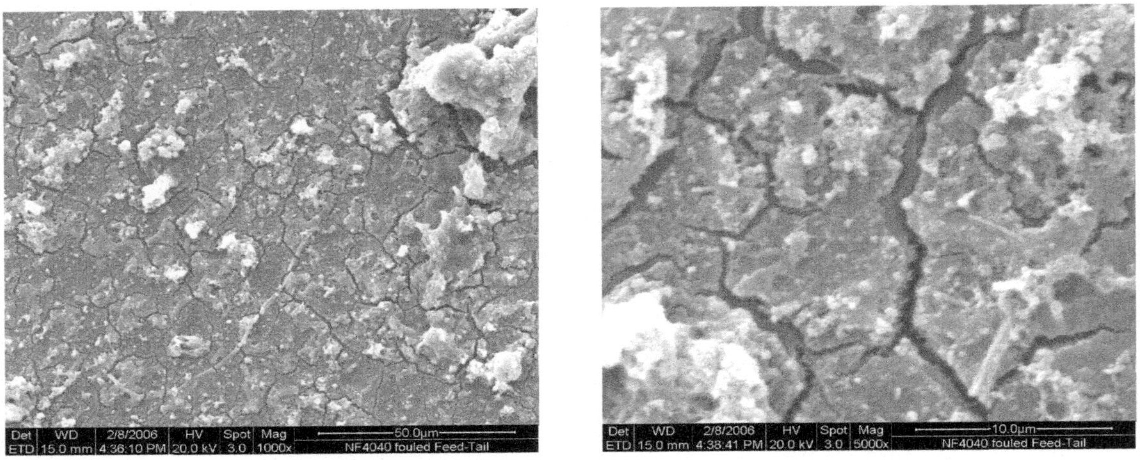

(a) (b)

Figure F.11 ESEM micrographs of the feed-end tail-element NF-4040 membrane specimen at different magnifications (a) x1000; (b) x5,000

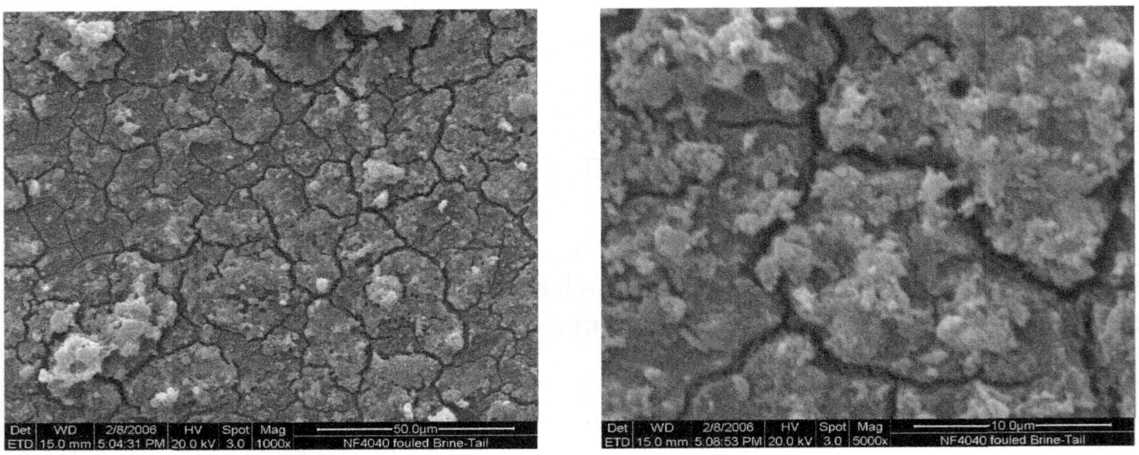

(a) (b)

Figure F.12 ESEM micrographs of the brine-end tail-element NF-4040 membrane specimen at different magnifications (a) x1000; (b) x5,000

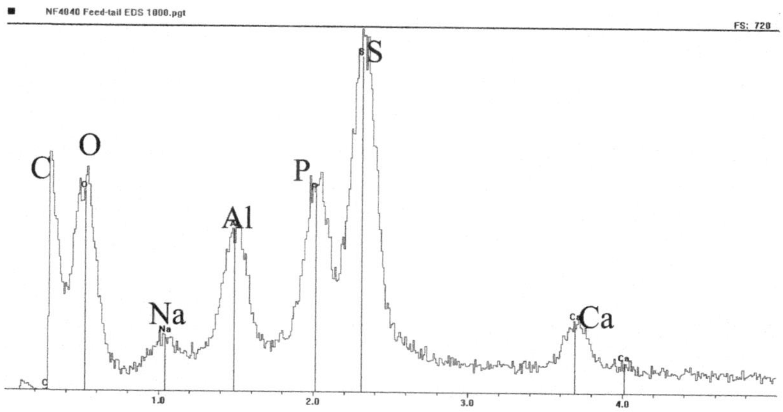

Figure F.13 EDS spectrum of the feed-end tail-element NF-4040 membrane specimen

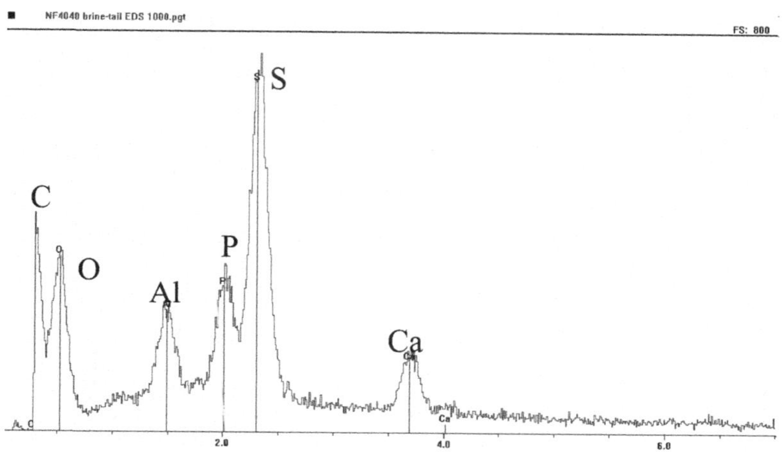

Figure F.14 EDS spectrum of the brine-end tail-element NF-4040 membrane specimen

ATR-FTIR Measurement

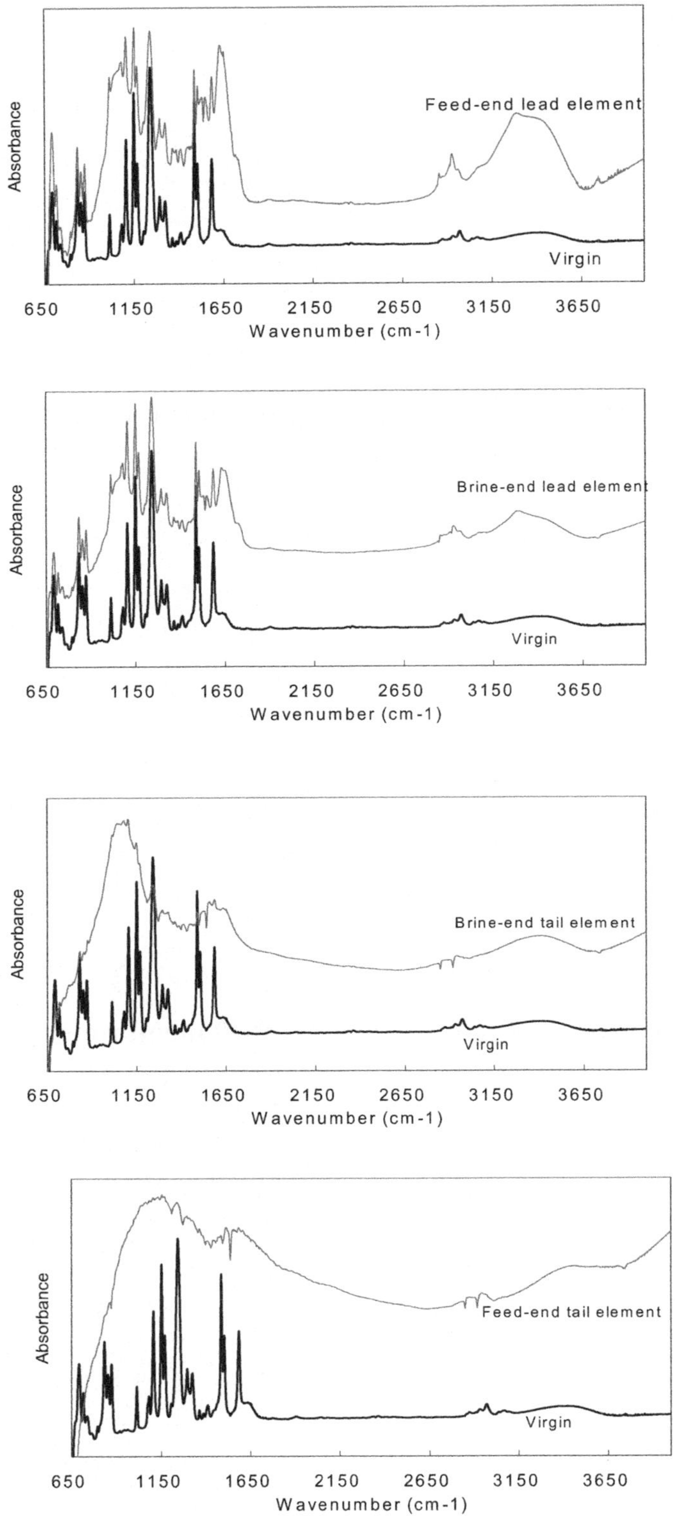

Figure F.15 ATR-FTIR spectra of virgin and fouled NF-4040 membrane specimen

APPENDIX G
SPIEGLER-KEDEM SOLUTE TRANSPORT MODELING

Spiegler-Kedem Model Development

This section will focus on the current strategies for applying the Spiegler-Kedem model to organic compounds. The fundamental component of the model is the separation of the diffusive and convective transport contributions for solute flux as presented in equation G.1 (Cussler 1997, Yoon et al. 2002, Jain et al. 2004).

$$J_S = \text{diffusion} + \text{convection} = \omega\Delta\Pi + (1 - \sigma)\, C_{avg} J_V \qquad \text{(G.1)}$$

where J_S = solute flux (mol/m^2-s)
σ = reflection coefficient (unitless) – modeled with log-normal distribution (equation G.12)
C_{avg} = average interfacial solute concentration between feed and permeate side (mol/m^3)

The molecular transport coefficient (ω, mol/m^2-s-Pa) is calculated with equation G.2.

$$\omega = \frac{D_P}{\delta RT} \qquad \text{(G.2)}$$

The membrane diffusion coefficient (D_P, m^2/s) is calculated with equation G.3.

$$D_P = D_W H_M \qquad \text{(G.3)}$$

where D_W = diffusion coefficient in water (m^2/s)
H_M = membrane partition coefficient (unitless)
δ = thickness of the membrane separating layer (m)
R = universal gas constant (m^3-Pa/mol-^{o}K);
T = absolute temperature (^{o}K)

The solvent flux of a particular membrane can be calculated using equation G.4.

$$J_V = \text{solvent flux (m/s)} = L_P(\Delta P - \sigma\Delta\Pi) \qquad \text{(G.4)}$$

where L_P = pure water permeability (m^3/m^2-s-Pa)
ΔP = transmembrane pressure (Pa)
σ = reflection coefficient (unitless)
$\Delta\Pi$ = osmotic pressure difference (Pa) = MRT

The osmotic pressure difference across the membrane can be calculated using the van't Hoff law which is given as equation G.5.

$$\Delta\Pi = RT(M_F - M_P) \qquad \text{(G.5)}$$

where M = molarity of feed and permeate (M_F and M_P) solution (mol/m^3)
R = universal gas constant (m^3-Pa/mol-°K)
T = absolute temperature (°K)

The Spiegler-Kedem equation is often presented in the differential form of equation G.6.

$$J_S = -P\Delta x \frac{dC}{dx} + (1-\sigma)J_V C \qquad \text{(G.6)}$$

where J_S = solute flux (mol/m^2-s)
P = solute permeability (m/s)
Δx = separating layer thickness (m)
C = concentration (mol/m^3)
σ = reflection coefficient (unitless)
J_V = solvent flux (m^3/m^2-s)

When equation G.6 is integrated over the membrane thickness and combined with the solvent flux equation (equation G.4), the rejection of a solute becomes a function of solute permeability (P), reflection coefficient (σ), and solvent flux (J_v) shown in equation G.7 (Van der Bruggen et al., 2002).

$$R = \frac{\sigma(1-F)}{1-\sigma F} \qquad \text{(G.7)}$$

where R = rejection (unitless)
σ = reflection coefficient (unitless)

Equation G.8 is used to calculate F

$$F = \exp\left(-\frac{1-\sigma}{P} J_V\right) \qquad \text{(G.8)}$$

The solute permeability (P, m/s) can be calculated using the solute diffusion parameter and the membrane thickness, but is commonly calculated as the diffusion parameter (ρ, m^2/s) divided by solute radius. Both forms of calculating P are given as equation G.9.

$$P = \frac{D_S}{\Delta x} \quad = \text{“}\rho / r\text{”} \qquad \text{(G.9)}$$

The solute diffusion coefficient (D_S, m^2/s) is commonly calculated using the Stokes-Einstein equation given as equation G.10.

$$D_S = \frac{kT}{6\pi\eta r} \quad = \text{“a / r”} \qquad \text{(G.10)}$$

where k = Boltzmann's constant (1.38 x 10^{-23} J/K
T = absolute temperature (K)
η = viscosity (Pa s)
r = solute size (m)
ρ = a / Δx (m^2/s)

The diffusion contribution is determined experimentally by investigating rejection at varying pressures. Increases in driving pressure increases the solvent flux (equation G.4), and increases the convective transport (equation G.6, Wang et al. 2005, Van der Bruggen et al. 2002). A range of pressures should be used ranging from slightly higher than the osmotic pressure where solvent flux begins to develop across the membrane to slightly below the maximum operating pressure for a membrane. By plotting rejection versus pressure and fitting equation G.7 and G.8 with the solute permeability (P) and reflection coefficient (σ), these model parameters can be determined for a particular membrane and solute. The solute permeability (P) is dependent upon the membrane thickness and the diffusion coefficient (equation G.9). The diffusion coefficient is related to solute size using the Stokes-Einstein relationship (equation G.10). If temperature and viscosity are constant, the diffusion coefficient is only a function of solute size. This relationship can be used in equation G.9 so that the solute permeability depends upon a membrane permeability constant (ρ) and the solute size. The membrane permeability constant includes the membrane thickness and the constants from the Stokes-Einstein relationship that can be used at constant temperature and viscosity, neglecting changes in Δx. Therefore, once ρ has been determined for the membrane, the solute permeability can be predicted from the solute size (Van der Bruggen et al. 2002).

Log Normal model for calculating the reflection coefficient

Once the membrane has been characterized with respect to the membrane permeability constant (ρ), the next step is to calculate the reflection coefficient (σ) based on solute size in order to predict the convective contribution of transport. One method assumes that each pore will reject or pass the solute depending on the size of the solute in relation to the pore size (e.g., each pore has a reflection coefficient of 1 or 0). When all the pores in the membrane are combined, the reflection coefficient is the percent of pores small enough to reject the solute (Kargol 2001). The pore size distribution has been described by the log-normal distribution (Van der Bruggen et al. 2002). The log-normal probability density function is given below by P(x) in equation G.11 and the log-normal cumulative distribution is given as D(x) in equation G.12.

$$P(x) = \frac{1}{S_P\sqrt{2\pi}}\frac{1}{r}\exp\left[-\frac{\left(\ln(r) - \ln(\bar{r})\right)^2}{2S_P^{\,2}}\right] \qquad \text{(G.11)}$$

$$D(x) = \frac{1}{2}\left[1 + erf\left(\frac{\ln(r) - \ln(\bar{r})}{S_P\sqrt{2}}\right)\right] \quad \text{(G.12)}$$

where S_P = log-normal standard deviation of pore sizes (m)
r = solute size (m)
$\bar{r}$ = mean membrane pore size (m)
erf = error function

The reflection coefficient (σ) must first be determined experimentally for some compounds in order to calibrate the log-normal model. Generally, 6-8 compounds are sufficient to calibrate the model (Van der Bruggen et al. 2002). This is accomplished by assuming the reflection coefficient equals the rejection at infinite pressure. Using the plots of rejection vs. pressure described above for the membrane permeability constant, the reflection coefficient is equal to the maximum rejection. The reflection coefficient is plotted vs. solute size. Then, the mean membrane pore size and the log-normal standard deviation of pore sizes are used to fit the cumulative distribution curve to the data.

The model involves characterizing the membrane with the pure water permeability (L_P), the membrane permeability constant (ρ), the mean membrane pore size ($\bar{r}$), and the log-normal standard deviation of pore sizes (S_P). Once these are known, the solvent flux can be calculated (equation G.4), and the solute size can be used to calculate the solute permeability (P) (equation G.9) and reflection coefficient (σ) (equation G.12). All of these parameters can be used in the system of equations (G.7 and G.8) to describe and ultimately predict rejection.

Characterization of Solutes and Membranes

In order to apply the Spiegler-Kedem model to make predictions of the rejection of organic compounds, solutes and membranes need characterization. The important parameters include: a descriptor of molecular size, solute/membrane permeability for a given membrane and solute, a mean effective pore radius for a given membrane, the pore size distribution of a membrane, and the solvent permeability of a given membrane.

Solute Geometry Determination. Chen et al. (2004) used Hyperchem software for solute modeling to get size parameters that could be used for rejection predictions. This study adopted a similar method where molecular width, length, and depth were determined using the software package Hyperchem G.5 (Gainesville FL). The compounds were drawn in 2-D using the structures obtained from Chemfinder.com and the program performed an energy minimization step using the MM+ method. The molecular length, width, and depth were found by fitting a box around the unhydrated molecule and selecting the longest dimension as the length, the shortest as the depth, and the middle dimension as the width. Additional QSAR data were available within the software including the molecular volume and surface area. These parameters were also considered as viable size indicators for further consideration during this study.

Membrane Permeability Constant (ρ). The membrane permeability constant (ρ) was determined by fitting equations G.7 and G.8 with the permeability coefficient (P) to rejection versus pressure data using the least squares method. Using equation G.9 and the solute size, the membrane permeability constant could be calculated.

Membrane Effective Pore Size and Molecular Weight Cut-Off. The reflection coefficients for all the compounds were plotted versus molecular size parameters including molecular weight, width, length, volume, surface area, and effective diameter. The data was fitted using equation G.12 by the least squares method. The effective pore size ($\bar{r}$) was initially chosen as the molecular size resulting in 50 percent rejection. Once the initial value was selected, the solver function in Microsoft Excel was used to fit the effective pore size and the pore size distribution parameters using the least squares method and the solver function.

Membrane Characterization

Several model parameters are required for the Spiegler-Kedem model. A membrane characterization procedure was performed in order to determine membrane specific model parameters including the effective pore size of the membrane, the membrane permeability constant, the pure water permeability, and the standard deviation between pores. The membrane characterization procedure requires rejection data for several compounds performed at several different feed pressures.

Selecting a Molecular Size Parameter

Size exclusion is generally considered the most important rejection mechanism for non-ionic hydrophilic organic compounds, but no single molecular size parameter is widely accepted for describing the size exclusion mechanism in membrane treatment (Kiso et al. 2001, Chen et al. 2004, Van der Bruggen et al. 2002, Agenson et al. 2002). Molecular size has been described by many parameters including the molecular weight, molecular width, molecular length, stokes radius, effective diameter and many other parameters (Kiso et al. 2001, Chen et al. 2004, Van der Bruggen et al. 2002, Agenson et al. 2002). Generally, molecular weight is used under the assumption that molecules generally increase in size as they increase in mass (Schäfer et al. 2002, Van der Bruggen et al. 2002).

The Spiegler-Kedem model requires the calculation of a given compounds reflection coefficient (σ, maximum rejection) and permeability (P) for a given membrane. For nonionic solutes, the reflection coefficient, permeability and rejection are a function of molecular size, solvent flux and membrane pore size and permeability. Selecting the appropriate size parameter for the solutes employed to characterize a membrane as well as solutes in which the model will be used to predict rejection becomes integral to applying the Spiegler-Kedem model. Model compound rejection plotted as a function of molecular weight for each TFC-HR, TMG10 and NF-90 membranes are plotted in Figures G.2, G.3, and G.4, respectively.

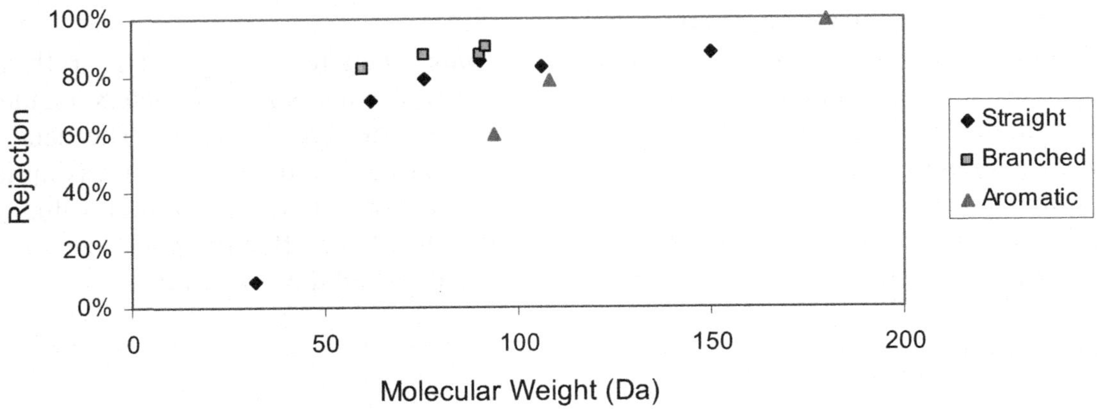

Figure G.2 Rejection of membrane characterization compounds by molecular weight for the TFC-HR membrane (flat-sheet experiments)

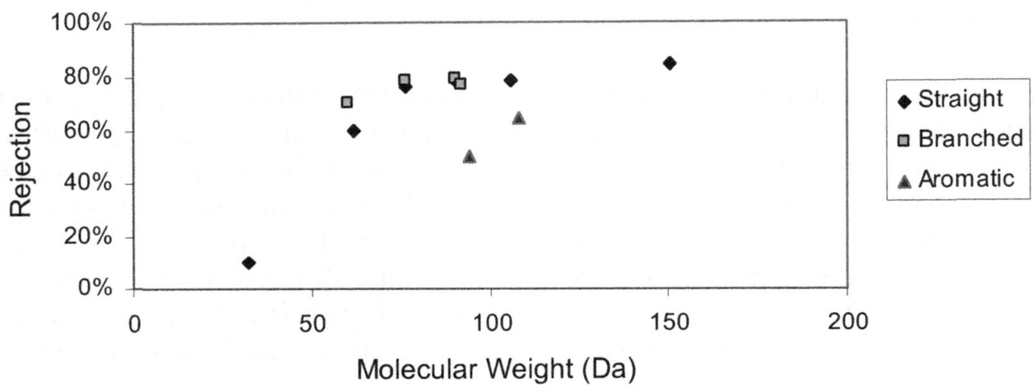

Figure G.3 Rejection of membrane characterization compounds by molecular weight for the TMG10 membrane (flat-sheet experiments)

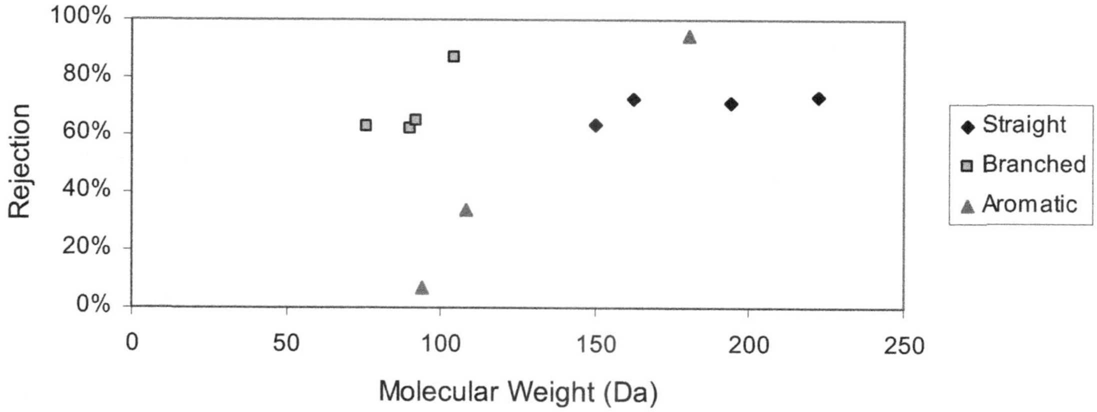

Figure G.4 Rejection of membrane characterization compounds by molecular weight for the NF-90 membrane (flat-sheet experiments)

Model compound rejection was observed to increase with molecular weight for the straight chain aliphatic and aromatic compounds for all three membranes tested (Figures G.2, G.3, G.4). With the exception of the NF-90 membrane, the rejection of the straight aliphatic compounds increased substantially with small increments of molecular weight, but leveled off at a maximum rejection that appears independent of molecular size. Additionally, for the TFC-HR and TMG10 membranes, rejection was observed to increase for compounds with similar molecular weights from the straight chain aliphatics to the branched aliphatics (Figures G.2 and G.3). Other researchers have concluded that rejection increases with molecular weight within a homologous series and with the amount of branching for compounds with similar molecular weights (Wiesner et al. 1996). Although molecular weight appears to adequately describe the increase of rejection due to increased size exclusion, the NF-90 membrane data is more scattered and does not show a clear trend between molecular weight and rejection. Therefore, a molecular size parameter that adequately represents the size of a molecule was required in order to pursue the Spiegler-Kedem approach.

Selecting the molecular size parameter based on the reflection coefficient

The maximum rejection (reflection coefficient, σ) for each of the membrane characterization compounds was plotted as a function of several size parameters including the molecular width, length, effective diameter, molecular surface area, and molecular volume. Due to the number of parameters, only the figures for TMG10 are shown below. The results of the TMG10 membrane are representative of the other membranes.

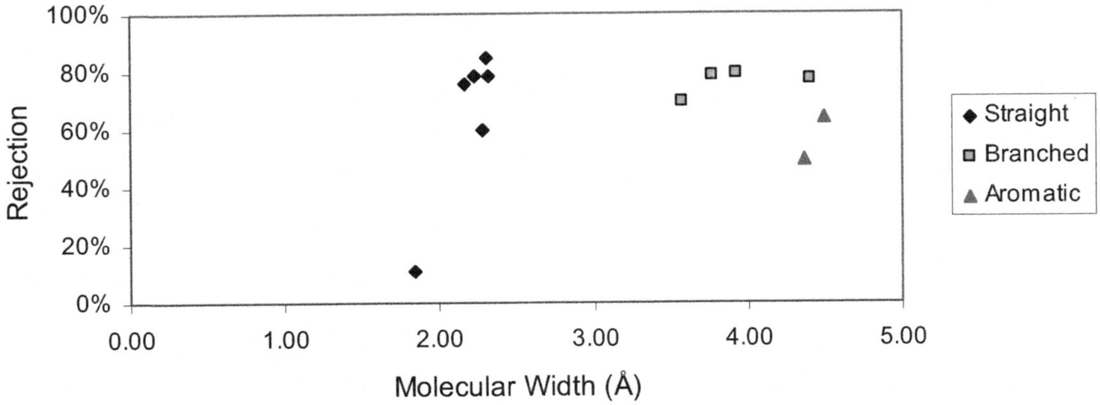

Figure G.5 Rejection of the membrane characterization compounds by TMG10 membrane as a function of molecular width

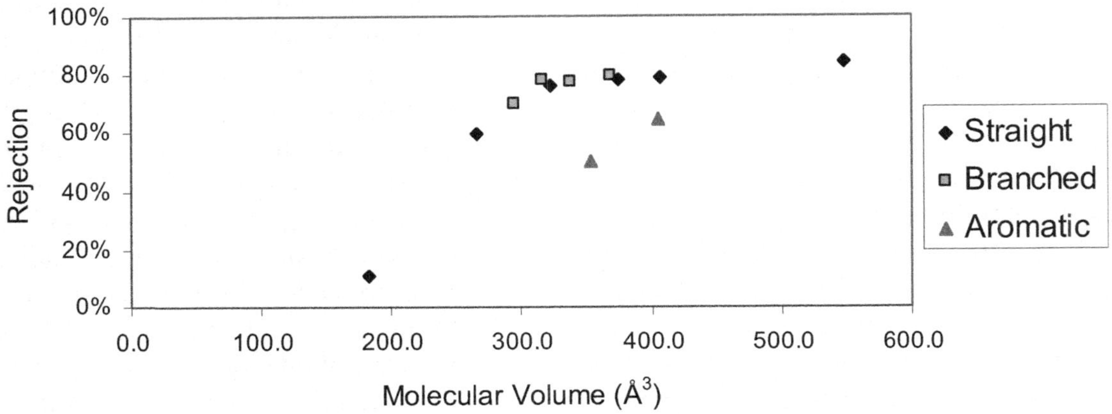

Figure G.6 Rejection of the membrane characterization compounds by the TMG10 membrane as a function of the molecular volume

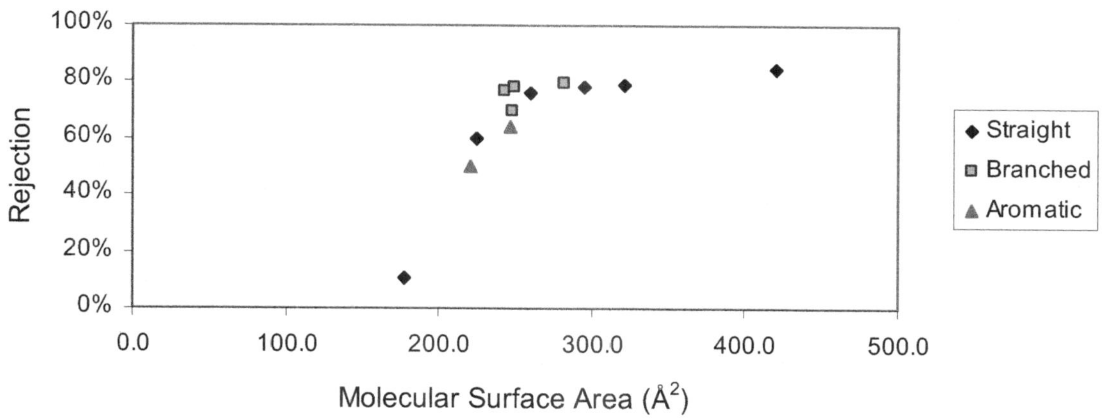

Figure G.7 Rejection of the membrane characterization compounds by TMG10 membrane as a function of the molecular surface area

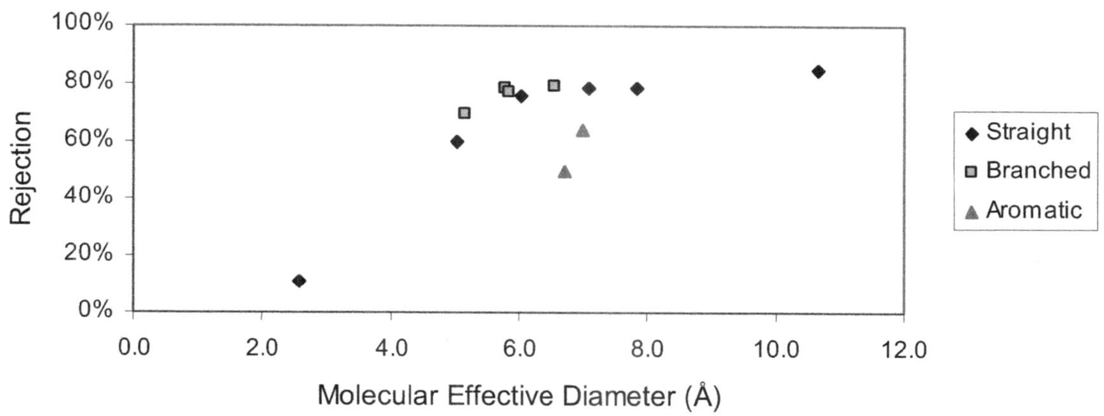

Figure G.8 Rejection of the membrane characterization compounds by TMG10 membrane as a function of the molecular effective diameter

Molecular width was not found to correlate well with TMG10 membrane rejection data, which is expected given the suite of compounds that was investigated. The straight chain aliphatics have nearly the same width although molecular weights ranged from 32 to 222 Daltons. The molecular volume and the effective molecular diameter did not appear to provide any substantial advantage over the molecular weight (Figure G.3) other than having length units. Of the parameters investigated, effective molecular diameter, molecular surface area and molecular volume were determined to be the best descriptors of molecular size.

Selecting the size parameter based on the permeability coefficient

Solute permeability (P) for a particular compound depends on the permeability constant of the membrane (ρ) and the size of the compound (given as radius). Therefore, the solute permeability coefficient was plotted with the various molecular size parameters for each membrane (TMG10 results shown in Figures G.9 through G.13).

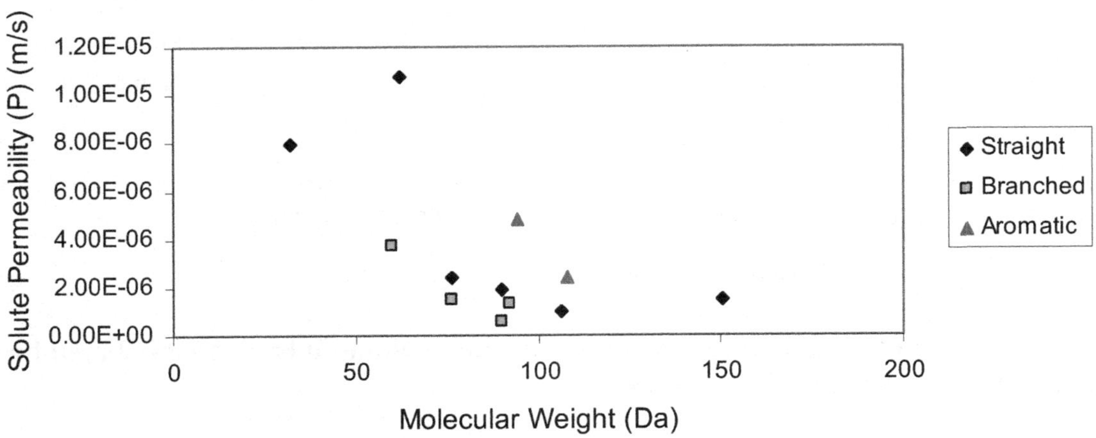

Figure G.9 Solute permeability of the membrane characterization compounds for TMG10 membrane as a function of molecular weight

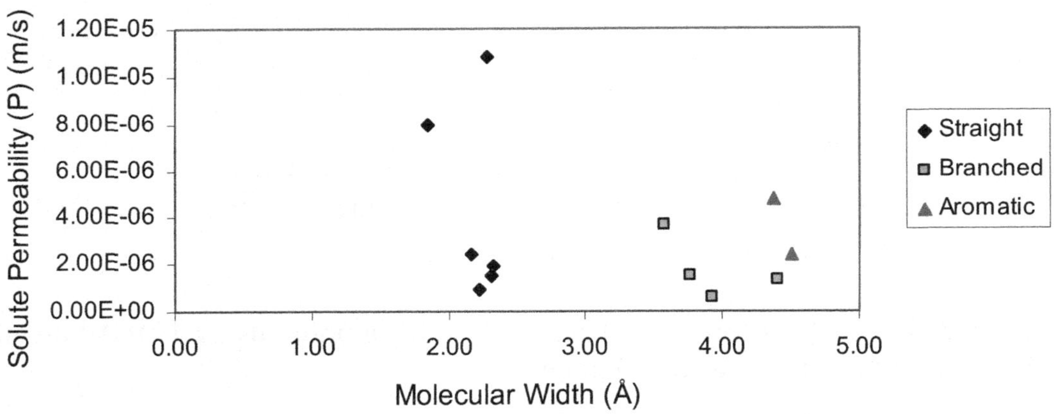

Figure G.10 Solute permeability of the membrane characterization compounds for TMG10 membrane as a function of molecular width

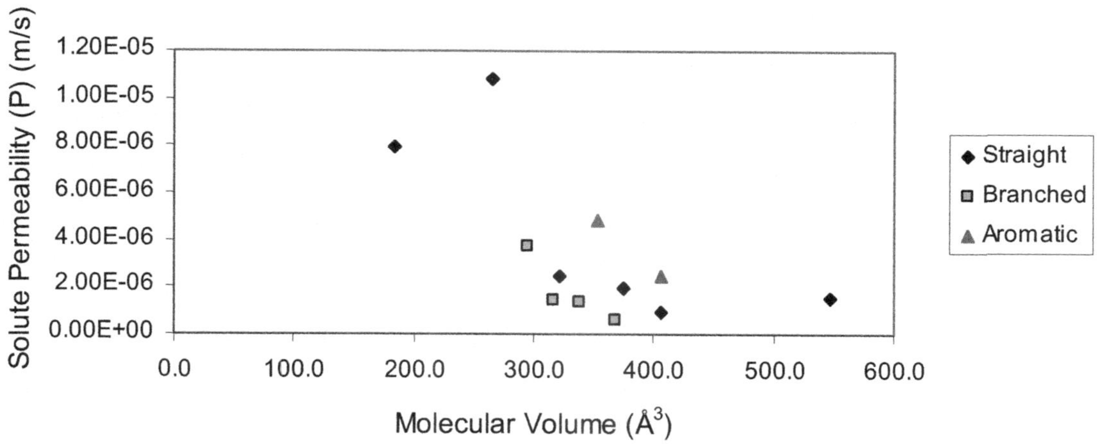

Figure G.11 Solute permeability of the membrane characterization compounds for TMG10 membrane as a function of molecular volume

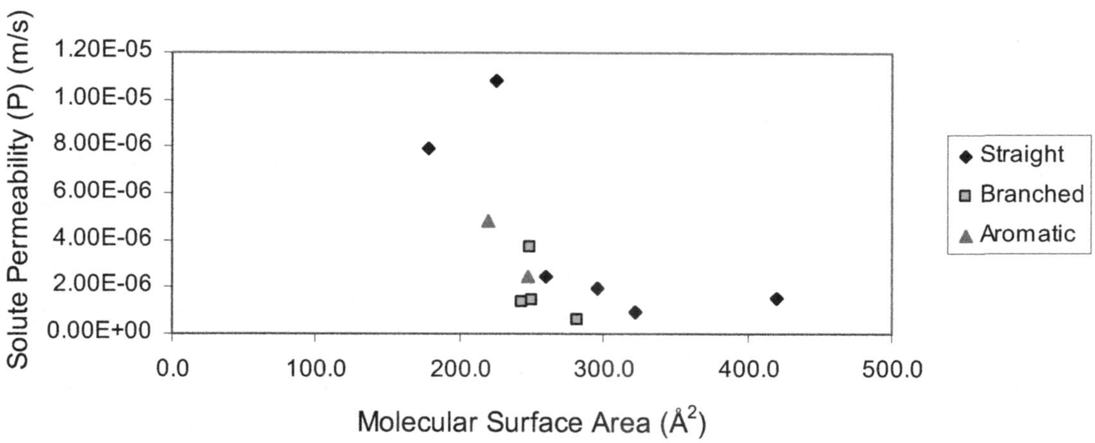

Figure G.12 Solute permeability of the membrane characterization compounds for TMG10 membrane as a function of molecular surface area

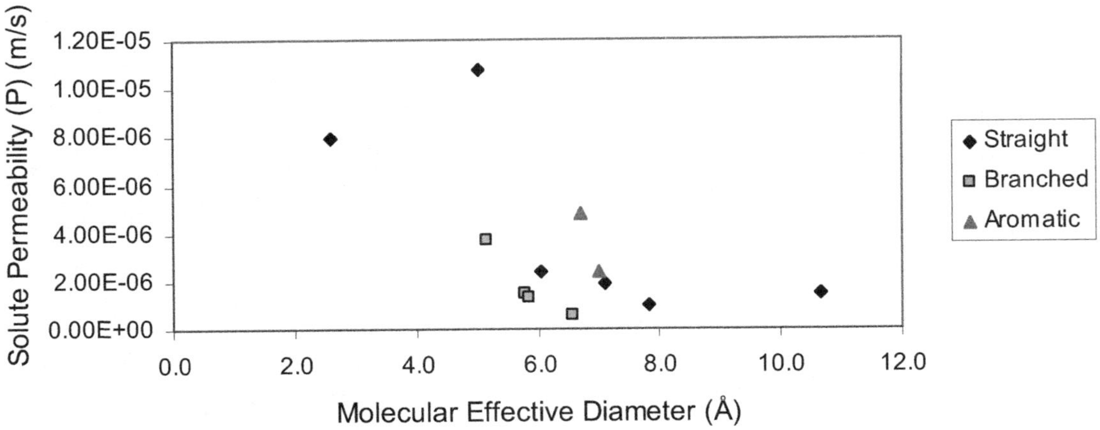

Figure G.13 Solute permeability of the membrane characterization compounds for TMG10 membrane as a function of molecular effective diameter

Similar to maximum rejection, molecular width did not correlate well with the solute permeability for the targeted molecular size parameter. Therefore, there is no clear advantage of selecting one molecular size parameter over another.

Selection of the molecular size parameter

The expected trend of increasing maximum rejection (reflection coefficient, σ) of the characterization compounds by the three membranes with increasing molecular size was best described by the molecular weight, molecular volume and molecular surface area. Due to the requirement of a length parameter for the Spiegler-Kedem model, molecular weight cannot be used directly. Since molecular weight is widely published in the literature, however, a molecular size parameter closely related to molecular weight would be ideal. Figure G.14 presents the molecular volume of all of the solutes employed in the modeling effort versus molecular weight while Figure G.15 shows the correlation between surface area and molecular weight. Molecular volume was found to closely correlate with molecular weight (Figure G.14, R^2 = 0.93), while surface area did not (Figure G.15, R^2 = 0.3G). Molecular volume, therefore, was subsequently used for the remaining Spiegler-Kedem modeling effort since it was found to allow the adequate description of increasing rejection with increasing molecular size for the three membranes and could be estimated using molecular weight and the regression equation presented in Figure G.14.

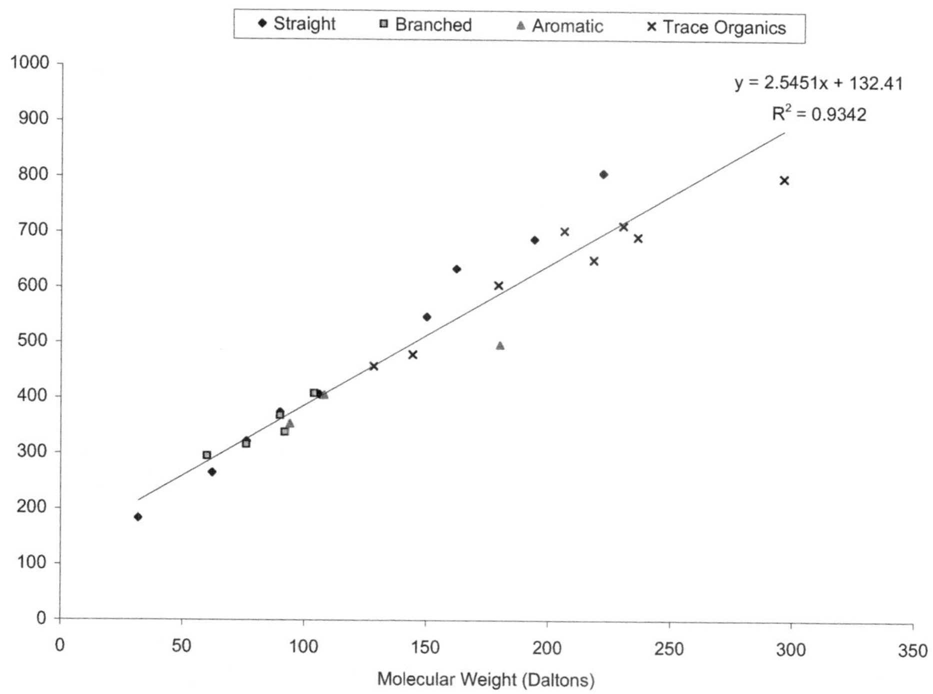

Figure G.14 Molecular volume versus molecular weight for three different structural configurations of compounds and trace organics with linear regression

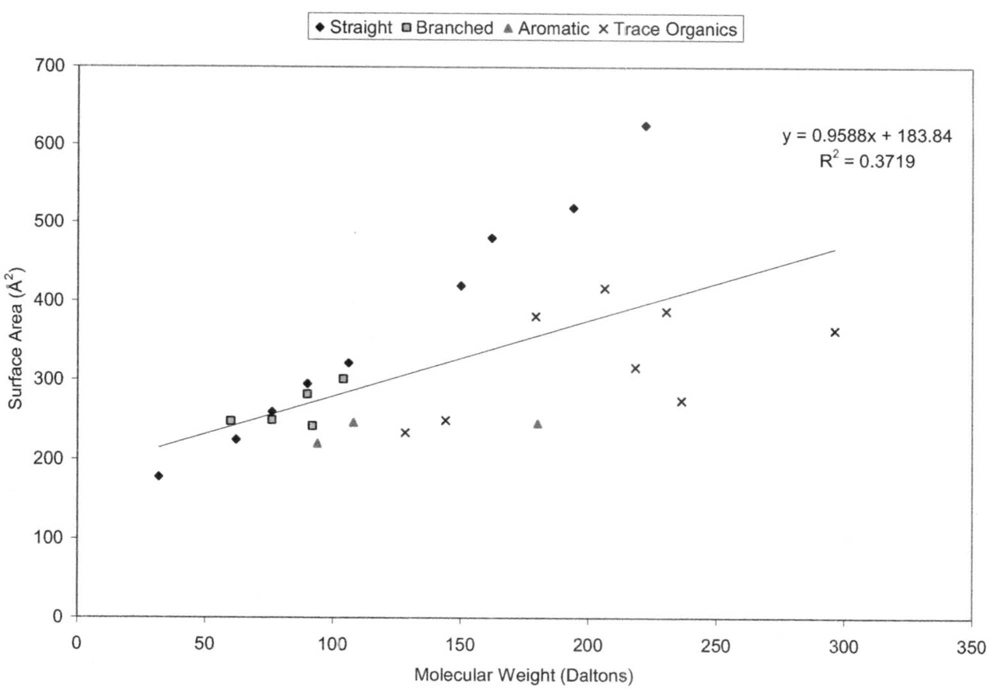

Figure G.15 Molecular surface area related to molecular weight for the membrane characterization compounds and the trace organics

In order to convert this parameter into a length dimension, the molecules are assumed to be spherical and the radius was calculated from the volume of a sphere as expressed through equation G.13.

$$r = \sqrt[3]{\frac{3 * V}{4 * \pi}} \qquad \text{(G.13)}$$

where r = the radius of the spheroid molecule
V = the volume of the spheroid molecule

While it is understood that all compounds are not spherical, the correlation shown in Figure G.14 indicates that this assumption is valid for the compounds selected and that it is also applicable to the trace organics selected for this study. The resultant molecular diameters calculated from molecular volume for the compounds employed in this study are listed in Table G.2.

Table G.2
Molecular diameter calculated from the molecular volume

Compound	Diameter from volume (Å)
Straight Chain Aliphatics	
Methanol	7.0
Ethylene Glycol	8.0
1,3-Propanediol	8.5
1,4-Butanediol	8.9
Diethylene Glycol	9.2
Triethylene Glycol	10.2
Diethylene Glycol Monobutyl Ether	10.7
Tetraethylene Glycol	11.0
Tetraethylene Glycol Dimethyl Ether	11.6
Branched Aliphatics	
Isopropanol	8.3
1,2 Propanediol	8.4
1,3 Butanediol	8.9
Glycerin	8.6
Neopentyl Glycol	9.2
Aromatics/Ringed	
Phenol	8.8
Benzyl Alcohol	9.2
dextrose (D-glucose)	9.8
Trace Organics (Aromatic/Ringed)	
Primidone	11.8
Phenacetine	11.6
Carbamazepine	12.1
2-Naphthol	10.7
Naproxen	12.2
Diclofenac	12.7
Ibuprofen	12.1
Naphthalene	10.5

Results of Membrane Solvent Permeability (L_P) Coefficient

The amount of water produced is dependent upon the applied pressure and the solvent permeability coefficient. The solvent permeability coefficient (L_P) was measured during the flat sheet cross flow experiments by measuring the permeate flow rate at 350, 490, 700, 1,050, and 1,400 kPa (50, 70, 100, 150, and 200 psi). The osmotic pressure was assumed negligible (calculated at < 1.4 kPa (0.2 psi)) during the experiments. Table G.3 presents the L_P data obtained experimentally and calculated from manufacturer test conditions.

Table G.3
Membrane pure water permeability coefficients

Membrane	Manufacturer Data (L_P) (L/m^2-day-kPa)	Experimental Data (L_P) (L/m^2-day-kPa)
TFC-HR	0.69	0.59 ± 0.07
TMG10	1.38	1.72 ± 0.17
NF-90	2.49	2.66 ± 0.23

The results indicate that pure water permeability in flat sheet experiments varied from the manufacturer specifications for the three membranes investigated (nearly 25 percent difference for the TMG10 membrane). The differences, however, most likely reflect the difficulty in replicating manufacturer test conditions and were considered insignificant. L_P is an important parameter in the model for predicting the solvent flux and subsequently the solute rejection.

Determining the Solute Permeability (P)

The solute permeability was determined by fitting rejection versus pressure data. The rejection for each compound was determined at 350, 490, 700, 1,050, and 1,400 kPa (50, 70, 100, 150, and 200 psi). The rejection of solutes generally increases with increasing pressure due the increase of solvent flux (Van der Bruggen et al., 2002). The fitting procedure is demonstrated in Figure G.16. An increase in solute permeability was found to decrease rejection. The value of the solute permeability highlighted in Figure G.16 is 2.4E-6 and was obtained by using the solver function in Microsoft Excel. This procedure was repeated for all of the membrane characterization compounds for each of the three membranes. Rejection, however, did not always increase with increasing pressure. Notably, the NF-90 exhibited a different rejection behavior for some solutes in which the rejection increased from 350 kPa (50 psi) to 700 kPa (100 psi) and then decreased as pressure increased to 1,400 kPa (200 psi) (Figure 7.17). This behavior is believed to be a result of the highly permeable NF-90 and the resultant beta values (permeate volume/feed volume) due to relatively high permeate flows compared to the other membranes.

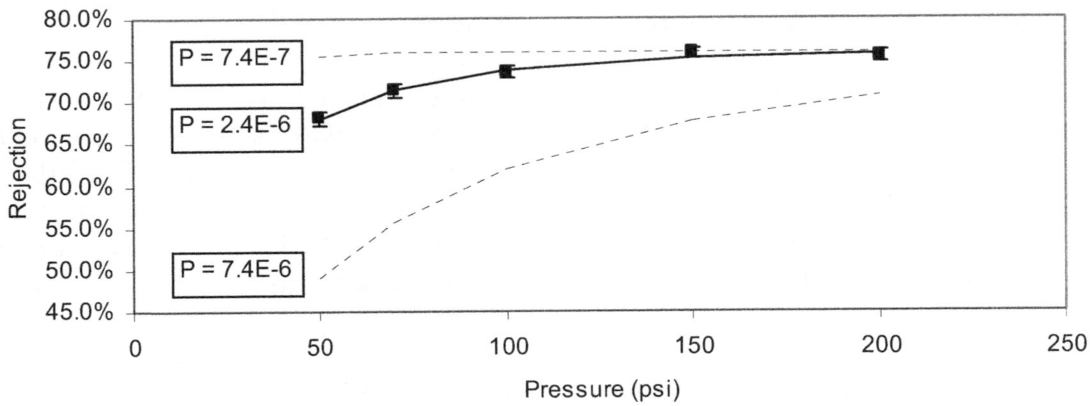

Figure G.16 Rejection of 1-3 propanediol by the TMG10 membrane as a function of feed pressure

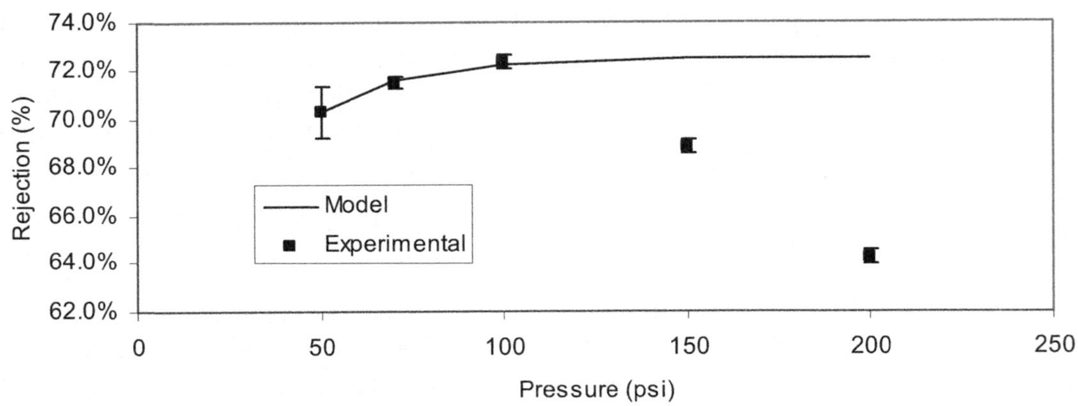

Figure G.17 Rejection of diethylene glycol butyl ether by NF-90 membrane at varying pressures

Membrane Permeability Constant (ρ)

The membrane permeability constant (ρ) was calculated using the solute permeability described in the previous section and the molecular radius calculated from molecular volume. The results for each compound for the TFC-HR membrane are listed in Table G.4. The summary results for all three membranes are listed in Table G.5.

Table G.4
Solute permeability and membrane permeability constant for the TFC-HR membrane

Compound	P (m/s)	ρ (m^2/s)
Straight Aliphatic		
Methanol	3.63E-06	2.56E-15*
Ethylene Glycol	4.13E-06	3.30E-15*
1,3-Propanediol	8.37E-07	7.12E-16
1,4-Butanediol	9.54E-07	8.53E-16
Diethylene Glycol	7.88E-07	7.25E-16
Triethylene Glycol	3.87E-07	3.93E-16
Branched Aliphatic		
Isopropanol	9.11E-07	7.52E-16
1,2-Propanediol	5.36E-07	4.53E-16
1,3-Butanediol	2.65E-07	2.36E-16
Glycerin	2.28E-07	1.97E-16
Aromatic/Ringed		
Phenol	6.53E-07	5.73E-16
Benzyl Alcohol	4.89E-07	4.49E-16
Glucose	9.27E-08	9.10E-17

*Considered an outlier and not included in statistical summary.

Table G.5
Membrane permeability coefficient

Membrane	Membrane permeability constant (ρ) (m^2/s)
TFC-HR	4.94E-16 ± 2.52E-16
TMG10	15.5E-16 ± 4.89E-16
NF-90	29.2E-16 ± 5.03E-16

The experimental standard deviation for TFC-HR, TMG10, and NF-90 were calculated with n=11, n=10, and n=6, respectively.

Membrane permeability for the NF-90 membrane could not be calculated using the full dataset since certain compounds did not follow the pattern of increasing rejection with increasing pressure (Figure G.17). Using the data that did follow the expected trend, a value for the membrane permeability constant was determined to be very near the values obtained for the TFC-HR and TMG10.

Membrane Pore Size and Distribution ($\bar{r}$, S_P)

The membranes were characterized with respect to the membrane pore size and distribution using the membrane characterization compound rejection data. The reflection coefficient was plotted as a function of the molecular size parameter. The log-normal cumulative distribution function (equation G.13) was used to fit the membrane parameters $\bar{r}$ and S_P. The modeled and experimental data are presented for the TFC-HR and TMG10 membranes in Figures G.18 and G.19, respectively.

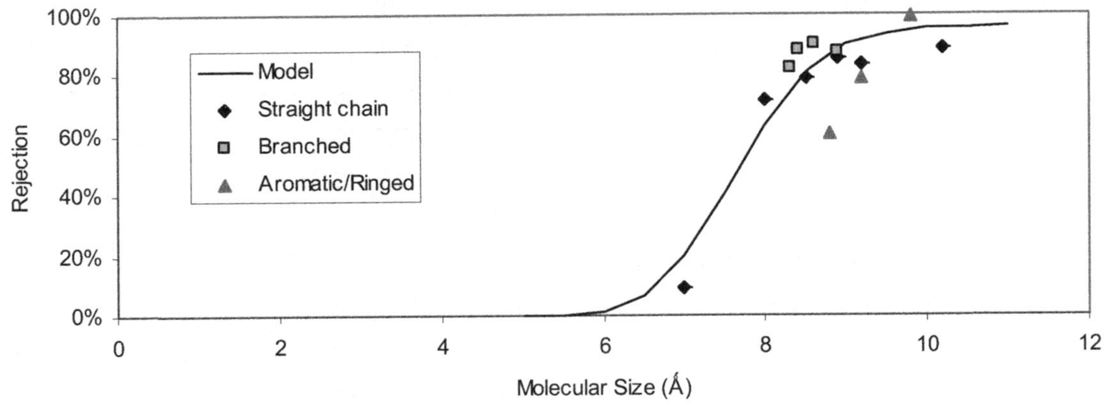

Figure G.18 Generation of TFC-HR model fitting parameters including average fictive pore radius (r, r = 3.85 angstroms), standard fictive pore radius deviation (S_p, S_p = 0.11 angstroms) and sum of error (sum of error = 0.14)

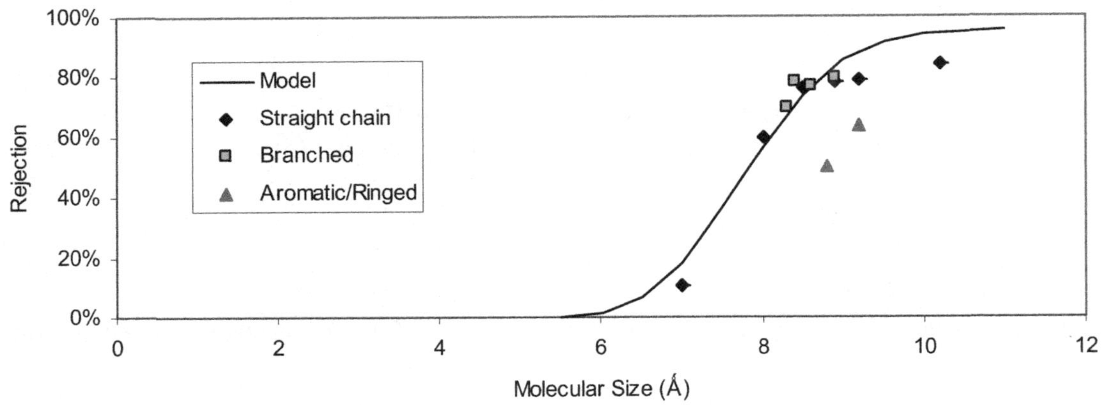

Figure G.19 Generation of TMG10 model fitting parameters including average fictive pore radius (r, r = 3.9 angstroms), standard fictive pore radius deviation (S_p, S_p = 0.124 angstroms) and sum of error (sum of error = 0.2)

Based on the sum of error values for the TFC-HR and TMG10 membranes (0.14 and 0.2, respectively) the model was adequate in describing the data. Unfortunately, there were no compounds with rejection values in the 20 to 50 percent rejection range for either of the membranes. This undoubtedly led to a better 'fit' since a large percentage of the dataset is clustered around 80 percent rejection. In addition, the aromatic/branched compounds are not as well described by the model for either membrane. The rejection of the straight chain aliphatic and branched compounds employed in the study; however, appear to be adequately described using the Spiegler-Kedem approach.

As mentioned previously, the NF-90 did not always exhibit the greatest rejection at the highest pressure. Since the reflection coefficient (maximum rejection at maximum pressure, σ) could not be experimentally determined for all the compounds using the NF-90 membrane, the maximum rejection observed during the experiment was used for the fitting exercise (Figure G.20).

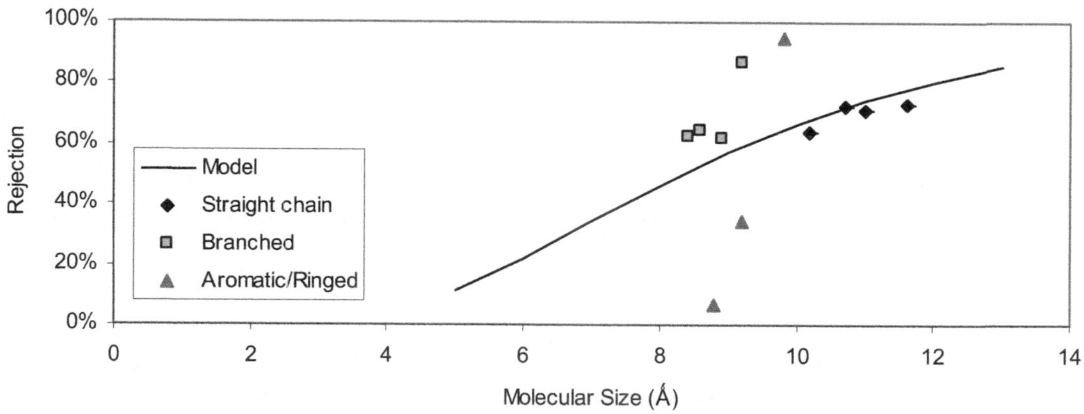

Figure G.20 Generation of NF-90 model fitting parameters including average fictive pore radius (r, r = 4.15 angstroms), standard fictive pore radius deviation (S_p, S_p = 0.42 angstroms) and sum of error (sum of error = 0.5)

The branched aliphatics (isopropanol, 2-propanediol, 3-butanediol, and glycerin) and one aromatic/ringed compound (glucose) were not well described by the model as shown by the sum of error calculation value of 0.50 (Figure G.20). The model was fitted a second time using only the straight aliphatic and the smaller two aromatic/ringed compounds and the results are presented in Figure G.21.

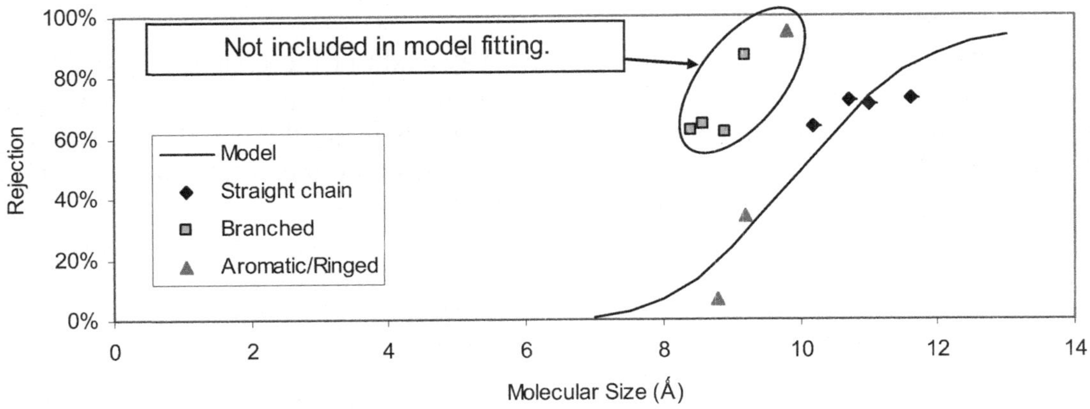

Figure G.21 Generation of NF-90 model fitting parameters without data outliers including fictive pore radius (r, r = 5 angstroms), standard fictive pore radius deviation (Sp, Sp = 0.15 angstroms) and sum of error (sum of error = 0.04)

The model is fitted with a much smaller sum of error (0.04). It is unfortunate that the model did not capture the rejection behavior for all of the compounds; however, the model simulates the rejection for those compounds lower than the experimental value and would therefore be conservative in its predicted rejection for other compounds. The pore size and distribution of pore sizes are summarized in Table G.6 for the three membranes employed in the study.

Table G.6
Membrane mean pore size and logarithmic distribution of pore sizes

Membrane	$\bar{r}$ (Å)	S_P
TFC-HR	3.85	0.111
TMG10	3.9	0.124
NF-90	5	0.148

The TFC-HR and TMG10 membrane have very similar fictive pore size characteristics. This is to be expected since they are both classified as RO membranes. The NF-90 has a slightly larger mean fictive membrane pore size and pore size distribution which is also expected because it is classified as an NF membrane

NONIONIC TRACE ORGANIC REJECTION PREDICTION

The results of bench-scale nonionic trace organic experiments conducted with the TFC-HR, TMG10 and NF-90 in terms of predicted and experimentally determined rejection are presented in Figures G.22, G.23, and G.24. The non-ionic trace organic solutes employed in bench-scale experiments include Tier 5 solutes (Table 2.1) primidone, phenacetine, carbamazepine, 2-napthol and naphthalene. Due to the relatively large size of the nonionic solutes (>10 Å in molecular diameter), the TFC-HR demonstrated rejection values greater than 88 percent (Figure G.22). In the size range of the solutes employed, the model predicted rejection values between 92 and 94 percent, which are similar to the observed rejection values for the TFC-HR (Figure G.21). A list of the predicted and observed rejection values for individual solutes is presented in Table G.7.

During laboratory-scale testing (see Chapter 3), the TFC-HR was observed to achieve approximately 90 percent rejection of phenacetine and reduce feed concentrations of primidone and carbamazepine to below detection limits (Figure 3.7). The compounds primidone and carbamazepine were detected/quantified in WBWRP and SWC feed water (Figures 4.6 and D.9) during sampling campaigns conducted on full-scale treatment trains employing the TFC-HR. Both primidone and carbamazepine were not detected/quantified in any permeate samples indicating that rejection is sufficient to reduce concentrations in the feed water below detection limits. The model, therefore, appears to sufficiently predict the rejection of nonionic trace organic solutes by the TFC-HR on bench-, laboratory- and pilot-scale.

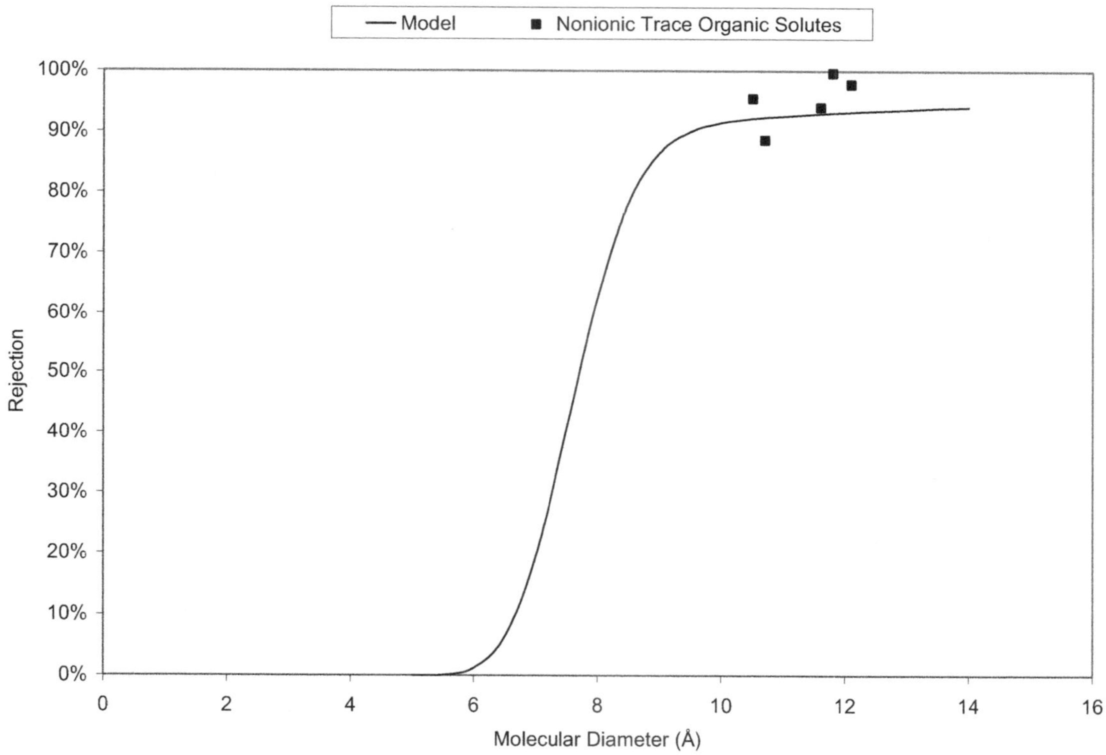

Figure 7.22 Rejection of nonionic trace organics (primidone, phenacetine, carbamazepine, 2-napthol, naphthalene) by the TFC-HR, experimental and modeled

Rejection values for non-ionic trace organic solutes during bench-scale TMG10 membrane experiments were between 80 and 99 percent and generally lower than the TFC-HR (Figure G.23). The model overpredicted the rejection of 2-napthol and phenacetine by approximately 10 percent (Table G.7), while underpredicting the rejection of naphthalene by 7 percent. The rejection of primidone and carbamazepine were observed to be well predicted by the model. Similarly to the TFC-HR, the rejection of all trace organic solutes employed in the study was greater than 78 percent, indicating that due to the relatively large molecular size of common trace organic pollutants that rejection by RO and ULPRO membranes is efficient. A list of the predicted and observed rejection values for individual solutes is presented in Table G.7.

The rejection of phenacetine during laboratory-scale experiments (Figure 3.3) was also over predicted by the Spiegler-Kedem model (80 percent experimentally versus 93 percent predicted). During spiking experiments conducted at the SWC and WBWRP however, phenacetine was reduced to below quantification concentrations in permeate samples indicating that phenacetine rejection was greater than 90 percent.

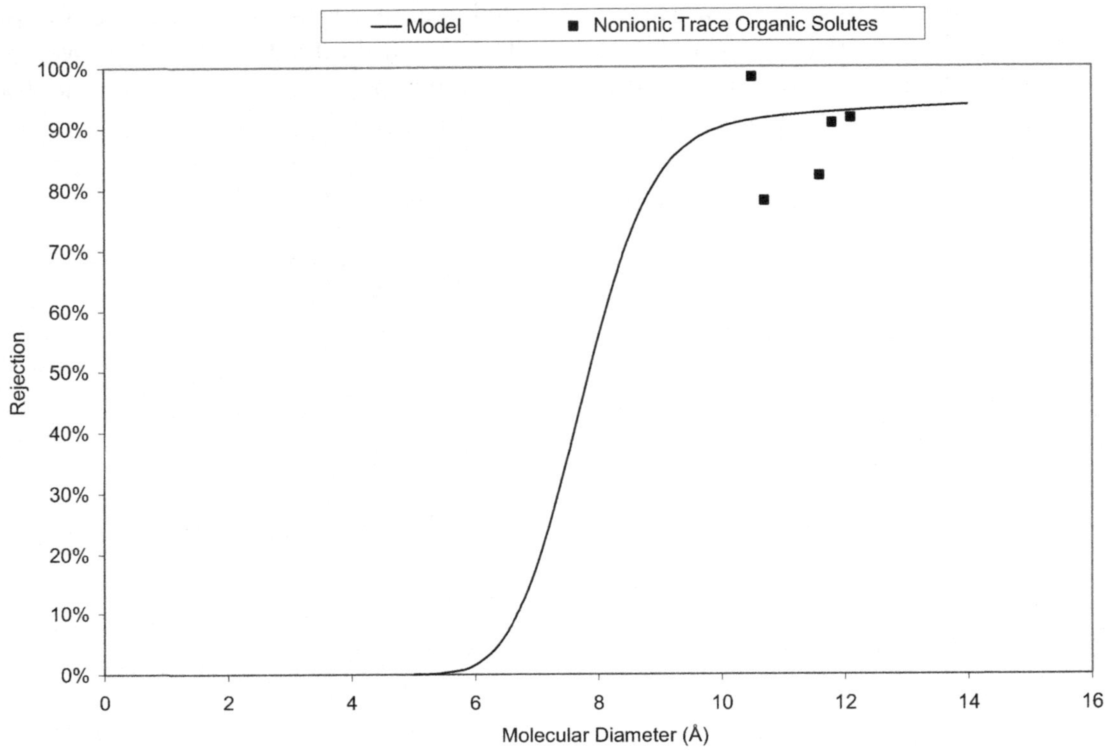

Figure G.23 Rejection of nonionic trace organics (primidone, phenacetine, carbamazepine, 2-napthol, naphthalene) by the TMG10, experimental and modeled

Nonionic trace organic rejection by the NF-90 was found to be highly variable and weakly dependent upon the size of the molecule during bench-scale experiments (Figure G.24). Phenacetine and 2-napthol rejection was poorly predicted by the model and experimental rejection for these solutes during bench-scale was 45 and 25 percent, respectively. Bench-scale NF-90 rejection experiments indicate that the NF-90 cannot effectively reject the trace organic solutes phenacetine, 2-napthol and naphthalene. Although laboratory-scale experiments were not conducted with naphthalene and 2-napthol, the NF-90 was observed to reject phenacetine at approximately 75 percent during laboratory-scale experiments (Figure 3.20), which is much closer to the predicted value of 81 percent (Table G.7). In addition, during spiking experiments conducted at the WBWRP during pilot-scale NF-90 testing, phenacetine was not detected in any permeate sample indicating efficient rejection.

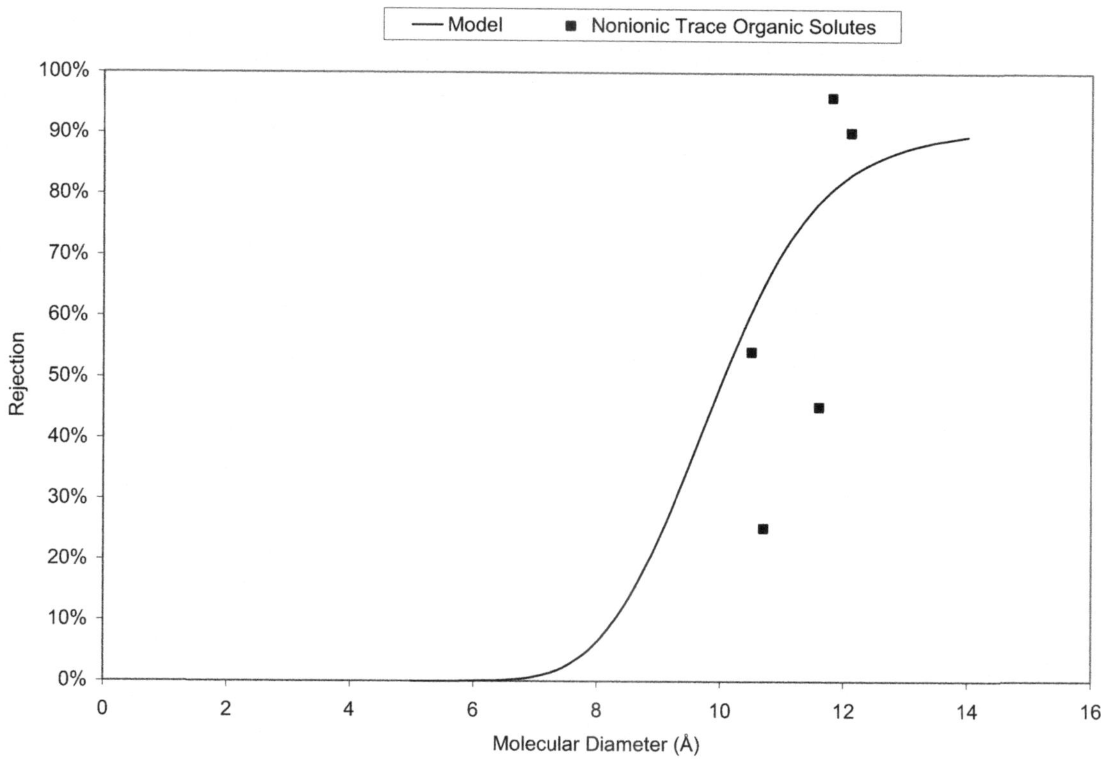

Figure G.24 Rejection of nonionic trace organics (primidone, phenacetine, carbamazepine, 2-napthol, naphthalene) by the NF-90, experimental and modeled

Table G.7
Predicted and observed rejection for the nonionic trace organic solutes

Compound	Predicted Rejection	Observed Rejection	Difference (%)
	TFC-HR		
Primidone	0.93	1.00	6.6
Phenacetine	0.93	0.94	1
Carbamazepine	0.93	0.98	4.5
2-Naphthol	0.92	0.89	-3.4
Naphthalene	0.92	0.96	3.8
	TMG10		
Primidone	0.93	0.91	-1.8
Phenacetine	0.93	0.82	-10.3
Carbamazepine	0.93	0.92	-1.2
2-Naphthol	0.92	0.78	-13.5
Naphthalene	0.91	0.98	7
	NF-90		
Primidone	0.83	0.96	13
Phenacetine	0.81	0.45	-35.6
Carbamazepine	0.86	0.90	4.7
2-Naphthol	0.66	0.25	-41.1
Naphthalene	0.62	0.54	-7.7

REFERENCES

Agenson, K.O., J.I. Oh, T. Kikuta, and T. Urase. 2002. Rejection Mechanisms of Plastic Additives and Natural Hormones in Drinking Water Treatment by NF. *Conference Proceedings of Membranes in Drinking Water and Industrial Water Production*. Mülheim an der Ruhr, Germany: (September, 2002) 323-331.

Alexander, K.L., S. Alt, E. Owens, M.V. Patel, L. McGovern. Low fouling reverse osmosis membranes: Evidence to the contrary on microfiltered secondary effluent. *Proceedings AWWA Membrane Technology Conference*, Atlanta, Georgia. 2003.

Balannec, B., G. Gesan-Guiziou, B. Chaufer, M. Rabiller-Baudry, and G. Daufin. 2002. Treatment of dairy process waters by membrane operations for water reuse and milk constituents concentration. *Desalination*, 147:89-94.

Bandini, S. and D.Vezzani. 2003. Nanofiltration modeling: the role for dielectric exclusion in membrane characterization. *Chemical Engineering Science*, 58: 3303-3326.

Berg, P., G. Hagmeyer, and R. Gimbel. 1997. Removal of pesticides and other micropollutants by nanofiltration. *Desalination*, 113 (2-3): 205-208.

Bellona, C., J.E. Drewes, P. Xu, and G. Amy. 2004. Factors affecting the rejection of organic solutes during NF/RO treatment – a literature review. *Water Research*, 38, 2795-2809.

Bellona, C. and J.E. Drewes. 2005. The role of membrane surface charge and solute physico-chemical properties in the rejection of organic acids by NF membranes. *Jour. of Membrane Science*, 249:227-234.

Beverly, S., S. Seal, and S. Hong. 2000. Identification of surface chemical functional groups correlated to failure of reverse osmosis polymeric membranes. *Jour. Vac. Sci. Technol. A.*, 18 (4):1107-1113.

Bowen, W.R. and H. Mukhtar. 1996. Characterization and prediction of separation performance of nanofiltration membranes. *Jour. of Membrane Science*, 112:263-274.

Bowen, W.R., A.W. Mohammad, and N. Hilal. 1997. Characterization of nanofiltration membranes for predictive purposes – use of salts, uncharged solutes and atomic force microscopy. *Jour. of Membrane Science*, 125:91-105.

Bowen, W.R. and A.W. Mohammad. 1998. Diafiltration by nanofiltration: prediction and optimization, *Jour. of American Institute of Chemical Engineers*, 44:1799-1812.

Bowen, W.R. and J.S. Welfoot. 2002a. Modelling the performance of membrane nanofiltration – critical assessment and model development. *Chemical Engineering Science*, 57:1121-1137.

Bowen, W.R. and J.S. Welfoot. 2002b. Modelling of membrane nanofiltration – pore size distribution effects. *Chemical Engineering Science*, 57:1393-1407.

Bowen, W.R., J.S. Welfoot, and P.M. Williams. 2002. Linearized transport model for nanofiltration: development and assessment. *Jour. of the American Institute of Chemical Engineering*, 48(4): 760-772.

Bowen, W.R., B. Cassey, P. Jones, and D.L. Oatley. 2004. Modelling the performance of membrane nanofiltration – application to an industry relevant separation. *Jour. of Membrane Science*, 242:211-220.

Braghetta, A., F.A. Digiano, and W.P. Ball. 1997 Nanofiltration of natural organic matter: pH and ionic strength effects. *Jour. Environ. Engineer.*, 123(7):628-640.

Braeken, L., R. Ramaekers, Y. Zhang, G. Maes, B. Van der Bruggen, and C. Vandecasteele. 2006. Influence of hydrophobicity on retention in nanofiltration of aqueous solutions containing organic compounds. *Jour. of Membrane Science*, 252:195-203.

California Department of Public Health (CDPH). 2007. *Groundwater Recharge Draft Regulations*. January 2007. Sacramento, California.

Chen, S.S., J.S. Taylor, L.A. Mulford, and C.D. Norris. 2004. Influences of Molecular Weight, Molecular Size, Flux, and Recovery for Aromatic Pesticide Removal by Nanofiltration Membranes. *Desalination*, 160(2):103-111.

Cho, J., G. Amy, and J. Pellegrino. 2000. Membrane filtration of natural organic matter: comparison of flux decline, NOM rejection, and foulants during filtration with three UF membranes. *Desalination*, 127: 283-298.

Crozes, G., and T. Seacord. 2001. *Evaluation of Membrane Technologies for Removal of Atrazine and Other SOCs*. Denver, Colo.: AwwaRF and AWWA.

Cussler, E.L. 1997. *Diffusion: Mass transfer in fluid systems.* 2nd edition. Cambridge.

Deeb, R., E. Hawley, Z. Chowdhury, and M.C. Kavanaugh. 2003. NDMA sources, formation and destruction during wastewater treatment: Current and future research. *Proceedings Annual Water Reuse Research Conference*. WateReuse Foundation. San Francisco, Cal.

DiGiano, F.A., S. Arweiler, and J.A. Riddick Jr. 2000. Alternative tests for evaluating NF fouling. *Jour. AWWA.*, 92(2):103-115.

Dresner, L. 1972. Some remarks on the integration of the extended Nernst-Planck equations in the hyperfiltration of multicomponent solutions. *Desalination*, 10: 27-46.

Drewes, J.E., Fox, P., and Reinhard, M. 2001. Removal of Organics in Indirect Potable Reuse Systems: A Comparison of Efficiencies of Long-term Soil-Aquifer Treatment (SAT) and Best Available Technologies (BAT). Final project report submitted to National Centers for Water Treatment Technologies (NCWT). Project No. NCWTT 699-799-005. March 2001.

Drewes, J.E., M. Reinhard, and P. Fox. 2003. Comparing microfiltration-reverse osmosis and soil-aquifer treatment for indirect potable reuse of water. *Water Research,* 37:3612-3621.

Drewes, J.E., C. Bellona, M. Oedekoven, P. Xu, T.-U.Kim, and G. Amy. 2005. Rejection of wastewater-derived micropollutants in high-pressure membrane applications leading to indirect potable reuse. *Environmental Progress,* 24(4):400-409.

Drewes, J.E., P. Xu, C. Bellona, M. Oedekoven, D. Macalady, G. Amy, and T-U. Kim. 2006. Rejection of wastewater-derived micropollutants in high-pressure membrane applications leading to indirect potable reuse. Final Project Report WRF-02-001. WateReuse Foundation, Alexandria, Virginia.

Duranceau, S.J., J.S. Taylor, and L.A. Mulford. 1992. SOC removal in a membrane softening process. *Jour. AWWA*, 84(1): 68-78.

Findlay, R.H., G.M. King, and L. Watling. 1989. Efficiency of phospholipids analysis in determining microbial biomass in sediments. *Applied Environ. Microbiol.*, 55(11):2888-2893.

Glater, J., S-K. Hong, and M. Elimelech. 1994. The search for a chlorine-resistant reverse osmosis membrane. *Desalination*, 95:325-345.

Goldstein, J., D. Newbury, D. Joy, C. Lyman, P. Echlin, E. Lifshin, L. Sawyer, and J.R. Michael. 2003. *Scanning electron microscopy and X-ray microanalysis*. Kluwer Academic/Plenum Publishers, 3rd Edition.

Hagmeyer G. and R.Gimbel. 1998. Modeling the salt rejection of nanofiltration membranes for ternary ion mixtures and for single salts at different pH values. *Desalination*, 117:247-256.

Heberer, T. 2002. Occurrence, fate, and removal of pharmaceutical residues in the aquatic environment: a review of recent data. *Toxicology Letters,* 131:5-17.

Helsel, D.R. 2004. *Nondetects and Data Analysis : Statistics for Censored Environmental Data (Statistics in Practice).* Wiley-Interscience Publisher.

Her, N., G. Amy, and C. Jarusutthirak. 2000. Seasonal variations of nanofiltration (NF) foulants: identification and control. *Desalination*, 132:143-160.

Hoek, E.M.V., S. Bhattacharjee, and M. Elimelech. 2003. Effect of membrane surface roughness on colloid-membrane DLVO interactions. *Langmuir*, 19:4836-4847.

Hofman, J.A., T.H. Noij, J.C. Kruithof, and J.C. Shippers. 1993. Removal of pesticides and other micropollutants with membrane filtration. *Water Supply,* 11(3/4):259-269.

Ivnitsky, H., I. Katz, D. Minz, E. Shimoni, Y. Chen, J. Tarchitzky, R. Semiat, and C.G. Dosoretz. 2005. Characterization of membrane biofouling in nanofiltration processes of wastewater treatment. *Desalination*, 185:255-268.

Jacangelo, J.G., J-M. Laîn, E.W. Cummings, A. Deutschmann, J. Mallevialle, and M.R.Wiesner. 1994. *Evaluation of Ultrafiltration Membrane Pretreatment and Nanofiltration of Surface Waters.* Denver, Colo. AwwaRF and AWWA.

Jain, S. and S.K. Gupta. 2004. Analysis of modified surface force flow model with concentration polarization and comparison with Spiegler-Kedem model in reverse osmosis systems. *Jour. of Membrane Science*, 232 (1-2):45-62

Jarusutthirak, C., G. Amy, and J-P. Croue. 2002. Fouling characteristics of wastewater effluent organic matter (EfOM) isolates on NF and UF membranes. *Desalination*, 145: 247-255.

Kargol, A. 2001. A Mechanistic model of transport processes in porous membranes generated by osmotic and hydrostatic pressure. *Jour. of Membrane Science*, 191: 61-69.

Kastelan-Kunst, L., K. Kosutic, V. Dananic and B. Kunst. 1997. FT30 membranes of characterized porosities in the reverse osmosis organics removal from aqueous solutions. *Water Research*, 31(11): 2878-2884.

Kimura, K., G. Amy, J.E. Drewes, and Y. Watanabe. 2003. Adsorption of hydrophobic compounds onto NF/RO membranes – an artifact leading to overestimation of rejection. *Jour. of Membrane Science*, 221: 89-101.

Kimura, K., S. Toshima, G. Amy, and Y. Watanabe. 2004. Rejection of neutral endocrine disrupting compounds (EDCs) and pharmaceutically active compounds (PhACs) by RO membranes. *Jour. of Membrane Science*, 245: 71- 78.

Kiso Y., Y. Nishimura, T. Kitao, and K. Nishimura. 2000. Rejection properties of non-phenylic pesticides with nanofiltration membranes. *Jour. of Membrane Science*, 171: 229-237.

Kiso, Y., T. Kitao, J. Kiyokatsu, and M. Miyagi. 1992. The effects of molecular width on permeation of organic solute through cellulose acetate reverse osmosis membrane. *Jour. of Membrane Science*, 74(1-2):95-103.

Kiso, Y., T. Kon, T. Kitao, and K. Nishimura. 2001. Rejection properties of alkyl pthalates with nanofiltration membranes. *Jour. of Membrane Science*, 182:205-214.

Kiso, Y., Y. Sugiura, T. Kitao, and K. Nishimura. 2001. Effects of hydrophobicity and molecular size on rejection of aromatic pesticides with nanofiltration membranes. *Jour. of Membrane Science*, 192:1-10.

Kosutic K., L. Kastelan-Kunst, and B. Kunst. 2000. Porosity of some commercial reverse osmosis and nanofiltration polyamide thin-film composite membranes. *Jour. of Membrane Science*, 168:101-108.

Koyuncu, I., D. Topacik, M. Turan, M.S. Celik, and H.Z. Sarikaya. 2001. Application of the membrane technology to control ammonia in surface water. *Water Science and Technology,* 1:117-124.

Koyuncu, I. 2002. Effect of Operating Conditions on the Separation of Ammonium and Nitrate Ions with Nanofiltration and Reverse Osmosis Membranes. *Jour. of Environmental Science & Health, Part A - Toxic/Hazardous Substances & Environmental Engineering,* 37(7):1347-1360.

Koyuncu, I., D. Topacik, and M.R. Wiesner. 2004. Factors influencing flux dcline during nanofiltration of solutions containing dyes and salts. *Water Research*, 38: 432-440.

Kurama H., J. Poetzschke, and R. Haseneder. 2002. The Application of Membrane Filtration for the Removal of Ammonium Ions from Potable Water. *Water Science,* 36(11):2905-2909.

Kwon, B., S. Lee, J. Cho, H. Ahn, D. Lee, and H.S. Shin. 2005. Biodegradability, DBP formation and membrane fouling potential of natural organic matter: characterization and controllability. *Environ. Sci. Technol.*, 39:732-739.

Kurama, H., J. Poetzschke, and R. Haseneder. 2002. The application of membrane filtration for the removal of ammonium ions from potable water. *Water Research*, 36 (11): 2905-2909.

Kutlan, D., P. Presits, and I. Molnar-Perl. 2002. Behavior and characteristics of amine derivatives obtained with *o*-phthaldialdehyde/3-mercaptopropionic acid and with *o*-phthaldialdehyde /*N*-acetyl-L-cysteine reagents. *Jour. of Chromatography A*, 949: 235-248.

Lee, S., J. Cho, and M. Elimelech. 2004. Influence of colloidal fouling and feed water recovery on salt rejection of RO and NF membranes. *Desalination*, 160:1-12.

Lee, S. and R. M. Lueptow. 2001. Membrane rejection of nitrogen compounds. *Environ. Sci. Technol.*, 35:3008-3018.

Levine B., K. Reich, P. Sheilds, I.H. Suffet, and V. Lazarova. 2000. Water quality assessment for indirect potable reuse: a new methodology for controlling trace organic compounds at the West Basin Water recycling plant (California, USA). *Water Sci Technol.*, 43(10):249–527.

Li, Q. and M. Elimelech. 2004. Organic fouling and chemical cleaning of nanofiltraion membranes: measurements and mechanisms. *Environ. Sci. Technol.*, 38(17): 4683-4693.

Lide, D. 2004-2005. *CRC Handbook of Chemistry and Physics*, 85th Edition, CRC Press, 10-148.

Mansell, J. and J.E. Drewes. 2004. Fate of steroidal hormones during soil-aquifer treatment (SAT). *Jour. Ground Water Monitoring and Remediation,* 24(2): 94-101.

Meng, F.G., H.M. Zhang, Y.S. Li, X.W.Zhang, F.L. Yang, and J.N. Xiao. 2005. Cake layer morphology in microfiltration of activated sludge wastewater based on fractal analysis. *Separation and Purification Technology*, 44(3):250-257.

Metcalf and Eddy. 2003. *Wastewater Engineering. Treatment and Reuse.* 4th Edition. Boston: McGraw- Hill.

Mitch, W.A., J.O. Sharp, R.R.Trussel, R.L.Valentine, L. Alvarez-Cohen, and D. Sedlak. 2003. N-Nitrosodimethylamine (NDMA) as a Drinking Water Contaminant: A Review. *Environmental Engineering Science,* 20(5): 389-404.

Najim, I. and R.R. Trussel. 2001. NDMA formation in water and wastewater. *Jour. AWWA*, 93(2):92-99.

National Research Council. 1998. *Issues in Potable Reuse: The Viability of Augmenting Drinking Water Supplies with Reclaimed Water*. National Academy Press, Washington, D.C.

National Research Council. 2004. *Review of the Desalination and Water Purification Technology Roadmap*.

Ng, H. Y. and M. Elimelech. 2004. Influence of colloidal fouling on rejection of trace organic contaminants by reverse osmosis. *Jour. of Membrane Science*, 244: 215-226

Nghiem, L.D., A.I. Schäfer, and M. Elimelech. 2005. Pharmaceutical Retention Mechanisms by Nanofiltration Membranes. *Environ. Sci. Technol.* 39(19):7698-7705.

Ozaki, H. and H. Li. 2002. Rejection of organic compounds by ultra-low pressure reverse osmosis membrane. *Water Research*, 36(1):123-130.

Park, N., B. Kwon, I.S. Kim, and J. Cho. 2005. Biofouling potential of various NF membranes with respect to bacteria and their soluble microbial products (SMP): characterizations, flux decline, and transport parameters. *Jour. of Membrane Science*, 258:43-54.

Rauch-Williams, T. and J.E. Drewes. 2006. Using soil biomass as an indicator for the biological removal of effluent-derived organic carbon during soil infiltration. *Water Research*, 40(5): 961-968

Reddersen, K. and T. Heberer. 2003. Multi-compound methods for the detection of pharmaceutical residues in various waters applying solid phase extraction (SPE) and gas chromatography with mass spectrometric (GC-MS) detection. *Jour. Sep. Sci.*, 26:1443-1450.

Reinhard, M., N.L. Goodman, P.L. McCarty, and D.G. Argo. 1986. Removing trace organics by reverse osmosis using cellulose acetate and polyamide membranes. *Jour. AWWA*, 78(4):163-174.

Ridgway, H.F., M.G. Rigby, and D.G. Argo. 1984. Adhesion of a Mycobacteriun sp. To cellulose diacetate membranes used in reverse osmosis. *Applied and Environmental Microbiology*, Jan, 61-67.

Roudman, A.R. and F.A. DiGiano. 2000. Surface Energy of Experimental and Commercial Nanofiltraion Membranes: Effect of Wetting and Natural Organic Matter Fouling. *Jour. of Membrane Science*, 175(1): 61-73.

Santos, J.L.C., P. de Beukelaar, I.F.J. Vankelecom, S. Velizarov, and J.G. Crespo. 2006. Effect of solute geometry and orientation on the rejection of uncharged compounds by nanofiltration. *Separation and Purification Technology*, In Press.

Schaep, J., C. Vandecasteele, A.W. Mohammad, and W.R. Bowen. 1999. Analysis of the salt retention of nanofiltration membranes using the Donnan-steric partitioning model. *Separation Science and Technology*, 34(15):3009-3030.

Schäfer, S., M. Mastrup, and R. Jensen. 2002. Particle Interactions and Removal of Trace Contaminants from Water and Wastewaters. *Desalination*, 147: 243-250.

Schäfer, A.I., L.D. Nghiem, and T.D. Waite. 2003. Removal of natural hormone estrone for aqueous solutions using nanofiltration and reverse osmosis. *Environmental Science and Technology*, 37: 182-188.

Schäfer, A.I., A.G. Fane, and T.D. Waite. 2005 (eds.). *Nanofiltration: Principles and Applications.* Elsevier Advanced Technology Publishing, Oxford, UK.

Schäfer, A.I. and T.D. Waite. 2002. Trace contaminant removal using hybrid processes in water recycling. In *Chemical Water and Wastewater Treatment VII*, eds. H.H. Hahn, E. Hoffman, H. Odegaard, IWA Publishing, London. p. 319-330.

Schippers, J. C., Kruithof, J. C., Nederlof, M. M., Hofman, J. A. M. H. and Taylor, J. 2004. *Integrated Membrane Systems.* Denver, Colo. AwwaRF and AWWA.

Schneider, R.P., L. M.Ferreira, P. Binder, and R. Ramos. 2005. Analysis of foulant layer in all elements of an RO train. *Jour. of Membrane Science.* 261: 152-162.

Schutte, C.F. 2003. *The Rejection of Specific Organic Compounds by Reverse Osmosis Membranes.* Desalination, 158, 285-294.

Seidel, A. and M. Elimelech. 2002. Coupling between chemical and physical interactions in natural organic matter (NOM) fouling of nanofiltration membranes: implications for fouling control. *Jour. of Membrane Science*, 203:245-255.

Shon, H.K., S. Vigneswaran, I.S. Kim, J. Cho, and H.H. Ngo. 2004. Effect of pretreatment on the fouling of membranes: application in biologically treated sewage effluent. *Jour. of Membrane Science*, 234:111-120.

Snyder, S., S. Adham, A.M. Redding, F.S. Cannon, J. DeCarolis, J. Oppenheimer, E.C. Wert, and Y. Yoon. 2006. Role of membranes and activated carbon in the removal of endocrine disruptors and pharmaceuticals. *Desalination*, 202:56-181.

Sorin, A., A. Favre-Reguillon, S. Pellet-Rostaing, G. Bernier, and M. Lemaire. 2006. Polycarboxylic acids rejection by charged nanofiltration membrane. *Jour. of Membrane Science*, 279:446-452.

Speth, T.F., R.S. Summers, and A.M. Gusses. 1998. Nanofiltration foulants from a treated surface water. *Environ. Sci. Technol.*, 32:3612-3617.

Spiegler, K.S. and O. Kedem. 1996. Thermodynamics of Hyperfiltration (RO): Criteria for Efficient Membranes. *Desalination,* 1:311-326.

Taylor, J.S., L.A. Mulford, S.S. Chen, and C.N. Norris. 2000. *Flat Sheet, Bench, and Pilot Testing for Pesticide Removal Using Reverse Osmosis.* Denver, Colo. AwwaRF and AWWA.

Tödthheide, V., G. Laufenberg, and B. Kunz. 1997. Waste water treatment using reverse osmosis: real osmotic pressure and chemical functionality as influencing parameters on the retention of carboxylic acids in multi-component systems. *Desalination*, 110:213-222.

Trussell, R. R. 2001. Endocrine disruptors and the water industry. *Jour. AWWA,* 93 (2):5865.

Tsuru, T., S-I. Nakao, and S. Kimura 1991a. Calculation of ion rejection by extended Nernst-Planck equation with charged reverse osmosis membranes for single and mixed electrolyte solutions. *Jour. of Chemical Engineering Japan*, 24:511-517.

Tsuru, T., M. Urairi, S-I. Nakao, and S. Kimura. 1991b. Reverse osmosis of single and mixed electrolytes with charged membranes: experiment and analysis. *Jour. of Chemical Engineering Japan*, 24:518-524.

United States Environmental Protection Agency (U.S. EPA) 2004. *Guidelines for Water Reuse.* EPA/625/R-04/108, September 2004.

Van der Bruggen, B., J. Schaep, W. Maes, D. Wilms, and C. Vandecasteele. 1998. Nanofiltration as a treatment method for the removal of pesticides from ground waters. *Desalination*, 117:139

Van der Bruggen, B., K. Everaert, D. Wilms, and C. Vandecasteele. 2001. Application of nanofiltration for removal of pesticides, nitrate and hardness from ground water: rejection properties and economic evaluation. *Jour. of Membrane Science*, 193:239-248.

Van der Bruggen, B., L. Braeken, and C. Vandecasteele. 2002. Evaluation of parameters describing flux decline in nanofiltration of aqueous solutions containing organic compounds. *Desalination*, 147:281-288.

Van der Bruggen, B. and C. Vandecasteele. 2002. Modelling of the retention of uncharged molecules with nanofiltration. *Water Research*, 36:1360-1368.

Vernon, W., J. Clune, and A. Nunez. 2001. Applying an integrated membrane system to meet water quality goals for the indirect potable reuse of tertiary effluent. *Proceedings American Water Works Association Membrane Technology Conference*, San Antonio, Texas.

Vestal, J.R. and D.C. White. 1989. Lipid analysis in microbial ecology. Quantitative approaches to the study of microbial communities. *BioScience*, 39(8):535-541.

Vrijenhoek, E.M., S. Hong, and M. Elimelech. 2001. Influence of membrane surface properties on initial rate of colloidal fouling of reverse osmosis and nanofiltration membranes. *Jour. of Membrane Science*, 188:115-128.

Wang, X. L., T. Tsuru, S. Nakao, and S. Kimura. 1995. Electrolyte transport through nanofiltration membranes by the space-charge model and the comparison with the Teorell-Meyer-Sievers model. *Jour. of Membrane Science*, 103:117-133.

Wang, K.Y. and T.S. Chung. 2005. The Characterization of Flat Composite Nanofiltration Membranes and Their Applications in the Separation of Cephalexin. *Jour. of Membrane Science,* 247(1-2): 37-50.

Wiesner, M.R. and P. Aptel. 1996. Mass Transport and Permeate Flux and Fouling in Pressure-Driven Processes. *Water Treatment: Membrane Processes*, eds. J. Mallevaille, P.E. Odendaal, and M.R. Wiesner. 9.1 - 9.7.

Wiesner, M.R. and C.A. Buckley. 1996. Principles of Rejection in Pressure-Driven Membrane Processes. *Water Treatment Membrane Processes,* eds. J. Mallevaille, P.E. Odendaal, and M.R. Wiesner.

Wijmans, J. G. and R. W. Baker. 1995. The solution-diffusion model: a review. *Jour. of Membrane Science*, 107: 1-21

Wilf, M. and S.Alt. 2000. Application of low fouling RO membrane elements for reclamation of municipal wastewater. *Desalination*, 132:11-19.

Xu, P., J.E. Drewes, C. Bellona, G. Amy, T-U. Kim, M. Adam, and T. Heberer. 2005. Rejection of emerging organic micropollutants in nanofiltration/reverse osmosis membrane application. *Water Environment Research*, 17(1):40-48.

Xu, P., Drewes, J.E., Bellona, C. and Amy, G. 2006a. Effect of Membrane Fouling on Transport of Emerging Organic Contaminants in NF/RO Membrane Applications. *Jour. of Membrane Science*, 279:165-175.

Xu, P., J.E. Drewes, M. Oedekoven, C. Bellona, and G. Amy. 2006b. Rejection of Non-ionic Organic Micropollutants by Nanofiltration Membranes: Effect of Membrane Fouling. *AWWA Membrane Best Papers Book*, in press.

Xu, P. and J.E. Drewes. 2006. Viability of Nanofiltration and Ultra Low Pressure Reverse Osmosis Membranes for Multi-Beneficial Use of Methane Produced Water. *Separation & Purification Technology*, 52:67-76.

Yoon, Y., and R.M. Lueptow. 2005. Removal of organic contaminants by RO and NF membranes. *Jour. of Membrane Science*, 261:76-86.

Yoon, Y., G. Amy, J. Cho, N. Her, J. Pellegrino. 2002. Transport of perchlorate (ClO_4^-) through NF and UF membranes. *Desalination*, 147:11-17.

Zander, A. and N.K. Curry. 2001. Membrane and solution effects on solute rejection and productivity. *Water Research,* 35(18): 4426-4434.

Zhao, Y. and J.S. Taylor. 2004. Modeling membrane performance over time. *Jour. AWWA*, 96: 90-97.

Zhu, X. and M.Elimelech. 1997. Colloidal fouling of reverse osmosis membranes: measurements and fouling mechanisms. *Environ. Sci. Technol.*, 31:3654-3662.

ABBREVIATIONS

Actual Flux	Observed permeate flow dived by membrane area
AFM	Atomic Force Microscopy
APECs	Alkylphenolpolyethoxycarboxylates
ATR	Attenuated total reflection
AWT	Advanced Water Treatment
AwwaRF	Awwa Research Foundation
AZ	Arizona
CA	California
CDPH	State of California Department of Public Health
C_f	Feed Concentration
C_p	Permeate Concentration
CCL	Contaminant Candidate List
CIP	Clean-in-place
CSM	Colorado School of Mines
CU	University of Colorado-Boulder
DBPs	Disinfection By-Products
DCAA	Dichloroacetic Acid
DI	Distilled water (type II water)
DOC	Dissolved Organic Carbon
DON	Dissolved Organic Nitrogen
DSPM	Donnan steric pore model
EDCs	Endocrine Disrupting Compounds
EDTA	Ethylenediaminetetraacetic Acid
EfOM	Effluent organic matter
ELISA	Enzyme Linked Immunosorbent Assay
EPA	Environmental Protection Agency
EPS	Extracellular polymer substances
ESEM	Environmental Scanning Electron Microscopy
FTIR	Fourier Transform Infrared Spectroscopy
GC	Gas Chromatography
GC/ECD	Gas Chromatography/Electron Capture Detector
GC/MS	Gas Chromatography/Mass Spectrometry
gfd	Gallons per Square Foot and Day
gfd/psi	Gallons per Square Foot per psi and Day
HAAs	Haloacetic Acids
HPLC-DAD	High Performance Liquid Chromatography/Diode Array Detector
HS	Humic Substances

IC	Ion Chromatograph
ICP	Inductive Coupled Plasma
IMS	Integrated Membrane System
J	Flux
J_o	Initial Permeate Flux Taken at Filtration Time of 30 Minutes
L	Liters
L/m^2	Liter per Square Meter
L/m^2-d	Liter per Square Meter per Day
L/min	Liter per Minute
LOD	Limit of Detection
LOQ	Limit of Quantification
log D	Octanol-water Partitioning Coefficient (log K_{OW}) at a Specific pH
log K_{OW}	Octanol-water Partitioning Coefficient
m^3/s	Cubic Meter Per Second
MCL	Maximum Contaminant Level
Metro	Denver Metro Wastewater Reclamation District's Plant
MF	Microfiltration
mgd	Million Gallon Per Day
mL	Milliliters
ML	Million Liters
MRL	Method Reporting Limit
MtBE	Methyl t-butyl ether
MTBSTFA	N-(t-butyldim ethylsilyl)-N-methyl-trifluoraacetamide
MW	Molecular Weight
MWCO	Molecular Weight Cut-off
NDMA	N-nitrosodimethylamine
NDP	Net Driving Pressure
NF	Nanofiltration
nm	nanometer
OPA/NAC	*o*-phthaldialdehyde/*N*-acetyl-L-cysteine
ΔP	Differential Pressure
PAHs	Polycyclic Aromatic Hydrocarbons
P_F	Feed pressure
PFBBr	Pentalfuorobenzylbromide
PhACs	Pharmaceutically Active Compounds
pK_a	Acid Disassociation Constant
PWP	Pure water permeability
P_P	Permeate Pressure.
$\Delta\Pi$	Aaverage Osmotic Pressure Differential

Q_f	Feed Flow
Q_p	Permeate Flow
QA/QC	QualityAssurance/Quality Control
QSAR	Quantitative Structure-Activity Relationship
QSPR	Quantitative Structure-Property Relationship
RO	Reverse Osmosis
SCADA	Supervisory Control and Data Acquisition System
SDI	Silt Density Index
SEM	Scanning electron microscopy
SF	Specific flux
SMP	Soluble microbial products
SPI	Separation Processes, Inc.
Stdev	Standard Deviation
SUVA	Specific UV absorbance
SWC	Scottsdale Water Campus
T	Temperature
T_a	Actual Temperature
T_s	Standard Temperature
TC	Total Carbon
TCAA	Trichloroacetic Aacid
TCEP	Tris(2-chloroethyl-)phosphate
TCPP	Tris(1-chloro-2-propyl)phosphate
TCF	Temperature Correction Factor
TCSF	Temperature Corrected Specific Flux
TDCPP	Tris(1,3-dichloro-2-propyl)phosphate
THM	Total Haloorganic Methane
TIC	Total Inorganic Carbon
TN	Total Nitrogen
TOC	Total Organic Carbon
TON	Total Organic Nitrogen
Type I Water	Ultrapure water (Milli-Q-Water)
Type II Water	Deionized water
µS/cm	MicroSimens per centimeter
U	Temperature-corrected flux coefficient
UF	Ultrafiltration
ULPRO	Ultra-low Pressure Reverse Osmosis
UVA	Ultraviolet absorbance
WBMWD	West Basin Municipal Water District
WBWRP	West Basin Water Recycling Plant
WRF	WateReuse Foundation
WRP	Water Recycling Plant

IWA Publishing's authorised EU representative for General Product Safety Regulations is Diane D'Arras, 15 rue Duret, 75116 Paris, France, e-mail: safety@iwap.co.uk.

Printed and bound by CPI Group (UK) Ltd, Croydon, CR0 4YY
17/09/2026
02218808-0006